GAL\

This book
the l?

083210

The Effects of Fishing on Non-target Species and Habitats

Biological, Conservation and Socio-economic Issues

M.J. Kaiser
School of Ocean Sciences, University of Wales – Bangor, United Kingdom

S.J. de Groot
Netherlands Institute of Fisheries Research (RIVO-DLO), IJmuiden, The Netherlands

Published in association with
The European Commission Fisheries,
Agriculture and Agroindustrial
Research Programme (FAIR)

Blackwell
Science

Copyright © 2000 by
Blackwell Science Ltd
Editorial Offices:
Osney Mead, Oxford OX2 0EL
25 John Street, London WC1N 2BL
23 Ainslie Place, Edinburgh EH3 6AJ
350 Main Street, Malden, MA 02148 5018,
  USA
54 University Street, Carlton, Victoria 3053,
  Australia
10, rue Casimir Delavigne, 75006 Paris, France

Other Editorial Offices:

Blackwell Wissenschafts-Verlag GmbH
Kurfürstendamm 57
10707 Berlin, Germany

Blackwell Science KK
MG Kodenmacho Building
7–10 Kodenmacho Nihombashi
Chuo-ku, Tokyo 104, Japan

First published 2000

Produced by and typeset in Times
by Gray Publishing, Tunbridge Wells, Kent
Printed and bound in Great Britain by
MPG Books Ltd, Bodmin, Cornwall

The Blackwell Science logo is a
trade mark of Blackwell Science Ltd,
registered at the United Kingdon
Trade Marks Registry

For further information on Blackwell Science,
visit our website:
www.blackwell-science.com

DISTRIBUTORS

Marston Book Services Ltd
PO Box 269
Abingdon
Oxon OX14 4YN
(*Orders:* Tel: 01235 465500
          Fax: 01235 465555)

USA
  Blackwell Science, Inc.
  Commerce Place
  350 Main Street
  Malden, MA 02148 5018
  (*Orders:* Tel: 800 759 6102
                    781 388 8250
            Fax: 781 388 8255)

Canada
  Login Brothers Book Company
  324 Saulteaux Crescent
  Winnipeg, Manitoba R3J 3T2
  (*Orders:* Tel: 204 837–2987
            Fax: 204 837–3116)

Australia
  Blackwell Science Pty Ltd
  54 University Street
  Carlton, Victoria 3053
  (*Orders:* Tel: 03 9347 0300
            Fax: 03 9347 5001)

A catalogue record for this title is available
from the British Library

ISBN 0–632–05355–0

Library of Congress Cataloging-in-Publication
Data
Kaiser, M.J.
   The effects of fishing on non-target species
and habitats: biological conservation and
socio-economic issues/M.J. Kaiser,
S.J. de Groot
       p.   cm.
   Includes bibliographical references.
   ISBN 0–632–05355–0 (hardback)
   1. Fisheries—Environmental aspects.
I. Groot, S. J. de.   II. Title
QH545.F53K35   1999
577.6′27– dc21   99-32458
                                                  CIP

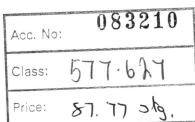

Contents

Introduction

The impact of human harvesting of marine organisms is manifested throughout the world and is most clearly seen in the resultant decreases in the population sizes of target species. Not surprisingly, this has led to concerted attempts to manage or sustain populations of commercially valuable species on which many local economies depend. Until recently, this preoccupation has deflected the majority of fisheries-related research towards improving management of the stocks of target species. The wider effects of fishing activities on marine ecosystems were eluded to prior to this century (de Groot, 1984), but it wasn't until 1955 that Michael Graham undertook the first scientific assessment of the likely effects of fishing on non-target species. Graham's paper summarily dismissed the possibility that fishing could adversely affect seabed communities and probably precluded any further research in this area for a further 15 years (Graham, 1955). Preliminary studies of the effects of bottom trawls on the seabed and its fauna were published in a number of International Council for the Exploration of the Sea (ICES) papers in the early 1970s (see de Groot, 1984, for review). However, these studies did not progress much further beyond quantifying by-catch organisms in trawl catches and the depth to which trawl gears penetrated the seabed. Once again, this issue did not attract sufficient attention to maintain its research momentum. Finally, in the 1980s, reports of the disappearance of once common structural biogenic reefs began to cause concern that fishing might have caused widespread alteration of the seabed and the wider marine ecosystem (Riesen & Reise, 1982). Towards the end of the 1980s, it became clear that large-scale changes had occurred in the distribution and abundance of scavenging seabirds in the North Sea. Furthermore, it was suggested that discarding activities associated with trawling were the source of these population changes (Hudson & Furness, 1988; Furness et al., 1988, 1992). Since these findings were reported, research interest in the ecosystem effects of fishing activities has increased substantially world-wide and now encompasses effects on benthic biota, non-target fish species, marine reptiles, marine mammals and seabirds (Jennings & Kaiser, 1998).

Europe has been a focus of research initiatives that have addressed the ecosystem effects of fishing on the marine environment for the last two decades. Hence, we considered it appropriate and timely to hold a workshop that gathered representatives of European research groups working on different aspects of this topic to update and review progress and future research needs. The structure of our meeting was biologically dominated. This was intentional, as the biology of the ecological processes that result from fishing form the baseline of facts against which additional concerns, such as methods to reduce fishing impacts, socio-economic and legalistic issues, can be assessed. In addition, while those from the social sciences have had to dip into the world of biology to understand the rationale behind conservation issues and their implications, biologists often are unaware of the feasibility or ramifications of the recommendations that emanate from their research.

The book is divided into distinct themes. The first theme describes the distribution of bottom-fishing disturbance and the direct physical process of sediment modification and the possible biological consequences of increased sediment resuspension (Jennings *et al.*; Fonteyne; Ardizzonne *et al.*). The second theme focuses on the direct effects of fishing on benthic fauna and a variety of habitats that vary in their stability and sensitivity to disturbance (Bergman & van Santbrink; Ball *et al.*; Bradshaw *et al.*; Hall-Spencer & Moore). The next theme considers the ecological consequences of redirecting energy in the marine ecosystem either by discarding material at the surface of the sea or through the animals and plants killed directly on the seabed (Demestre *et al.*; Fonds & Groenewold; Ramsay *et al.*; Camphuysen & Garthe). The long-term consequences of the quantifiable short-term effects of fishing (themes 1, 2 and 3) are investigated in the fourth theme (Craeymeersch *et al.*; Frid & Clark; Greenstreet & Rogers; Jennings & Reynolds). Then, consideration is given to important conservation issues (theme 5) such as those species vulnerable to the effects of fishing outwith the target species and the importance of adopting an ecosystem approach to management of the marine environment. Methods for reducing the impacts of fishing for target, non-target species and habitats are also investigated (van Marlen; Tregenza; Tasker *et al.*; Lindeboom; Horwood). Finally, the economic incentives for improving fishing practice, the priorities that society attaches to conservation of the marine habitat and the reality of trying to implement ecosystem approaches to fisheries management are explored (Pascoe; McGlade & Metuzals; Jones; Symes).

These themes are brought together in a final concluding chapter, which distils and integrates the major points that emerge from each study. Throughout the text, the reader will find reports of similar research findings and conclusions that are expressed by authors from a wide range of institutes with different professional backgrounds. This is a source of great encouragement, as it gives one confidence in the reported results and trends and that the opinions expressed are not merely emotive outbursts, rather that they are founded on well-conducted science that addresses an increasingly important ecological issue.

References

Graham, M. (1955) Effect of trawling on animals of the seabed. *Deep Sea Research*, **3** (Suppl.), 1–16.
De Groot, S.J. (1984) The impact of bottom trawling on the benthic fauna of the North Sea. *Ocean Management*, **10**, 21–36.
Furness, R.W., Hudson, A.V. & Ensor, K. (1988) Interactions between scavenging seabirds and commercial fisheries around the British Isles. In: *Seabirds and other Marine Vertebrates: Competition, Predation and other Interactions* (ed. J. Burger), pp. 240–68. Columbia University Press, New York.

Furness, R.W., K. Ensor, K. & Hudson, A. (1992) The use of fishery waste by gull populations around the British Isles. *Ardea*, **80,** 105–13.

Furness, R. W., A. V. Hudson & Ensor, K. (1988) Interactions between scavenging seabirds and commercial fisheries around the British Isles, In: *Seabirds and other Marine Vertebrates: Competition, Predation and other Interactions* (ed. J. Burger), pp 240–68. Columbia University Press, New York.

Hudson, A. V. & Furness, R.W. (1988) Utilization of discarded fish by scavenging seabirds behind whitefish trawlers in Shetland. *Journal of Zoology, London.* **215,** 151–66.

Jennings, S. & Kaiser, M.J. (1998) The effects of fishing on marine ecosystems. *Advances in Marine Biology*, **34,** 201–352.

Riesen, W. & Reise, K. (1982) Macrobenthos of the subtidal Wadden Sea: revisited after 55 years. *Helgolander Meeresuntersuchungen*, **35,** 409–23.

Michel J. Kaiser
Bas de Groot

Acknowledgements

This workshop was partly funded by the European Commission, DGXIV FAIR programme of accompanying and support measures contract number MAC/06/98, and by the Ministry of Agriculture, Fisheries and Food, UK, under contract MF0716. Myti Mussel Ltd and Deep Dock Ltd kindly provided seafood for the conference dinner. Local organisation was greatly assisted by the help of Ms Fiona Spence, Ms Sam Vize and the staff of The Bulkeley Hotel, Beaumaris. We gratefully acknowledge the help of the anonymous referees whose suggestions have improved many of the manuscripts presented within this volume. We extend our thanks to Richard Miles of Blackwell Science for his help in putting this book into publication. Finally, MJK would like to express his thanks to Diane for her support and patience throughout this project when she was busy producing our daughter Holly.

Contributors

G D Ardizzone	Dipartimento di Scienze dell'Uomo e dell'Ambiente, Università di Pisa, Via A. Volta 6, 56100 Pisa, Italy.
B Ball	Martin Ryan Institute for Marine Science, National University of Ireland, Galway, Ireland. E-mail: brendan.ball@nuigalway.ie
A Belluscio	Dipartimento di Scienze dell'Uomo e dell'Ambiente, Università di Pisa, Via A. Volta 6, 56100 Pisa, Italy.
M J N Bergman	Netherlands Institute for Sea Research, NIOZ, PO Box 59, 1790 AB Den Burg, Texel, The Netherlands. E-mail: magda@nioz.nl
C Bradshaw	Port Erin Marine Laboratory, Port Erin, Isle of Man, IM9 6JA, UK. E-mail: cbrads@liv.ac.uk
A R Brand	Port Erin Marine Laboratory, Port Erin, Isle of Man, IM9 6JA, UK. E-mail: arbrand@liv.ac.uk
J Buijs	Netherlands Institute for Fisheries Research, RIVO-DLO, Haringkade 1, PO Box 68, NL-1970 AB, IJmuiden, The Netherlands.
C J Camphuysen	Netherlands Institute for Sea Research, NIOZ, PO Box 59, 1790 AB Den Burg, Texel, The Netherlands. E-mail: camphuys@nioz.nl
R Clark	Dove Marine Laboratory, University of Newcastle, Cullercoats, North Shields, NE30 4PZ, UK. E-mail: robin.clark@ncl.ac.uk
A J R Cotter	CEFAS, Lowestoft Laboratory, Pakefield Road, Lowestoft, NR33 0HT, UK.
J A Craeymeersch	RIVO-DLO, Postbus 77, 4400 AB Yerseke, The Netherlands. E-mail: johan@rivo.dlo.nl
M Demestre	Consejo Superior de Investigations Cientificas, Instituto de Ciencas del Mar, Passeig Joan de Borbo, 08039 Barcelona, Spain.
J R Ellis	CEFAS, Lowestoft Laboratory, Pakefield Road, Lowestoft, NR33 0HT, UK. E-mail: j.r.ellis@cefas.co.uk
M Fonds	Netherlands Institute for Sea Research, NIOZ, PO Box 59, 1790 AB Den Burg, Texel, The Netherlands.

R Fonteyne
Rijksstation voor Zeevisserij RVA, Ankerstraat 1, 8400 Oostende, Belgium.
E-mail: rfonteyne@unicall.be

C L J Frid
Dove Marine Laboratory, University of Newcastle, Cullercoats, North Shields, NE30 4PZ, UK.
E-mail: c.l.j.frid@ncl.ac.uk

S Garthe
Institut für Meereskunde, Abt. Meereszoologie, Düsternbrooker Weg 20, 24105 Kiel, Germany

S P R Greenstreet
Fisheries Research Institute, Marine Laboratory, PO Box 101, Victoria Road, Torry, Aberdeen, AB9 8DB, UK.

S Groenewold
Netherlands Institute for Sea Research, NIOZ, PO Box 59, 1790 AB Den Burg, Texel, The Netherlands.

S J de Groot
Brederoodseweg 49, 2082 BS, Santpoort-Zuid, The Netherlands.
E-mail: grootres@hetnet.nl

J M Hall-Spencer
University Marine Biological Station, Millport, Isle of Cumbrae, KA28 0EG, UK.

A S Hill
Port Erin Marine Laboratory, Port Erin, Isle of Man, IM9 6JA, UK.

J W Horwood
CEFAS, Lowestoft Laboratory, Pakefield Road, Lowestoft, NR33 0HT, UK.
E-mail: j.w.horwood@cefas.co.uk

S Jennings
CEFAS, Lowestoft Laboratory, Pakefield Road, Lowestoft, NR33 0HT, UK.
E-mail: s.jennings@cefas.co.uk

P J S Jones
Jackson Environment Institute, University College London, 5 Gower Street, London, WC1E 6HA, UK.
E-mail: p.j.jones@ucl.ac.uk

M J Kaiser
School of Ocean Sciences, University of Wales – Bangor, Menai Bridge, Gwynedd, LL59 5EY, UK.
E-mail: m.j.kaiser@bangor.ac.uk

P A Knapman
English Nature, Northminster House, Peterborough, PE1 1UA, UK.
E-mail: paul.knapman@english-nature.org.uk

D Laffoley
English Nature, Northminster House, Peterborough, PE1 1UA, UK.
E-mail: dan.laffoley@english-nature.org.uk

H J Lindeboom
Netherlands Institute for Sea Research, NIOZ, PO Box 59, 1790 AB Den Burg, Texel, The Netherlands.
E-mail: hanl@nioz.ac.nl

R van Marlen
Netherlands Institute for Fisheries Research, RIVO-DLO, Haringkade 1, PO Box 68, NL-1970 AB, IJmuiden, The Netherlands.
E-mail: b.vanmarlen@rivo.dlo.nl

J M McGlade
Centre for Coastal and Marine Science, Plymouth Marine Laboratory, Prospect Place, West Hoe, Plymouth, PL1 3DH, UK.
E-mail: jmgl@ccms.ac.uk

K I Metuzals
Centre for Coastal and Marine Science, Plymouth Marine Laboratory, Prospect Place, West Hoe, Plymouth, PL1 3DH, UK.

P G Moore
University Marine Biological Station, Millport, Isle of Cumbrae, KA28 0EG, UK.
E-mail: pmoore@udcf.gla.ac.uk

B Munday
Martin Ryan Institute for Marine Science, National University of Galway, Galway, Ireland.
E-mail: b.w.munday@ucg.co

S Pascoe
University of Portsmouth, Department of Economics, Centre for the Economics and Management of Aquatic Resources, Locksway Road, Portsmouth, PO4 8JF, UK.
E-mail: sean.pascoe@port.ac.uk

G J Piet
Netherlands Institute for Fisheries Research, RIVO-DLO, Haringkade 1, PO Box 68, NL-1970 AB, IJmuiden, The Netherlands.

K Ramsay
School of Ocean Sciences, University of Wales – Bangor, Menai Bridge, Gwynedd, LL59 5EY, UK.
E-mail: k.ramsay@bangor.ac.uk

A D Rijnsdorp
Netherlands Institute for Fisheries Research, RIVO-DLO, Haringkade 1, PO Box 68, NL-1970 AB, IJmuiden, The Netherlands.

S I Rogers
CEFAS Lowestoft Laboratory, Pakefield Road, Lowestoft, NR33 0HT, UK.
E-mail: s.i.rogers@cefas.co.uk

J D Reynolds
School of Biological Sciences, University of East Anglia, Norwich, NR4 7TJ, UK.

P Sanchez
Consejo Superior de Investigations Cientificas, Instituto de Ciencas del Mar, Passeig Joan de Borbo, 08039 Barcelona, Spain.

J W van Santbrink
Netherlands Institute for Sea Research, NIOZ, PO Box 59, 1790 AB Den Burg, Texel, The Netherlands

A Somaschini
Dipartimento di Scienze dell'Uomo e dell'Ambiente, Università di Pisa, Via A. Volta 6, 56100 Pisa, Italy.

D Symes
University of Hull, School of Geography and Earth Sciences, Cottingham Road, Kingston upon Hull, HU6 7RX, UK.

M L Tasker
Joint Nature Conservation Committee, Dunnet House, 7 Thistle Place, Aberdeen, AB10 1UZ, UK.

N J C Tregenza   Cornwall Wildlife Trust, Beach Cottage, Long Rock,
Penzance, Cornwall, TR20 8JE, UK.
E-mail: nick@chelonia.demon.co.uk

P Tucci   Dipartimento di Scienze dell'Uomo e dell'Ambiente,
Università di Pisa, Via A. Volta 6, 56100 Pisa, Italy.

I Tuck   Fisheries Research Institute, Marine Laboratory, PO
Box 101, Victoria Road, Torry, Aberdeen, AB9 8DB,
UK.
E-mail: tucki@marlab.aberdeen.ac.uk

L O Veale   Port Erin Marine Laboratory, Port Erin, Isle of Man,
IM9 6JA, UK.

K J Warr   CEFAS, Lowestoft Laboratory, Pakefield Road,
Lowestoft, NR33 0HT, UK.

Part 1
Distribution of fishing effort and physical interaction with the seabed

(A) Dutch 12-m wide commercial beam trawl rigged with tickler chains for fishing on a sandy seabed for sole and plaice. (B) Gang of Newhaven spring-toothed scallop dredges for fishing on stony grounds.

Chapter 1
Spatial and temporal patterns in North Sea fishing effort

S. JENNINGS[1,2], K.J. WARR[1], S.P.R. GREENSTREET[3] and A.J.R. COTTER[1]

[1]*The Centre for Environment, Fisheries and Aquaculture Science, Lowestoft Laboratory, Pakefield Road, Lowestoft, NR33 0HT, UK*
[2]*School of Biological Sciences, University of East Anglia, Norwich, NR4 7TJ, UK*
[3]*Fisheries Research Services, Marine Laboratory, Aberdeen, AB9 8DB, UK*

Summary

1. An understanding of spatial and temporal patterns in fishing effort is a prerequisite for investigating the effects of fishing in the marine environment. We describe patterns in demersal (bottom) fishing effort by trawlers and seine netters in the North Sea.
2. Total international demersal fishing effort (vessels largely > 12 m) was relatively constant in the period 1990–1995. In 1994, total effort was 2.5 million h year^{-1} of which 51% and 44% was due to beam and otter trawling, respectively.
3. Spatial analyses indicate that the proportion of beam trawling effort increases from north to south, while the proportion of otter trawling and seine net effort increases from south to north.
4. Plots of annual fishing effort by International Council for the Exploration of the Sea statistical rectangle (211 boxes of 0.5° latitude by 1° longitude) indicate that the majority of fishing effort is concentrated in relatively few rectangles.
5. Mean annual total trawling effort (1990–1995) exceeded 40 000 h in 4% of rectangles. Effort was less than 2000 h in 29% of rectangles and 10 000 h in 66% of rectangles.
6. Most seine netting is conducted by Scottish vessels in the northern and central North Sea. Seine netting effort has declined between 1991 and 1994 to 125 000 h year^{-1}.
7. The validity of reported trawling effort in the southern and central North Sea was tested by comparing official effort reports with calculated trawler sightings per unit effort (SPUE) from fisheries enforcement boats and overflights.
8. Reported effort provides a good general picture of the areas where fishing effort is concentrated. However, the large-scale resolution of reported effort (by rectangle) provides a poor indication of the specificity of fishing grounds. SPUE data suggest that parts of rectangles are heavily fished while others are unfished.
9. If fishing effort data are needed to examine the wider ecological effects of fishing, the resolution at which data are collected will have to be refined. Satellite tracking of vessels may help to achieve this.

Keywords: trawling effort, overflight data, spatial resolution.

Introduction

Most bottom fishing in the North Sea is conducted using beam trawls, otter trawls and seine nets (Anon., 1997). These gears have direct and indirect impacts on benthic fauna and habitats (Gislason, 1994; Anon., 1995; Jennings & Kaiser, 1998; Lindeboom & de Groot, 1998), so knowledge of the spatial and temporal distribution of fishing effort is fundamental to many ecological studies. Eight countries (Belgium, Denmark, England, France, Germany, Netherlands, Norway, Scotland) border the North Sea and have significant fishing interests. Assessments of fishing effort have often been based on national data and cannot provide a complete picture of the spatial and temporal distribution of effort. In this chapter, we compile international fishing effort data for beam trawling, otter trawling and seine netting in the North Sea.

Trawls and seine nets have important effects on the North Sea ecosystem because they remove a large proportion of the biomass of target and by-catch species and because they have direct impacts on the substratum and associated biota. Beam and otter trawls cause most disturbance because they are in direct contact with the seabed and because water movements induced by the gear lead to turbulent resuspension of surface sediments. The magnitude of impact is determined by the speed of towing, physical dimensions and weight of the gear, type of substratum and strength of currents or tides in the area fished. The effects may persist for a few hours in shallow areas with strong tides or for decades in deeper areas subject to less natural disturbance (Hall, 1994; Lindeboom & de Groot, 1998; Fonteyne, this volume, Chapter 2).

Methods

Data sources

The collation of fishing effort data is described in detail by Jennings *et al.* (1999). In this chapter, we report data for the 6-year period from 1990 to 1995. For the purposes of this compilation, otter trawlers were defined as any vessels that towed bottom-fishing nets held open by trawl doors. This simplistic definition was required to allow the aggregation of data for different countries and to ensure that we did not mask general trends in the dataset by introducing too many variables. Thus, otter trawling included gears such as *Nephrops* and prawn trawls. The smallest vessels for which data were included in this analysis were 12 m (Denmark), 12 m (England and Wales), 13 m (beam trawler, Germany) and 15 m (otter trawler, Germany), 50 hp (Holland), 21 m (Norway) and 17 m (Scotland). Polet *et al.* (1994), Jennings *et al.* (1999) and Lindeboom & de Groot (1998) provide more information on vessels in the North Sea and the regulations for reporting effort data. Effort was recorded as hours fishing in each International Council for the Exploration of the Sea (ICES) statistical rectangle (boxes of 0.5° latitude by 1° longitude). Fishing effort data were not available from France and Belgium. Both countries have significant fishing interests in the southern North Sea (ICES Area IVc). Demersal seine net effort data are presented for Scotland, the only country to use this gear extensively in the North Sea (Greenstreet *et al.*, 1999).

Effort data are collected on a coarse scale (ICES rectangle) and may be inaccurate as a result of misreporting. In order to investigate the finer scale distribution of fishing effort, and to see if patterns in official effort data correlate with those from an alternative source, we used data collected by British Fishery Protection flights, the Royal Navy and Fishery Protection vessels during patrols in the southern and central North Sea. They record the location, type and identification number of all vessels sighted in this area and whether they are fishing or steaming. We have used these data to calculate sightings per unit of effort (SPUE) as an index of fishing intensity by assigning sighted vessels of all nationalities to ICES subrectangles (four areas of 0.25° latitude by 0.5° longitude within each rectangle) and dividing by surveillance effort (number of visits by ship or aircraft) in those subrectangles. To compare reported effort with SPUE, we calculated mean SPUE by rectangle (mean SPUE for the four subrectangles) and tested the significance of any correlation with reported effort. We have also plotted actual positions of all sighted vessels in 1997, by gear type, to assess the spatial distribution of fishing effort.

Results

Patterns in fishing effort

Total demersal fishing effort by Danish, Dutch, English, German, Norwegian, Scottish and Welsh vessels was relatively constant in the period 1990–1995 (Table 1.1): an increase in beam trawling, particularly in IVc, being balanced by reductions in the amount of otter trawling and seine netting. Otter trawl effort decreased in all areas except the northern North Sea (ICES Area IVa). Seine effort was highest in 1991 and least in 1994 (Table 1.1). Between 1990 and 1994, seine net effort decreased from 10% to 5% of total bottom fishing effort. No seine effort data were available for 1995. Otter trawling accounted for a greater proportion of total effort than beam trawling and seine netting in area IVa, where beam trawls were rarely used, but accounted for less than 50% of total effort in IVc and the central North Sea (Area

Table 1.1   Fishing effort in the North Sea (1000 h): trawl (otter and beam) data for Danish, Dutch, English, German, Norwegian, Scottish and Welsh vessels; seine net data for UK vessels landing in Scotland

Area	IVa			IVb			IVc			Total			
Year	Beam	Otter	Seine	Beam	Otter	Seine	Beam	Otter	Seine	Beam	Otter	Seine	All gears
1990	6.84	407.10	177.29	608.56	604.63	51.44	346.02	89.75	0	961.43	1101.48	228.73	2291.64
1991	10.77	469.59	239.67	687.06	624.49	82.00	355.41	82.66	0	1053.24	1176.74	321.67	2551.65
1992	16.66	477.49	226.33	754.89	550.13	61.53	366.93	74.06	0	1138.48	1101.70	287.86	2528.04
1993	17.47	448.08	201.65	768.29	556.35	50.05	411.84	74.75	0	1197.62	1079.17	251.70	2528.49
1994	16.80	465.53	102.95	823.85	541.97	22.54	427.06	69.33	0	1267.72	1076.83	125.49	2470.04
1995	20.38	466.30	no data	733.56	487.49	no data	477.19	72.59	no data	1231.13	1026.38	—	—

IVb) (Table 1.1). Total fishing effort in 1994 was approximately 2470 million h, of which approximately 51% could be attributed to beam trawling, 44% to otter trawling and 5% to seine netting. It is notable that the majority of trawling effort in the North Sea is confined to a small proportion of the total area. Thus, mean annual total trawling effort (1990–1995) was less than 2000 h in 29% of rectangles and less than 10 000 h in 66% of rectangles. Total effort exceeded 40 000 h in 4% of rectangles (Jennings *et al.*, 1999).

The spatial distribution of annual otter trawling effort by ICES statistical rectangle (Fig. 1.1) indicates that the grounds subject to the greatest fishing intensity have not changed in recent years. Most effort is confined to the north-east coast of the UK and east of the Shetland Islands. Parts of the Norwegian Deep and the central (offshore) North Sea are relatively lightly fished. Beam trawl effort (Fig. 1.2) is largely confined to the southern and eastern North Sea. Highest levels of fishing effort are recorded inshore off the Dutch coast. There was little change in the gross spatial distribution of effort from 1990 to 1995, although there are marked annual fluctuations in the intensity of fishing on the east coast of the North Sea. The spatial distribution of seine net effort indicates that most fishing takes place in the central

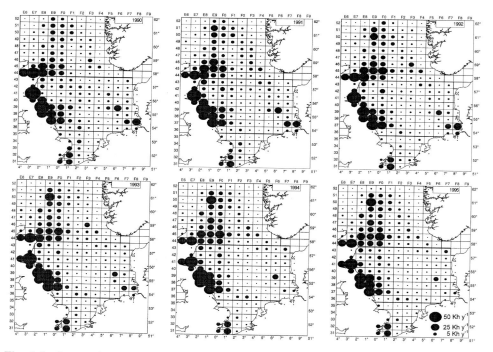

Fig. 1.1 International otter trawl effort in the North Sea 1990–1995. Fine lines denote borders of ICES rectangles and thicker lines indicate borders between ICES areas IVa (northern North Sea), IVb (central North Sea) and IVc (southern North Sea). From Jennings *et al.* (1999).

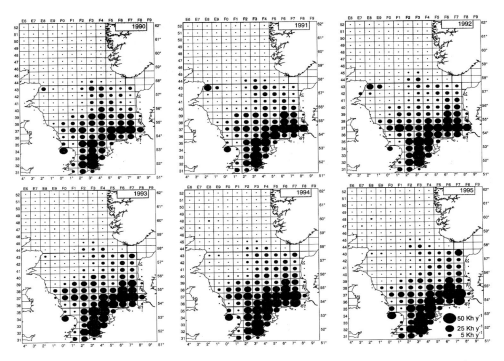

Fig. 1.2 International beam trawl effort in the North Sea 1990–1995. Fine lines denote borders of ICES rectangles and thicker lines indicate borders between ICES areas IVa (northern North Sea), IVb (central North Sea) and IVc (southern North Sea). From Jennings *et al.* (1999).

and northern North Sea (Fig. 1.3). The area to the east of the Shetland Islands was the most heavily fished.

In the southern and central western North Sea, patterns of effort shown by the official data (Figs 1.1 and 1.2) broadly correspond to those shown by the British Fishery Protection and Royal Navy SPUE data for both otter (Fig. 1.4) and beam (Fig. 1.5) trawling. The frequency of visits to subrectangles in Fig. 1.6 suggests that SPUE data are most likely to be unreliable in subrectangles adjacent to the coast and to the eastern border of the area patrolled, because few visits were made by patrols.

SPUE data also provide a useful indication of the location of favoured fishing grounds. For example, plots of all sightings during 1997 show the distribution of favoured beam trawling grounds in the southern and central North Sea and otter trawling grounds off the north-east coast of England (Fig. 1.7).

Relationships between reported effort and mean SPUE by rectangle indicate that the correlations between effort and SPUE are significant for otter trawling in all years from 1990 to 1995 (Fig. 1.8, Table 1.2). The correlations between beam trawling effort and SPUE in the same years were not significant on three occasions, and, in some rectangles, there were clear differences between reported effort and SPUE (Fig. 1.9, Table 1.2). This may be due to the inclusion of Belgian and French

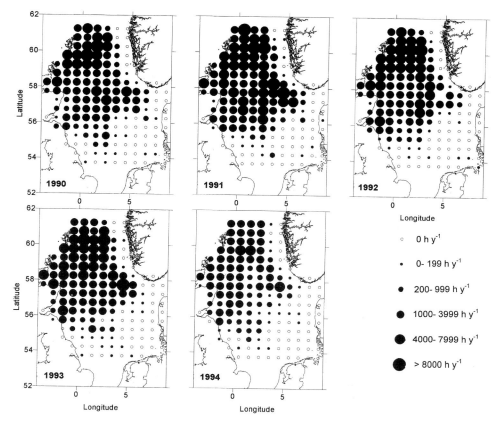

Fig. 1.3 Demersal seine effort in the North Sea 1990–1994 for UK vessels landing in Scotland. Data presented by rectangle.

Table 1.2 Correlation between reported international beam or otter trawling effort and SPUE by fisheries patrol vessels and aircraft

Gear	Beam			Otter		
	r	n	p	r	n	p
1990	0.64	42	<0.001	0.68	42	<0.001
1991	0.39	38	<0.01	0.67	39	<0.001
1992	0.76	37	<0.001	0.70	39	<0.001
1993	0.20	40	>0.1	0.79	40	<0.001
1994	0.25	40	>0.1	0.91	38	<0.001
1995	0.15	38	>0.1	0.89	40	<0.001

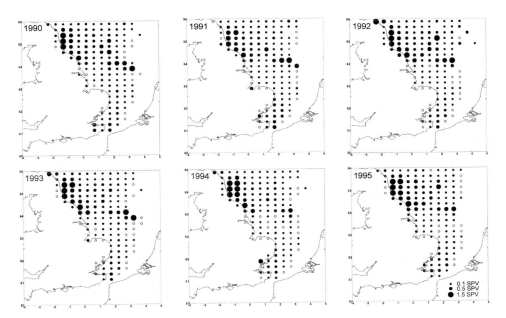

Fig. 1.4 SPUE of otter trawlers by fishery protection flights and vessels in the southern and central North Sea. Data are presented as mean sightings per visit (SPV) by ICES subrectangle. Open circles indicate that the subrectangle was visited but that no vessels were sighted.

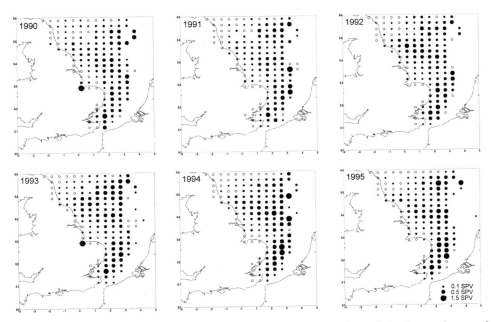

Fig. 1.5 SPUE of beam trawlers by fishery protection flights and vessels in the southern and central North Sea. Data are presented as mean SPV by ICES subrectangle. Open circles indicate that the subrectangle was visited but that no vessels were sighted.

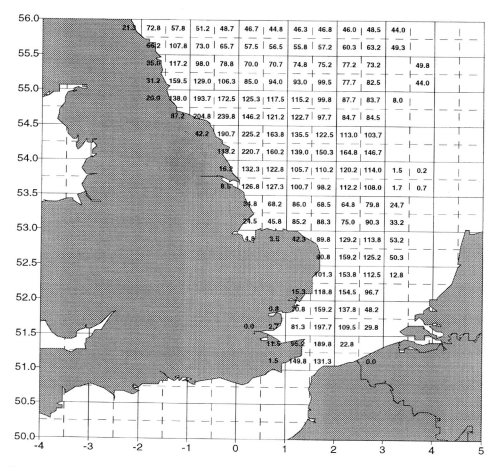

Fig. 1.6   Surveillance effort by fishery protection flights or vessels in the southern and central North Sea. Values indicate the mean number of visits per year to each subrectangle in the period 1990–1995.

beam trawlers in the SPUE statistics when they were not included in international effort data.

Discussion

No effort data were available for Belgian or French trawlers, but previous studies allow us to determine how these data might affect our estimates of the spatial distribution of effort and total effort. Lindeboom & de Groot (1998) calculated total fishing effort by Belgian, Dutch, English, German and Scottish beam and otter trawlers in the North Sea during 1994. Their compilation suggests that the Belgian beam fleet, for which we had no data, fish predominantly in the eastern part of IVc

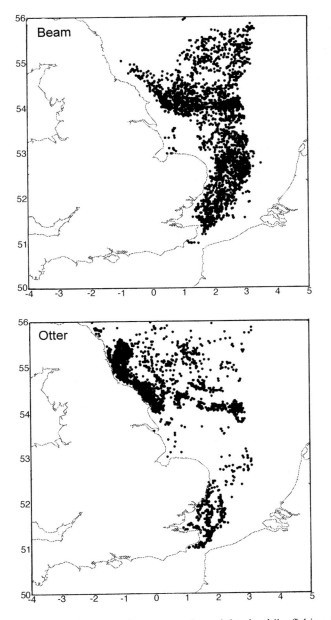

Fig. 1.7 Locations of all beam and otter trawlers sighted while fishing by UK fishery protection flights and vessels during 1997.

and that total effort in this area may be 20% higher than our estimate. The Belgian beam trawl fleet consists of fewer and smaller vessels than the Dutch fleet. In the early 1990s, the Belgian fleet consisted of around 170 vessels, and less than 20% of these were over 1100 hp. The Dutch fleet, conversely, averaged 550 vessels, and

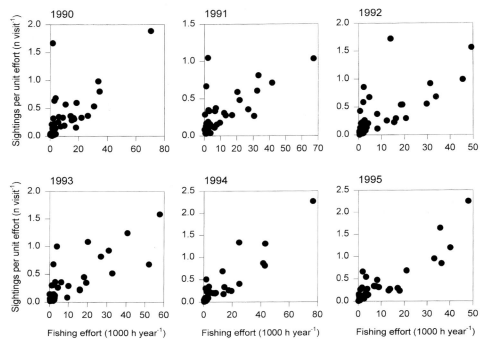

Fig. 1.8 Annual relationships between reported otter trawling effort and SPUE of otter trawlers by fishery protection vessels and flights. Each data point represents one rectangle.

approximately 80% of these were larger than 1100 hp (Lindeboom & de Groot, 1998). Otter trawling by Belgian vessels is limited and would have little effect on the patterns of total effort that we report.

French data were also not available. French vessels predominantly fish outside the North Sea, and significant effort in the North Sea is confined to Area IVc. Catch data show that the French fleet landed 26% of plaice, 15% of sole and 30% of the total international cod catches from IVc (Anon., 1990). This suggests that effort attributable to French vessels in IVc may increase our measures of international beam and otter trawl effort by 20% and 30%, respectively.

The present compilation of international effort is based on hours fished and takes no account of developments in gear design, the size and power of vessels or the use of different areas by different fleets (either by choice or because they are regulated). Subsequent compilations of international effort should account for the relative fishing power of different vessels. We recognise this is a challenging undertaking but, with improvements to national effort databases and better records of the gears used in North Sea fisheries, it should be possible in the future.

Reported international effort data provide a useful impression of gross spatial and temporal patterns in effort within the North Sea. However, their use in studies of fishing effects is limited, because they do not allow scientists to ascertain whether a given sample is taken from an area that has or has not been fished. This is because

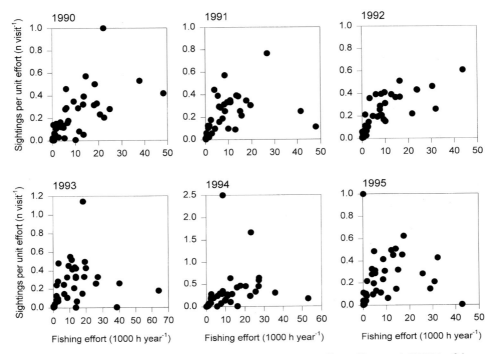

Fig. 1.9 Annual relationships between reported beam trawling effort and SPUE of beam trawlers by fishery protection vessels and flights. Each data point represents one rectangle.

effort data are collected on scales of tens of kilometres, while benthic samples are taken on scales of metres and tens of metres. As such, reported effort data are best used for larger scale studies of fishing effects on fish communities.

Reported effort and SPUE provide a similar picture of the spatial and temporal distribution of fishing effort in the southern and central North Sea. Since the resolution of the overflight data is better, these data may be useful when attempting to link spatial or temporal variation in benthic fish and invertebrate communities to fishing effort. However, given that few subrectangles are visited by patrol aircraft at mean intervals of less than 2 days (Fig. 1.6), SPUE data will exclude a proportion of fishing activity. In order to provide an accurate picture of spatial and temporal variations in fishing effort in the southern North Sea, it will be necessary to track vessels continuously during fishing operations. This can be achieved by continuous satellite monitoring (Rijnsdorp *et al.*, 1991), which will be implemented for most large fishing vessels in the North Sea from 1 January 2000.

Acknowledgements

We thank J. Alvsvåg, M.J. Brown, S. Ehrich, B. Gloerfeldt-Tarp, A. Jarre-Teichmann, N. Mergardt, B.D. Rackham, A.D. Rijnsdorp and O. Smedstad for helping to provide and analyse data presented in this chapter, and the Sea Fisheries Inspectorate (UK) for permission to use the sightings data. This project was funded by the EC under the FAIR shared cost contract 'Monitoring biodiversity in the North Sea using groundfish surveys'.

References

Anon. (1990) Data for 1987. *International Council for the Exploration of the Sea, Bulletin Statistique des Pêches Maritimes*, **72**.

Anon. (1995) *Report of the study group on ecosystem effects of fishing activities. ICES Co-operative Research Report*, 200, 120 pp.

Anon. (1997) Report of the working group on the assessment of demersal stocks in the North Sea and Skagerrak. *International Council for the Exploration of the Sea*, C.M. 1997/Assess 6.

Gislason, H. (1994) Ecosystem effects of fishing activities in the North Sea. *Marine Pollution Bulletin*, **29**, 520–7.

Greenstreet, S.P.R., Spence, F.E., Shanks, A.M. & McMillan, J.A. (1999) Fishing effects in northeast Atlantic shelf seas: patterns in fishing effort, diversity and community structure. II. Trends in fishing effort in the North Sea by U.K. registered vessels landing in Scotland. *Fisheries Research*, **40**, 107–24.

Hall, S.J. (1994) Physical disturbance and marine benthic communities: life in unconsolidated sediments. *Oceanography and Marine Biology Annual Review*, **32**, 179–239.

Jennings, S. & Kaiser, M.J. (1998) The effects of fishing on marine ecosystems. *Advances in Marine Biology*, **34**, 201–32.

Jennings, S., Alvsvåg, J., Cotter, A.J., Ehrich, S., Greenstreet, S. P. R., Jarre-Teichmann, A., Mergardt, N., Rijnsdorp, A.D. & Smedstad, O. (1999) Fishing effects in northeast Atlantic shelf seas: patterns in fishing effort, diversity and community structure. III. International fishing effort in the North Sea: an analysis of temporal and spatial trends. *Fisheries Research*, **40**, 125–34.

Lindeboom, H.J. & de Groot, S.J. (eds) (1998) *The effects of different types of fisheries on the North Sea and Irish Sea benthic ecosystems. NIOZ Report* 1998-1/*RIVO-DLO Report* C003/98, 404 pp. Netherlands Institute for Sea Research, Den Burg, Texel, The Netherlands.

Polet, H., Blom, W. & Thiele, W. (1994) An inventory of vessels and gear types engaged in the Belgian, Dutch and German bottom trawling. In: *Environmental impact of bottom gears on benthic fauna in relation to natural resources management and protection of the North Sea* (eds S.J. de Groot & H.J. Lindeboom), pp. 7–20. *NIOZ Report* 1994-11/*RIVO-DLO Report* C026/94. Netherlands Institute for Sea Research, Den Burg, Texel, The Netherlands.

Rijnsdorp, A.D., Groot, P. & van Beek, F.A. (1991) The micro distribution of beam trawl effort in the southern North Sea. *International Council for the Exploration of the Sea*, C.M. 1991/G:49.

Chapter 2
Physical impact of beam trawls on seabed sediments

R. FONTEYNE

Agricultural Research Centre – Ghent, Sea Fisheries Department, Ankerstraat 1, B-8400 Oostende, Belgium

Summary

1. The first data on the physical impact of beam trawling on the seabed were obtained during the 1970s, and consequently relate to rather light gears compared with those currently used. This paper deals with the impact on the seabed of modern, heavy beam trawls. It concentrates on the pressure exerted by the gears and on the changes to the seabed topography and sediment characteristics. A 4-m beam trawl equipped with a chain matrix was used in all experimental work. This gear is typical for 'Eurocutters' operating in coastal areas.
2. An instrumented trawl head was developed to measure directly the pressure of the trawl heads on the seabed. This device also allowed a description of the mechanical behaviour of the gear in contact with the seabed. The effect of gear and vessel size on gear pressure was modelled. The changes to the seabed topography were observed by side-scan sonar, and changes in sediment characteristics were measured using the RoxAnn seabed classification system.
3. The pressure exerted on the seabed by beam trawls is strongly related to the towing speed. As the speed increases, the lift of the gear increases and the resultant pressure force decreases. At higher speeds, the weight of the gear is fully compensated, and the trawl lifts off the bottom.
4. For the 4-m beam trawl studied, the pressure exerted by the trawl heads varied from 17 to 32 hPa at towing speeds of 4–6 kn. Bottom contact was lost at a towing speed of 7 kn.
5. Although larger vessels use heavier gears, this is compensated for by larger sole-plate dimensions and higher towing speeds, hence the pressure exerted is roughly equal to the 4-m beam trawl.
6. Beam trawls leave detectable marks on the seabed. The length of time that the beam trawl marks remain visible depends on the upper sediment layer. On a seabed consisting of mainly coarse sand, the tracks remained visible for up to 52 h, whereas on sediments with mainly finer particles, the tracks had completely faded after 37 h. The penetration depth could not be deduced from the side-scan sonar recordings, since the traces were too weak.
7. The movement of the gear causes the resuspension of the lighter sediment fraction. The changes are most pronounced in areas with finer sand. The suspended particles, however, settle down within a few hours.

Keywords: beam trawl, physical impact, pressure, seabed, sediment, track, penetration depth, sediment suspension.

Introduction

Systematic research on the physical effects of trawling dates mainly from 1970 onwards, when the International Council for the Exploration of the Sea (ICES) requested information on the effects of trawls and dredges on the seabed (ICES, 1971: Council Resolution 1970/5/1). In 1988, the ICES Study Group on the Effects of Bottom Trawling was convened in response to Council Resolution 1987/2/7 (ICES, 1988) to collect information available since 1972, and to report on the developments in bottom trawling gear, existing literature, national research and proposals for coordinated research (Anon., 1988). Most of the experiments carried out in the early 1970s dealt with rather light beam-trawl gears. Almost all beam trawls in the experiments were equipped with tickler chains, and only in one case was the beam trawl equipped with a chain matrix (De Clerck & Hovart, 1972).

Owing to the pressure of the gear on the seabed, parts of the gear penetrate to some extent into the sea bottom. The penetration depth largely depends on the nature of the seabed (Margetts & Bridger, 1971; Bridger, 1972; de Groot, 1972; Anon., 1973). Penetration of demersal gear has been studied in different ways. Direct observations have been made by divers (Bridger, 1970; Margetts & Bridger, 1971) and by using equipment such as underwater television cameras (Margetts & Bridger, 1971; Sydow, 1990) and side-scan sonar (de Groot, 1972; Sydow, 1990). Bridger (1972) implanted markers into the seabed and determined which part had been touched by the tickler chains of a beam trawl passing over them. Other researchers estimated the penetration depth from the benthic species caught by the gear (Houghton *et al.*, 1971; Bergman & Hup, 1992).

Depending on the sediment type, weight of the beam and trawl heads, weight per unit length, number and spacing of tickler chains, towing speed and tidal conditions, a beam trawl will produce a more or less distinct track, estimated to persist for up to 16 h (Margetts & Bridger, 1971; de Groot, 1972; Bergman *et al.*, 1990). The disturbance is most distinct on muddy or soft sandy grounds. On hard sandy grounds, the tracks are difficult to detect, since the path is more smoothed. On very soft grounds, the tracks are ill-defined and are soon erased. The most visible tracks are made by the sole plates. Margetts & Bridger (1971) observed sole-plate marks 80–100 mm deep on muddy sand but only 15 mm deep on sandy ridged ground. The tickler chains are generally not in firm contact with the bottom and exert only a limited pressure on the seabed. Successive layers of sediment are brought into suspension, but settle again after the gear passes. This is not likely to cause a problem in areas where natural sediment movement due to the effect of tidal action and gales is high (Anon., 1973, 1988; de Groot, 1984). Based on measurements made with implanted markers in the seabed, Bridger (1972) concluded that only the surface of the soil would be disturbed by a tickler chain. Even with an array of 15 tickler chains (1478 kg) operating on mud at a low speed of 2.2 kn, the penetration depth did not exceed 30 mm. Houghton *et al.* (1971) judged from the quantities of *Acanthocardia* and *Echinocardium* caught by a 9.5-m beam trawl fitted with 17 tickler chains that the gear disturbed the seabed to a depth of 10–20 cm. From the presence of benthic infauna (*Arctica islandica* and *Echinocardium cordatum*) in the catches of a 12-m beam trawl fishing on a hard sandy bottom, Bergman *et al.* (1990) concluded that the tickler chain penetration may be limited to part of the trawled area, and may extend to a depth of at least 6 cm.

The aim of the present study was to update the knowledge of the physical aspects of the impact of modern beam trawls on benthic habitat. The research was carried out in the period 1992–1997 in the frame of the EU IMPACT projects (Lindeboom & de Groot, 1998). A first series of experiments focused on the measurement of the forces exerted by the gear and its components on the seabed. Instrumentation was developed to measure directly the pressure force from the sole plates, since these cause the most distinct tracks. Gear characteristics and data from gear performance measurements allowed the gear pressure in relation to vessel size to be modelled. The seabed disturbance caused by a 4-m beam trawl equipped with a chain matrix was observed by hydroacoustic techniques. The persistence of trawl marks was estimated from side-scan sonar observations. The seabed classification system RoxAnn proved to be useful for evaluating the effect of fishing gear on the superficial sediments. As the measurements and observations were made on a beam-trawl type used by 'Eurocutters' for fishing in coastal waters, the test sites were chosen on representative fishing grounds off the Belgian and Dutch coast.

Methods

Fishing gear and vessel

The gear used in the experiments was a 4-m beam trawl equipped with a chain matrix (Fig. 2.1). The net is kept open horizontally by means of a steel beam, which is supported at both ends by the trawl heads. Flat steel plates, the sole plates, are welded to the bottom of the trawl head. During fishing, the sole plates are in direct contact with the seabed and generally slightly tilted. To reduce the wear of the sole plates, a heel is welded to the aft end. Beam trawls are normally provided with tickler chains to disturb the flatfish from the seabed. On rough grounds, the tickler chains are replaced by a chain matrix to prevent boulders from being caught by the net. A full description of beam trawls and their operation is given in Polet *et al.* (1994) and in Lindeboom & de Groot (1998). The gear employed in the experiments is used by the 'Eurocutters', beam trawlers with a maximum engine power of 221 kW (300 hp), allowed to fish within the 12-mile limits in certain coastal areas of the European Community with a pair of 4.5-m beam trawls. The target species of these vessels are plaice and sole. The weights of the gear and its components are given in Table 2.1.

The trawl was towed by the RV *Belgica*. The vessel is 50.9 m long, weighs 765 GRT and has an engine power of 1154 kW (1576 hp).

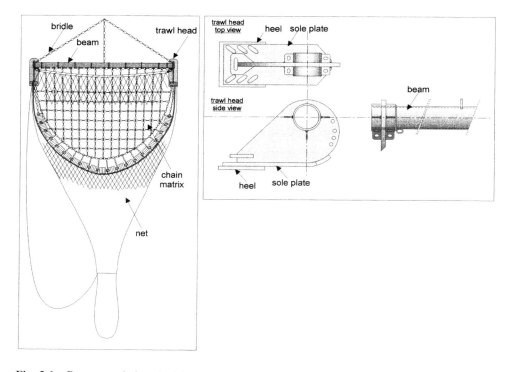

Fig. 2.1   Beam trawl rigged with a chain matrix. (Diagrams by H. Polet).

Table 2.1   Weight of the 4-m beam trawl and its components

Component	Weight in air (N)[a]	Weight under water (N)[a]
Complete trawl	26 925	20 440
Beam + trawl head + bridles	14 573	13 290
Beam + trawl head + bridles + chain mat	23 925	19 850
Net	3000	590

[a]1 N = 0.102 kg.

Pressure measurements

Forces exerted by the sole plates

In order to make direct measurements of the forces exerted by the sole plates on the bottom, an instrumented trawl head was developed and built (Fig. 2.2). The loose sole plate is connected to the trawl head by means of measuring axles 1 and 2. Strain gauges on the axles measure the forces generated in the x- and y-directions. The forces in the y-direction are a measure for the pressure exerted by the sole plate on

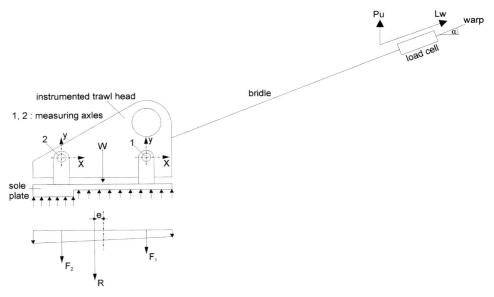

Fig. 2.2   Principle of the instrumented trawl head (see text for details).

the seabed, whereas the forces in the *x*-direction are a measure for the friction between the sole plate and the seabed. By measuring the seabed reactions at two different points, the eccentricity *e* of the resultant *R* of these forces can be determined. The eccentricity results mainly from the difference between the forces F_1 and F_2. This difference depends on the difference in load on each axle as well as on the tilt angle between the sole plate and the seabed profile. The resultant *R* is the total pressure force exerted by the trawl head on the seabed. The pressure is defined as the pressure force per unit of sole-plate area in contact with the seabed.

The measured values of the forces acting on the axles are averaged over a pre-set time interval and stored in an internal RAM memory for later read-out. The time interval between the two recordings can be chosen as 1, 2 or 4 s. In the present experiments, readings were made at 1-s time intervals.

Total gear pressure

The pressure of the complete gear on the seabed can be calculated from the weight *W* of the gear and the upwards pull P_u of the warp (Fig. 2.2). For a heavy load at the lower end of the trawl warp, as is the case in beam trawling, the warp curvature can be neglected and the upwards pull exerted by the warp on the gear is determined by:

$$P_u = L_w \sin\alpha = L_w(D/L) \tag{1}$$

in which P_u is the upwards component of the warp load, L_w is the warp load, α is the warp angle, D is the depth and L is the warp length.

The pressure force P of one trawl head on the seabed is

$$P = (W - P_u)/2 \qquad (2)$$

in which W is the underwater weight of trawl heads + beam + bridles.

The warp load was measured by an underwater load cell inserted between the bridles and the warp. The range of the load cell was 200 kN. The measured values are averaged over 4-s time intervals and stored in an internal RAM memory for later read-out.

For both the pressure exerted by the sole plates and the total gear pressure, two series of measurements were made: (i) with a constant warp length/depth ratio, to assess the influence of towing speed on the pressure exerted by the gear; and (ii) at constant towing speed, to assess the influence of the warp length/depth ratio on the pressure exerted. Towing speed was measured by the vessel's Doppler log and speed through the water by a SCANMAR speed log attached to the bridles.

Relation between total gear pressure and engine power

In order to obtain an insight into the variation of the total gear pressure with vessel and gear characteristics, data from a former series of gear performance measurements and from a detailed survey of vessel and gear characteristics (particularly the weight of the different gear components and actually measured on a number of vessels) were analysed to model the gear pressure against vessel engine power and gear weight.

Seabed disturbance

The seabed disturbance was studied by two hydroacoustic survey techniques: side-scan sonar to detect changes in seabed topography and the seabed classification system RoxAnn to detect changes in sediment characteristics.

Side-scan sonar

The side-scan sonographs recorded during the survey were obtained with a Klein 595 dual-frequency side-scan sonar. This instrument enables registrations to be carried out simultaneously with 100- and 500-kHz transducers. The 100-kHz has the highest range, but the resolution of the 500-kHz transducer is superior.

Prior to the fishing operations, side-scan sonar recordings were made along a number of lines parallel to the reference fishing track to check for possible earlier trawling activities. Superficial seabed sediment samples were taken with a Van Veen grab at regular intervals on the tracks for grain-size analysis. The reference track was fished a number of times and, following this, sonographs of the fished area were obtained at regular time intervals until the tracks could no longer be detected.

RoxAnn

The RoxAnn system enables real-time seabed classification through the analysis of the first and second echo of the ship's echo sounder. Linking of the system to the navigational system of the vessel permits rapid detailed survey work to be achieved on a chart on the computer screen. The integration E1 of the gated tail of the first echo provides information on the roughness of the seabed, whereas the integration E2 of the whole of the second bottom echo is a measure for the hardness. These data are then presented in an E1 vs. E2 scatter plot. Different clusters in the scatter plot correspond to different bottom types. The operator applies a coloured box to each cluster representing a bottom type. The bottom types are defined by ground-truth grab samples. Details of the theoretical operation of this technique are given by Burns *et al.* (1985), Chivers & Burns (1991) and Schlagintweit (1993).

The RoxAnn system was connected to an Atlas DESO 20 echo sounder operating at 210 kHz and to the vessel's DGPS navigation system. As for the side-scan sonar, the frequency used was relatively high, ideal for identifying changes to the surface of the sediment. Lower frequencies would get more penetration into the softer seabed, and may not detect superficial changes so well. Calibration and ground-truthing of the system was accomplished during several earlier cruises. Additionally, Van Veen samples were taken on the test sites during the RoxAnn surveys for later analysis ashore.

The survey methodology was the same as for the side-scan sonar observations. A blank recording was obtained of the reference track and on lines 20, 40 and 50 m to the sides of the track. After fishing, these lines were surveyed again at regular time intervals.

Test sites

Two series of side-scan sonar observations were made. The first series, in April 1992 and March 1993, was made on the Flemish Banks, in the Goote Bank area at locations I, II, IV in Fig. 2.3. The selection of the test areas was based on the occurrence of sandy rather than silty sediments. The pressure measurements were also made in the Goote Bank area in the period 1992–1993.

A second series of surveys was made in 1996. Side-scan sonar observations were made on the Goote Bank and on a site on the Dutch coast, off Scheveningen (Fig. 2.3). RoxAnn surveys were made on the Goote Bank and on Negenvaam, also on the Flemish Banks, and on the Scheveningen test site.

The depth at all test sites varied between 20 and 30 m. Although sand waves occur frequently on the Flemish banks, relatively flat areas were chosen for the reference fishing tracks.

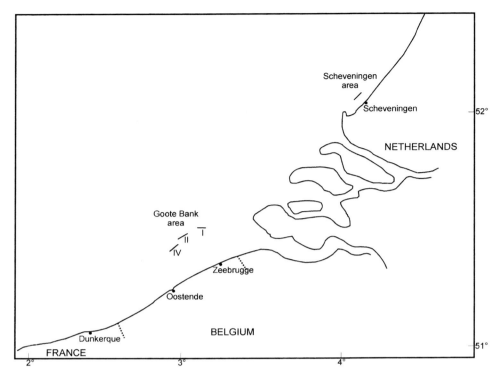

Fig. 2.3   Test sites used for experimental work off the coast of Belgium.

Results

Pressure exerted by a beam trawl

Pressure exerted by the sole plates

Figure 2.4 gives a typical representation in graphical form of pressure measurements made with the instrumented trawl head (Fig. 2.2). It is the result of a series (1 h) of measurements during a particular haul under normal trawling conditions and allows for the description of the mechanical performance of the gear. Figure 2.4(A) shows the vertical forces acting on the two measuring axles in the instrumented trawl head. When fishing with the current at a speed of 6 kn over the ground, situation (1), the load on the aft axle (cell 2) has a positive value of about 3.5 kN (385 kg), whereas the load on the front axle (cell 1) has a negative value. This indicates that the aft part of the sole plate is in firm contact with the bottom, whereas there is no bottom contact at all at the front part. The negative value measured in cell 1 is due to the moment of both the pressure and friction forces with regard to cell 2. After changing the course of the ship by 180° the gear is towed against the current. As the speed over the ground is kept constant at 6 kn, the speed of the gear relative to the water increases

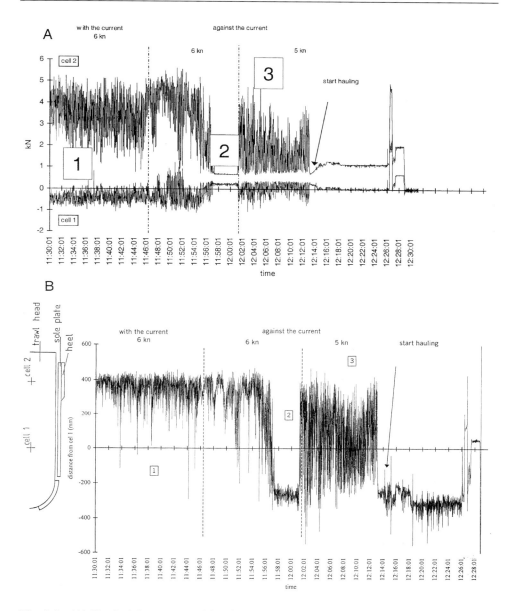

Fig. 2.4   (A) Vertical forces exerted by the sole plate; (B) centre of pressure.

and the gear tends to lift off the ground. This occurs in situation (2) in which the sole plate has completely lost bottom contact. The forces now acting on the load cells are determined by the water pressure on the sole plate only. When the speed is lowered to 5 kn, bottom contact is restored – situation (3). The fact that both the loads in cell 1 and cell 2 are positive indicates that the sole plate touches the seabed more horizontally than in situation (1). The decrease in speed is, however, not enough to keep the sole plates on the ground all the time. The soles plates regularly lose bottom contact, which is indicated by the forces in cells 1 and 2 being equal to the water pressure, as in situation (2).

The position of the centre of pressure can be calculated from the equilibrium equation $\sum$(moments) $= 0$. The result is graphically represented in Fig. 2.4(B) and confirms the conclusions drawn above:

- in situation (1), the centre of pressure is near to the middle of the heel, which is in firm contact with the bottom;
- in situation (2), bottom contact is lost and the centre of pressure is no longer located in the sole-plate area;
- in situation (3) the centre of pressure is located around the middle of the sole plate, which indicates that the whole sole plate is in contact with the bottom.

Figure 2.5a shows the average total pressure force acting on the sole plate as calculated from data obtained during four valid hauls. The warp length was kept at 90 m for depths varying from 23 m to 30 m. If the gear is towed against the current, the pressure force decreases from 5300 N (540 kgf) at a towing speed (over the ground) of 3 kn to 1000 N (102 kgf) at 6 kn. If the gear is towed with the current, an increase in towing speed from 4 to 7 kn results in a decrease in the pressure force from 4500 N (459 kgf) to 1000 N (102 kgf). At low speeds, the pressure force will normally act on the full surface of the sole plate, while, at high speeds, only the heel of the sole plate will be in contact with the seabed. The pressure can be calculated as: pressure $=$ pressure force/sole-plate area.

The sole-plate dimensions are 750 mm $\times$ 350 mm, the heels measure 250 mm $\times$ 350 mm. The average pressures in the present experiments were:

- fishing at 3 kn against the current with full sole-plate contact: 5300 N/75 $\times$ 35 cm^2 $= 2.019$ N cm^{-2} (0.206 kgf cm^{-2}) or 20.19 kPa
- fishing at 5 kn against the current with heel contact only: 2280 N/25 $\times$ 35 cm^2 $= 2.606$ N cm^{-2} (0.266 kgf cm^{-2}) or 26.06 kPa
- fishing at 4 kn with the current and with full sole-plate contact: 4500 N/75 $\times$ 35 cm^2 $= 1.71$ N cm^{-2} (0.174 kgf cm^{-2}) or 17.1 kPa
- fishing at 6 kn with the current and with heel contact only: 2750 N/25 $\times$ 35 cm^2 $= 3.14$ N cm^{-2} (0.320 kgf cm^{-2}) or 31.4 kPa.

At towing speeds of 6 kn (against the current) and 7 kn (with the current) the downward force on the gear is not sufficient to keep the trawl heads in contact with the bottom (identical to situation 2 in Fig. 2.4). In general, it can be stated that the 4-m beam trawl exerted average pressures on the seabed varying from 17 kPa to 31 kPa. An inquiry among Belgian skippers showed that, in commercial fishing, 'Eurocutters' tow 4-m beam trawls at an average speed of 3 or 4 kn, depending on whether the gear is towed against or with the tide. Under these circumstances, the sole-plate pressure will be 20 kPa and 17 kPa, respectively.

The pressures obtained are mean values at the given towing speeds. In general, the pressure forces are not constant but vary continuously, as can be seen in Fig. 2.4.

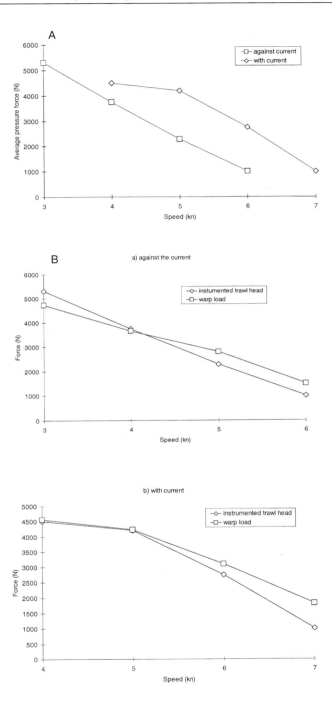

Fig. 2.5 (A) Average pressure force; (B) comparison of the pressure forces measured by the instrumented trawl head and calculated from the warp load either (a) against or (b) with the current.

These variations consist of more or less regular undulations of the average pressure and numerous peak values. The maximum and minimum values of the pressure force undulation at each towing speed all fall within the limits of the forces given above. The peak values, recorded at 1-s intervals, did not exceed 6000 N (612 kgf).

The periodic variations in the pressure forces correspond well with the heave of the ship as indicated by the echo sounder's heave compensator. The same periodic changes can be recognised in all other force measurements as well as in the shift of the pressure centre. Variations in the seabed morphology, however, are also superimposed on the variations due to the ship's movements. It is clear that the transmission of vessel movements to the gear may cause lifting of the gear in circumstances with light bottom contact. This will cause the sole plates to bounce over the seabed.

Total gear pressure

The average pressure force calculated from the warp load as compared with the pressure force measured with the instrumented trawl head is presented in Fig. 2.5b. The difference in pressure obtained by both measurement methods is not substantial. However, when fishing with the current, the difference in pressure load seems to increase with increasing speeds (Fig. 2.5b). At these higher speeds, the trawl heads have lost bottom contact, but this is not yet the case for most of the chain matrix and the bobbin rope. These gear components are attached to the trawl head at positions up the sole plate. The weight of ground gear affects the warp load, and hence the upwards component, but not the values recorded by the instrumented trawl head.

The pressure of the gear on the seabed also depends on the ratio between the depth and the length of the warp. The shorter the warp length, the larger the upward pull exerted by the warp on the gear as indicated by equation (1). The total gear pressure was calculated from the warp load for four hauls at a constant towing speed of 4 kn, but for warp lengths varying between 2.5 and 5.7 times the water depth. A linear relationship exists between the pressure force and the ratio warp length/depth, expressed by:

$$\text{pressure force} = 776.4(L/D) + 5708.2 \quad (r^2 = 0.9309).$$

Relation between gear pressure and engine power

Calculated from former sea trials on commercial vessels

Figure 2.6 shows the warp load against ground speed for five Belgian beam trawlers involved in former sea trials (Anon., 1996). The vessel and gear characteristics are given in Table 2.2. The pressure force exerted by the gear on the seabed can be calculated from:

$$\text{pressure force} = \text{gear weight} - \text{lift force}.$$

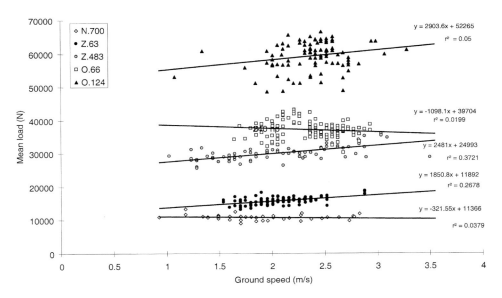

Fig. 2.6 Mean warp loads vs. ground speed of different commercial trawlers referred to by their registered port and number.

Table 2.2 Commercial vessel and gear characteristics

Vessel[a]	Engine class (hp)	Engine power (hp)	GRT (tons)	LOA (m)	Beam length (m)	Gear weight (kg)
N.700	<270	240	29.91	16.80	4	986
Z.63	270–300	297	68	20.04	4	1635
Z.483	600–899	600	187	34.21	9.5	3180
O.66	900–1100	900	224	30.58	10	3821
O.124	>1100	1200	259	32.23	10.3	4642

[a]Belgian vessel registration numbers.

The lift force, at an average towing speed, can be calculated from the warp load data presented in Fig. 2.6. A linear relationship exists between the lift force and the vessel hp:

$$\text{lift force} = 18.356 \times (\text{vessel hp}) - 337 \quad (r^2 = 0.9918).$$

The underwater weight of the gears is given by:

$$\text{gear weight} = 16810 \times \ln(\text{vessel hp}) - 83\ 912 \quad (r^2 = 0.995).$$

Finally, the resulting pressure force can be expressed as:

$$\text{pressure force} = 3052.4 \times \ln(\text{vessel hp}) - 14\,040 \quad (r^2 = 0.8674).$$

It can be concluded that the pressure force exerted on the bottom increases with vessel hp but on less than a proportional basis.

From fleet survey data

The pressure exerted by the sole plates depends on the weight of beam + trawl heads + bridles + block rather than on the weight of the complete gear. The net and the ground gear rest on the bottom and do not participate in the pressure generated by the sole plates. In case of a beam trawl rigged with a chain matrix, most of the chains rest on the seabed, but some are attached to the beam and will participate in the pressure generated by the sole plates. Furthermore, the pressure on the seabed will depend on the surface area of the sole plates.

The weight of the above-mentioned gear components and the sole-plate surface area, as obtained from a survey of 16 Belgian beam trawlers, increase proportionally with vessel hp:

$$\text{weight} = 1.5296 \times (\text{vessel hp}) + 699.04 \quad (r^2 = 0.6508)$$

$$\text{sole-plate surface area} = 1.6693 \times (\text{vessel hp}) + 1866.5 \quad (r^2 = 0.4169).$$

The slope of the trendlines differ only slightly, the sole-plate surface area being the steepest. The weight per cm^2 of the sole plate, which can be considered as a measure for the pressure exerted on the seabed, increases slowly with vessel hp:

$$\text{weight cm}^{-2} = 61.587 \times \ln(\text{vessel hp}) - 140.3 \quad (r^2 = 0.3749).$$

The pressure value calculated from this expression varies between 20 kPa and 30 kPa for 300-hp and a 1400-hp vessels, respectively. These values are remarkably close to the pressures measured during the field studies (17–31 kPa).

The results on the pressure/vessel hp relationship show that, under normal conditions, light and heavy beam trawls exert pressure forces of the same magnitude. The reason is that the difference in weight will be compensated by a higher towing speed and a larger sole-plate surface area.

Sea-floor disturbance

Side-scan sonar observations

This section gives the results of the side-scan sonar observations on tracks fished with the 4-m beam trawl. Different possible fishing procedures were simulated.

1992 observations on the Goote Bank

The sediment on the Goote Bank test sites consisted mainly of medium and coarse sand with variable amounts of silt. The grain-size parameters in each individual zone varied considerably along the sampled reference tracks. The main characteristics are given in Table 2.3.

On test zone IV, four parallel tracks about 3 km long and a distance of about 40 m apart were fished with the 4-m beam trawl. In total 10 side-scan sonar observations of the trawl marks were made between 15 min and 52 h after fishing. After 52 h, only very vague marks along 41% of the track could be identified. On test zone II, four parallel lines were fished in an area 40 m wide. Nine observations were made, up to 32 h after fishing. At that time, the complete track was still clearly visible. On test zone I, three parallel lines 10 m apart were fished. Three side-scan sonar observations were made. The last observation, made 20 h after fishing, showed relatively clearly visible marks on 70% of the track.

The penetration depths of the beam trawl in the superficial sediment could not be deduced from the sonographs as the traces on the recordings were too weak for this purpose. The depth of penetration was probably not very pronounced. No clear correlation could be made between the visibility of the trawl marks and the grain size of the sediment. Tracks were visible for the longest, however, on the coarse sand area of zone IV.

1993 observations on the Goote Bank

Zone I was fished twice on approximately the same track. The tracks were observed 21 h 25 min after fishing. Vague trails on 20.4% of the reference track could be detected. An attempt was made to correlate the visibility of the track on zone I with the sediment features. Sediment samples were taken with a Van Veen grab at six locations along the track and allowed a crude division of the test area into different sediment types. The gear marks were visible on 85% of the section covered with mud, on 18% of the section covered with coarse sand and on 21% of the section covered with coarse sand with shells. No tracks could be seen on the sections covered with coarse sand with superficial mud or mainly mud with some sand.

Within a period of 9 h, nine successive hauls were made on the reference track on zone II. Six side-scan sonar observations were made between 6.5 h and 44 h after the end of the fishing operations. Track visibility was best 7 h after fishing and, at that

Table 2.3 Sediment characteristics of samples taken on the Goote Bank test site

Track	Mean grain size (µm)	Size classification	% Silt
I	293–530	Medium (–coarse) sand	4.70–21.2
II	439–591	(Medium–) coarse sand	1.00–1.37
IV	374–>884	Medium–coarse sand	1.43–12.10

time, marks could be seen along 87% of the track. Afterwards, the visibility of the tracks decreased gradually but, after 44 h, imprints could still be detected on 23% of the track. The visibility of the marks at that time was again different for the different sediment types: 100% for coarse sand with shell debris, 36% for coarse sand with superficial mud, 10% for coarse sand with superficial mud and gravel debris, and 0% for coarse sand with some gravel elements.

The results indicate that the type of sediment is an important factor for the visibility of the trawl marks, although the results obtained in zone I and II are not in complete agreement. However, the results from zone II should be regarded as being more reliable, as the tracks could be more clearly detected than in zone I. As in 1992, the penetration depth could not be deduced from the recorded sonographs.

1996 observations on the Scheveningen area and on the Flemish Banks

Scheveningen area. Side-scan sonar observations were made on three parallel tracks, on each of which five consecutive fishing operations had taken place. The observations of the tracks were made just before fishing and several times after fishing, up to almost 37 h later. The sediment composition in the area was very homogenous and consisted mainly of medium and fine sand (Table 2.4). The tracks were clearly visible up to 22 h after fishing, though not over their complete length. One track was observed up to 37 h after fishing. At that time, only a short section was still visible, while most of the tracks could only be vaguely detected.

Goote Bank. The superficial sediment on the track consisted of nearly equal parts of fine and very fine sand and of shell debris and gravel (Table 2.4). The track observed was fished four times and side-scan sonar recordings were made up to

Table 2.4 Grain-size analysis (%) on the 1996 test sites (Udden–Wentworth classification (Leeder, 1988))

Sediment classification			Goote Bank samples			Negenvaam samples			
Description	Fraction (μm)	Scheve-ningen	1	2	3	1	2	3	4
Gravel	> 2000	0.40	43.96	32.18	47.92	1.03	4.60	1.73	8.49
Very coarse sand	> 1000	0.06	5.22	4.73	6.83	1.27	0.78	3.08	1.48
Coarse sand	> 500	1.83	3.26	3.68	3.26	5.13	4.10	17.10	3.11
Medium sand	> 250	67.86	28.94	30.38	16.99	32.21	16.21	47.89	12.89
Fine sand	> 125	27.19	15.44	26.42	22.24	57.26	67.99	28.22	65.57
Very fine sand	> 63	1.55	1.47	1.16	1.40	1.32	2.96	0.79	3.66
Mud	< 63	1.11	1.71	1.45	1.36	1.78	3.36	1.19	4.80

22.5 h after fishing. Even at the end of the observations, the tracks were still clearly visible.

Results from RoxAnn surveys

RoxAnn surveys were performed on one location on the Scheveningen area and on two locations on the Belgian coast (Table 2.4). The values of the RoxAnn parameters, E1 and E2, before fishing (t_0), are given in Table 2.5.

Scheveningen area. The low variation in the E1 (roughness) and E2 (hardness) parameters, and the low E1 value (Table 2.5 and Fig. 2.7 at t_0) confirms that the sediment in this area is homogenous and consist of mainly medium and fine sand (Table 2.4). Five consecutive hauls were made over the same track. The track was observed by RoxAnn 0.5, 3, 5 and 15 h after fishing. The E1 and E2 parameters along the track are compared with the values before disturbance (t_0) in Fig. 2.7 for each observation. Immediately after fishing ($t_0 + 0.5$), the seabed disturbance is very clear. The E1 value dropped, indicating that the 'roughness' had decreased. The E2 value, on the contrary increased considerably, indicating a harder bottom. Both the changes in E1 and E2 can be explained by the lighter sediment fractions being suspended by the gear. Note that the increase in E2 is variable. This is probably due to the fact that, during the RoxAnn surveys, not all sea bottom surfaces sampled by the echo sounder were equally affected by the fishing gear. It appears from the time series that the suspended sediment particles deposited rather quickly. After 3 h, the parameters E1 and E2 were again close to the t_0 value and, after 15 h, no difference could be distinguished. It should be noted, however, that the disturbance by the gear as recorded by the side-scan sonar became only less visible at the end of the time series.

Goote Bank. The presence of shell debris and gravel in the sediment is reflected by the higher values of both E1 and E2 (Table 2.5), indicating that the bottom was

Table 2.5 Values of RoxAnn parameters at the different locations

Area	E1/E2	Average	Min.	Max.	SD
Scheveningen	E1	0.1305	0.109	0.146	0.00782
	E2	0.2141	0.175	0.252	0.01557
Goote Bank	E1	0.4037	0.36	0.453	0.01943
	E2	0.4692	0.379	0.611	0.04403
Negenvaam	E1	0.2178	0.143	0.316	0.05719
	E2	0.2248	0.101	0.527	0.08492

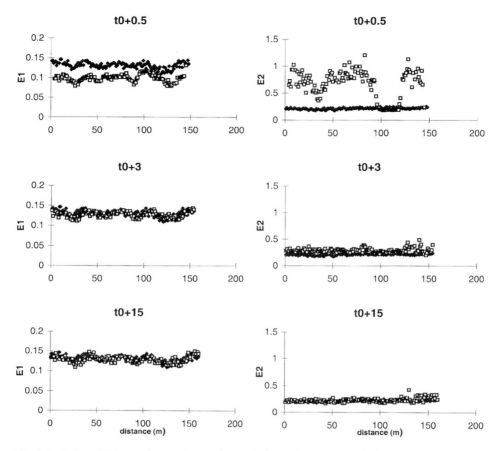

Fig. 2.7   E1 and E2 RoxAnn values estimated along the same trawled track at t_0 (◆) and at time intervals (h) after the initial trawling disturbance occurred (□).

rougher and harder than other locations examined here. The track was fished four times. RoxAnn surveys took place before fishing (t_0) and at 3, 10, 20 and 24 h after fishing. Owing to the relatively lower content of light particles, the changes in the sediment characteristics were less pronounced than on the Scheveningen area. The E1 and E2 values were dominated by the presence of large particles, and the suspension of the lighter particles after passage of the trawl caused only minor changes.

Negenvaam. The sediment on the Negenvaam area was not as homogenous as on the two other locations (Tables 2.4 and 2.5). Although it contained a lot of fine sand, there was also a considerable amount of larger particles. The effect of trawling on E1 and E2 was limited and soon faded away completely.

Discussion

Pressure

The pressure force exerted by a 4-m beam trawl is strongly related to the towing speed. As the speed increases, the lift of the gear increases and the resultant pressure force decreases. At the same time however, the tilt of the sole plates increases and a smaller surface of the sole plate will remain in contact with the bottom. The resultant pressure, expressed as force per unit area, tends to increase. At higher speeds, the weight of the gear will be fully compensated by the greater upward pull, and the beam will lift off the bottom. From the present experiments, it appears that the pressure exerted by the sole plates varies from 17 kPa to 32 kPa when fishing against the current at towing speeds (over the ground) of 4 kn and 6 kn, respectively. In Belgian commercial fishing with this gear, towing speeds are 3 kn when fishing against the current and 4 kn when fishing with the current. At these speeds, the sole-plate pressures are 20 kPa and 17 kPa, respectively. Bottom contact was lost at 6 kn or 7 kn, depending on whether the gear was towed against or with the current. These values are for warp lengths equal to three times the depth, the standard warp length/depth ratio. With shorter warp lengths, e.g. on soft grounds, the pressure will be lower, since the warp lift force will increase. Vessel movements are transmitted to the gear, even at low amplitudes. This may cause the gear to bounce on the bottom. From the comparison between the total gear pressure force and the pressure force exerted by the sole plates, it appears that the chain matrix and the bobbin gear exert only a limited pressure on the seabed.

Since larger vessels use heavier gears, the pressure force exerted by a beam trawl increases with vessel engine power, but the increase is less than proportional. The increase in gear weight, however, is compensated by larger sole-plate dimensions and a higher towing speed. As a result, heavy and light beam trawls will exert sole-plate pressures of the same magnitude. This is in agreement with the results of measurements made by van der Hak & Blom (1990). They calculated that for a 12 m/7000 kg beam trawl, the sole plates exerted a pressure of 0.15 kgf cm^{-2} (14.7 kPa). Taking into account that the entire sole plate was presumed to be in contact with the bottom, this value is quite close to the results of the present study. Even earlier studies point in the same direction. Margetts & Bridger (1971) calculated that the pressure of a 9-m beam plus trawl heads with a total weight of 324 kg (283 kg in water) is 0.1 kgf cm^{-2} (9.8 kPa). For comparison, a 556-kg (in water) otter board exerts a pressure of 0.236 kgf cm^{-2} (23.1 kPa). These otter board values, however, do not take account of the upwards pull of the warps.

Seabed disturbance

REMOTS and video observations made during the IMPACT projects (Lindeboom & de Groot, 1998) revealed that the passage of a beam trawl flattens the seabed and exposes shell debris at the surface. While differences between treatments and control were very clear, differences between treatments were not. Only a 300% fished area with a 12-m beam trawl could be distinguished from a 200% fished area with a 4-m beam trawl. From the penetration depth of the REMOTS prism, it can be estimated that on densely packed fine sand overlaid with a silt layer, heavy beam trawling will result in the removal of the upper 1 cm sediment layer.

The longer-term effects of fishing with a 4-m beam trawl can be judged from the present side-scan sonar recordings and the RoxAnn surveys. The movement of the trawl over the seabed causes the suspension of the lighter sediment fractions. The RoxAnn surveys indicated that the bottom becomes harder and less rough. The changes were most pronounced in an area with a lot of fine sand. The original situation, however, was quickly restored. On the most disturbed areas the 'hardness' and 'roughness' characteristics regained their original values in less than 15 h.

The duration that beam-trawl marks remained visible after fishing also depended on the upper sediment layer. On a seabed consisting of mainly coarse sand, the tracks remained visible for up to 52 h, whereas on sediments with mainly finer particles the tracks were completely faded after 37 h. The penetration depths of the beam trawl in the superficial sediment could not be deduced from the sonographs, as the traces on the recordings were too weak. Sydow (1990) also used side-scan sonar to study the tracks made by 5000-kg, 12-m beam trawls but could not detect the penetration depth from the recordings either. For the moment, the most reliable estimates of the penetration depth have been obtained indirectly from the presence of certain species of benthic infauna in the catch. From the presence of *Arctica islandica* and *Echinocardium cordatum* in the catches of a 12-m/7000-kg beam trawl operating on a hard sandy bottom, Bergman *et al.* (1990) concluded that the tickler chains, in a limited part of the trawled area, penetrated to a depth of at least 6 cm. It seems realistic to estimate the penetration depth into the seabed of beam trawls at 1–8 cm, depending on the sediment characteristics and the rigging of the gear (BEON, 1991).

Information on the seabed disturbance by towed gears other than beam trawls is available from the IMPACT projects (Lindeboom & de Groot, 1998). The passage of a *Nephrops* trawl was found to have a generally minor physical and visual impact on the soft sedimentary seabed, represented by a flattening of the normally mounded sediment surface and some disturbance of the sessile epifauna. Fewer openings of *Nephrops* burrows were seen in the trawled area, which suggests that the delicate and complex structure of the burrow system may be severely damaged by the action of the gear.

The main physical effect of otter trawling appears to be the tracks left in the sediment by the trawl doors, as indicated by the experiments in a Scottish loch. Both side-scan sonar and RoxAnn survey results were in general agreement on the time scale over which the effects were noticeable at this sheltered muddy site. Both indicate clear physical effects while trawling is ongoing, and suggest that, after 18 months, these effects are almost indistinguishable (no effect noticeable from RoxAnn but very faint tracks identified from side-scan sonar). Such effects would not be as large or long lasting in less sheltered areas with harder substrate.

Areas with a soft muddy bottom or with low tidal currents are likely to be more severely affected by bottom trawling than areas with a firm substratum (compacted sand) and those subject to periodic strong tidal currents or wave turbulence caused by gales in shallow coastal areas (Anon., 1988).

Acknowledgements

This study was carried out with financial support from the Commission of the European Communities, Agriculture and Fisheries (FAR & AIR) specific RTD programmes, FAR MA 2–549, 'Environmental Impact of Bottom Gears on Benthic Fauna in Relation to Natural Resources Management and Protection of the North Sea' and AIR2-CT94–1664 'IMPACT-II The Effect of Different Types of Fisheries on the North Sea and Irish Sea Benthic Ecosystems'. It does not necessarily reflect its views and in no way anticipates the Commission's future policy in this area.

References

Anon. (1973) Effects of trawls and dredges on the seabed. *International Council for the Exploration of the Sea*, C.M. 1972/B:2, 4 pp.

Anon. (1988) Report of the study group on the effects of bottom trawling. *International Council for the Exploration of the Sea*, C.M. 1988/B:56, 30 pp.

Anon. (1996) *Investigation of the relative fishing effort exerted by towed demersal gears on North Sea human consumption species. Final Report Project* AIR1-CT92-0445. Danish Institute for Fisheries Technology and Aquaculture, Hirshals, Denmark.

BEON (1991) *Effects of beamtrawl fishery on the bottom fauna of the North Sea. BEON Report* 13, 85 pp. Netherlands Institute for Sea Research, Den Burg, Texel, The Netherlands.

Bergman, M.J.N. & Hup, M. (1992) Direct effects of beam trawling on macrofauna in a sandy sediment in the Southern North Sea. *ICES Journal of Marine Science*, **49**, 5–11.

Bergman, M.J.N., Fonds, M., Hup, M., Lewis, W., van der Puyl, P. Stam, A. & den Uyl, D. (1990) Direct effects of beamtrawl fishing on benthic fauna in the North Sea – a pilot study. In: *Effects of beamtrawl fishery on the bottom fauna of the North Sea. BEON Report* 8, pp. 33–57. Netherlands Institute for Sea Research, Den Burg, Texel, The Netherlands.

Bridger, J.P. (1970) Some effects of the passage of a trawl over the seabed. *International Council for the Exploration of the Sea*, C.M. 1970/B:10, 10 pp.

Bridger, J.P. (1972) Some observations on the penetration into the sea bed of tickler chains on a beam trawl. *International Council for the Exploration of the Sea*, C.M. 1972/B:7, 9 pp.

Burns, D., Queen, C.B. & Chivers, R.C. (1985) Ground and fish discrimination in underwater acoustics. In: *Proceedings of the Ultrasonics International Conference*, Vol. 85, 49–55.

Chivers, R.C. & Burns, D. (1991) Processing of echo-sounder signals for real-time surveying. In: *Proceedings of the Meeting on Sonar Signal Processing*, Vol. 10, pp. 206–11.

De Clerck, R. & Hovart, P. (1972) On the effects of tickler chains. *International Council for the Exploration of the Sea*, C.M. 1972/B:15, 11 pp.

De Groot, S.J. (1972) Some further experiments on the influence of the beam trawl on the bottom fauna. *International Council for the Exploration of the Sea*, C.M. 1972/B:6, 7 pp.

De Groot, S.J. (1984) The impact of bottom trawling on benthic fauna in the North Sea. *Ocean Management*, **9**, 177–90.

van der Hak, W. and Blom, W.C. (1990) Cruise report experiments on the interacting fishing gear (beam trawl) – benthos with a commercial beam trawler, 21 August–1 September 1989. In: *Effects of beamtrawl fishery on the bottom fauna of the North Sea. BEON Report* 8, pp. 23–31. Netherlands Institute for Sea Research, Den Burg, Texel, The Netherlands.

Houghton, R.G., Williams, T. & Blacker R.W. (1971) Some effects of double beam trawling. *International Council for the Exploration of the Sea*, C.M. 1971/B:5, 16 pp.

ICES (1971) Procès-verbal de la Réunion 1970. Council Resolution 1970/S/1.

ICES (1988) Procès-verbal de la Réunion 1987. Council Resolution 1987/2:7.

Leeder, M.R. (1988) *Sedimentology, Process and Product*. Unwin Hyman, London.

Lindeboom, H.J. & de Groot, S.J. (eds) (1998) *IMPACT-II: the effect of different types of fisheries on the North Sea and Irish Sea benthic ecosystems. NIOZ Report* 1998-1/*RIVO-DLO Report* C003/98, 257 pp. Netherlands Institute for Sea Research, Den Burg, Texel, The Netherlands.

Margetts, A.R. & Bridger, J.P. (1971) The effect of a beam trawl on the sea bed. *International Council for the Exploration of the Sea*, C.M. 1971/B:8, 9 pp.

Polet, H., Blom, W. & Thiele, W. (1994) An inventory of vessels and gear types engaged in the Belgian, Dutch and German bottom trawling. In: *Environmental impact of bottom gears on benthic fauna in relation to natural resources management and protection of the North Sea* (eds S.J. de Groot & H.J. Lindeboom), pp. 7–19. *NIOZ Report* 1994-11/*RIVO-DLO Report* C026/94. Netherlands Institute for Sea Research, Den Burg, Texel, The Netherlands.

Schlagintweit, G.E.O. (1993) *Real-Time Acoustic Bottom Classification for Hydrography: A Field Evaluation of Roxann*. Canadian Hydrographic Service, Department of Fisheries and Oceans.

Sydow, J.S. (1990) Cruise report experiments on the interaction fishing gear (beamtrawl)–benthos with the R.V. *Mitra*. In: *Effects of beamtrawl fishery on the bottom fauna of the North Sea. BEON Report* 8, pp. 7–21. Netherlands Institute for Sea Research, Den Burg, Texel, The Netherlands.

Chapter 3

Is bottom trawling partly responsible for the regression of *Posidonia oceanica* meadows in the Mediterranean Sea?

G.D. ARDIZZONE, P. TUCCI, A. SOMASCHINI and A. BELLUSCIO

Dipartimento di Biologia Animale e dell'Uomo, Università di Roma 'La Sapienza', Viale dell'Università 32, 00185 Rome, Italy

Summary

1. The seagrass *Posidonia oceanica* is a marine angiosperm that is undergoing regression along Mediterranean coasts. Research in the last few years has demonstrated two possible main sources of damage: anthropogenic modification of sediment characteristics and the physical impacts of fishing gear. Trawl fisheries are considered to be one of the major factors leading to the deterioration of seagrass meadows. The aim of this study was to determine the physical and biological parameters that can be used to identify the reason for regression in different *Posidonia* meadows.
2. A total of 103 stations were sampled in two different areas in the Central Tyrrhenian Sea. The seagrass meadows in both areas are undergoing regression. The first area is strongly influenced by sedimentation and is untrawlable because of the presence of a hard and irregular seabed. In the second area, illegal trawling is known to have occurred for almost 20 years.
3. Regression analysis of environmental parameters on seagrass shoot density revealed that, in the untrawled area, the density of seagrass shoots is inversely proportional to the silt and clay content of the sediment, but independent of the depth gradient within the study area. At the same time, the percentage of dead 'matte' (a mat of dead seagrass roots and rhizomes) increases with higher proportions of silt and clay. This suggests that elevated levels of fine sediment may be one cause of the regression of *Posidonia*. Levels of silt and clay that exceed 10% of the sediment composition will cause a decline in seagrass beds. No relationship between sediment characteristics and meadow regression was found in the area that is trawled illegally. Thus, we conclude that fishing activities are the main cause of seagrass regression in this area.
4. While it is difficult to identify the possible sources of fine sediment inundation and thus ameliorate its effects on seagrass beds, illegal trawling can be controlled more readily through physical protection of the seabed using protective reefs or artificial seabed obstacles.

Keywords: environmental impact, illegal trawling, sediment, sedimentation, seagrass.

Introduction

Posidonia oceanica (L.) Delile is a marine angiosperm that is endemic to the Mediterranean Sea. This seagrass inhabits large areas of the coastal seabed down to depths of 40 m in optimal conditions, and occurs in dense aggregations called meadows or beds. The estimated total area of *Posidonia* beds in the Mediterranean is approximately 20 000 square nautical miles, which corresponds to 2% of the surface area of the littoral sea (Bethoux & Copin-Montegut, 1986). In common with other seagrasses, it is a highly productive species and acts as an important nursery ground for many fish species (Short & Wyllie-Echeverria, 1996).

Decrease in the area of seabed colonised and seagrass shoot density, known as 'regression', is a widespread phenomenon throughout the Mediterranean basin (Peres, 1984). Recent research has demonstrated two main sources of damage to *Posidonia* beds: changes in sediment composition and structure, and the physical impacts of fishing gears. Shifts in sediment structure are linked to changes in the use of the coastal zone (e.g. coastal construction projects) that cause a reduction in the transport of suspended solids along the coast. Concomitantly, increases in the mud content of riverine input have caused an ecologically significant increase in sea water turbidity (Peres & Picard, 1975; Blanc *et al.*, 1980; Boudouresque & Jeudy de Grissac, 1983; Peres, 1984). These changes have resulted in a reduction in the depth limit of the plant from 40 m to 20 m or less and have led to lower plant density on the seabed.

The regression of seagrass meadows is also possibly caused by mechanical stress from fishing activities that disturb the seabed. Coastal trawling is illegal within 3 n miles of the shore in most Mediterranean countries; nevertheless, it still occurs frequently and has negative effects on *Posidonia* meadows and their associated communities (Augier & Boudouresque, 1970; Peres, 1977; Ardizzone & Pelusi, 1984; Guillen *et al.*, 1994; Ardizzone & Belluscio, 1995).

While many studies have mapped or described the regression of *Posidonia* meadows in the Mediterranean Sea, few data are available on the physical and biological processes of damaged meadows or on their possible restoration rate (Jones, 1992). The regression of *Posidonia* is often explained as a 'natural phenomenon linked to unidentified pollution problems'. For effective coastal management, it is important to identify the source of *Posidonia* bed regression so that action can be taken to arrest this environmental damage. The identification of the cause of regression in *Posidonia* beds is complicated, as there are few clues that distinguish regression caused by sediment inundation or fishing.

This study examined physical and biological parameters that can be used to understand the real cause of regression in *Posidonia* by means of a comparison of two areas subjected to different types of environmental stress. The first area is affected by sediment transport from nearby rivers and construction projects along the coastline that have lowered water transparency (Ardizzone & Belluscio, 1995). Regression of the seagrass meadows has occurred throughout this area (Ardizzone & Belluscio, 1995). The second seagrass meadow has been studied for the last 15 years with reference to the impact of trawl fisheries and is not affected by riverine input (Ardizzone & Pelusi, 1984). Most of the meadow is characterised by a very low shoot density and the occurrence of dead 'matte'. In addition, the lower limit of *Posidonia* has altered from around 30 m to 22–24 m in depth (Ardizzone & Migliuolo, 1982; Ardizzone & Pelusi, 1984).

Methods

Two different areas of the Central Mediterranean Sea (Latium, Italy) with damaged *Posidonia* beds were studied. The first area (P1) is in the northern part of Latium bordering Tuscany and includes a 14-km sandy shoreline and substratum that extends down to a depth of 30 m. In this area, the seabed is composed of a mosaic of soft sediment, *Posidonia* and dead matte. Dead matte is composed of the layer of dead rhizomes that remains on and below the surface of the seabed. Rocks occur intermittently on the seabed such that the area is considered untrawlable. The second area (P2) in southern Latium includes 18 km of mixed rocky and sandy substrata that extends down to a similar depth (Fig. 3.1). The maximum depth sampled is greater than the depth to which *Posidonia* is found in this region. A small area with a rocky seabed was also considered (P24, P25) to compare its characteristics with the actively trawled *Posidonia* meadow. This area (P2) has been sampled to monitor the effects of fishing for the last 15 years.

The geographic distribution of *Posidonia* in the two areas has been mapped previously. The seagrass at P1 is affected by changes in sediment composition due to the influence of the River Tiber outflow, and of four minor rivers (Mignone, Marta, Arrone and Fiora) that flow directly into the area itself. Over the past 10 years,

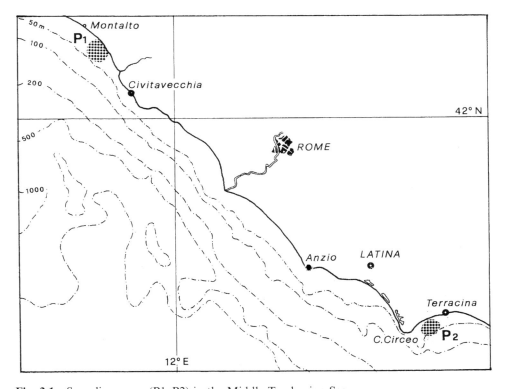

Fig. 3.1 Sampling areas (P1–P2) in the Middle Tyrrhenian Sea.

important construction projects have been carried out along the same coastline, producing large amounts of sediment-laden run-off (Civitavecchia harbour and Montalto di Castro electric power station).

A total of 78 sampling stations in P1 and 25 in P2 were located at depths between 5 and 30 m, some of which were collected from areas directly around *Posidonia* meadows. A higher number of stations were sampled to improve knowledge of the spatial distribution of *Posidonia* in that area. Sampling at each point was carried out by scuba divers, who measured shoot density within three randomly placed quadrants (33 × 33 cm), collected five *Posidonia* shoots (when present) and a 750-cm^3 sample of sediment. In this study, we considered the following variables:

- the percentage of seagrass covering the seabed;
- the percentage covering of dead matte;
- the shoot density (n m^{-2}), an important index of healthy conditions in *Posidonia* meadows (Giraud, 1977);
- the leaf area index (LAI), the leaf surface per bottom unit area (m^2 m^{-2}).

Sediment samples were analysed by both dry sieving and laser granulometry (for the finest fraction below 1000 μm) to obtain their grain-size distribution. The following percentage values from the Wentworth grain scale were used: pebble and coarse sand (>2 mm), sand (2 mm–63 μm), silt (63–3.9 μm) and clay (<3.9 μm). The standard deviation (SD) of the sample distribution was used as a sorting index of the sediment. Linear regression was used to describe the relationship between variables, log transforming the data when appropriate.

Table 3.1   Shoot density m^{-2} in P1 at different depths

Station	5 m	10 m	15 m	20 m	25 m	30 m
P11	–	267	240	–	20	–
P12	–	–	138	114	20	–
P13	–	–	114	111	–	–
P14	–	–	66	–	–	–
P15	–	–	99	–	20	42
P16	–	–	162	141	57	42
P17	–	–	198	90	33	20
P18	–	–	174	96	42	20
P19	–	–	120	126	36	–
P110	–	–	–	42	–	–
P111	–	–	135	81	–	–
P112	–	–	96	123	–	–
P113	–	–	–	93	–	–

Table 3.2   Percentage of dead matte in P1 at different depths

Station	5 m	10 m	15 m	20 m	25 m	30 m
P11	–	15	35	–	99	–
P12	–	–	65	75	99	–
P13	–	–	60	60	–	–
P14	–	–	75	–	–	100
P15	–	–	70	–	99	90
P16	–	–	15	65	75	90
P17	–	–	20	65	95	99
P18	–	–	85	80	95	99
P19	–	–	35	65	80	100
P110	–	–	–	99	–	–
P111	–	–	70	85	–	–
P112	–	–	60	70	–	–
P113	–	–	–	70	100	–

Results

In area P1, *Posidonia oceanica* first occurs at a depth of 10 m and maximum shoot density (267 m^{-2}) is found at the same depth. From this depth onward, density decreases steadily (Table 3.1). Concomitantly, the percentage of dead matte observed at different depths is inversely related to the shoot density, and thus confirms the active regression of the meadow from deep towards shallower depths (Table 3.2). Accordingly, the maximum values for the LAI were found at a depth of 15 m ($8.15 \text{ m}^2 \text{ m}^{-2}$), while values between 2.5 and 0.5 $\text{m}^2 \text{ m}^{-2}$ were found to a depth of 30 m.

Figure 3.2 shows the grain distribution of the sediment at different depths. It is clear that sediment is homogeneous at a depth of 5 m (sand), with increasing sediment particle size in deeper water; in particular, silt and clay increase with depth. The relationship between shoot density and percentage of silt and clay is inversely proportional ($r^2 = 0.704$ for silt and $r^2 = 0.789$ for clay). This relationship is independent of depth overall (Fig. 3.3). The LAI also decreases with increasing values of silt and clay, while the percentage values of dead matte increase with increasing proportions of silt and clay content.

The same analyses were carried out for area P2. *Posidonia* first occurred at a depth of 15 m and maximum shoot density (258 m^{-2}) occurred at a depth of 25 m (Table 3.3). The percentage of dead matte always increased with increasing depth; in some cases, samples reached very high levels (Table 3.4). The sediment grain composition of samples in P2 (Fig. 3.2) is more homogeneous than in P1, as most samples are characterised by a sandy component, while neither the silt nor the clay content ever exceeds 20%. In P2, *Posidonia* generally occurs deeper than in P1, and the regression

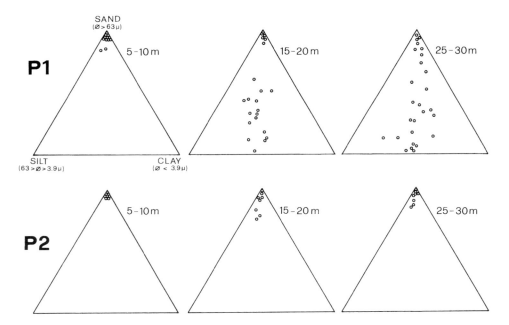

Fig. 3.2 Grain size distribution of the sediment at different depths at P1 and P2.

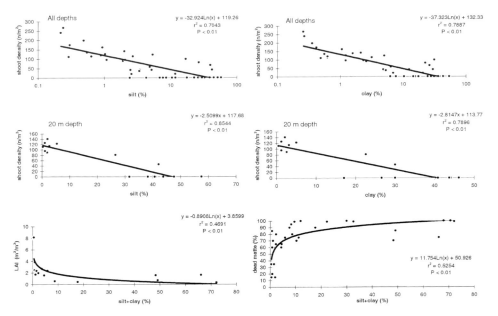

Fig. 3.3 Regression analyses of meadow and sediment parameters at site P1, showing the relationship between shoot density and percentage of silt and clay in the sediment for all depths and then at one depth (20 m) only. Variation in LAI and percentage dead matte with percentage silt and clay is also shown.

Table 3.3   Shoot density m^{-2} at P2 at different depths

Station	5 m	10 m	15 m	20 m	25 m	30 m
P21	–	–	–	–	3	0
P22	–	–	171	20	3	0
P23	–	–	180	–	–	–
P24	–	–	–	200	258	99
P25	–	–	–	171	180	63

Table 3.4   Percentage of dead matte at P2 at different depths

Station	5 m	10 m	15 m	20 m	25 m	30 m
P21	–	–	–	–	100	100
P22	–	–	75	99	100	100
P23	–	–	30	–	–	–
P24	–	–	–	15	15	25
P25	–	–	–	30	40	75

does not always tend to increase regularly with depth. Regression of *Posidonia* is absent or lower where rocky bottoms occur and prevent access to trawling (Tables 3.3 and 3.4, P24–P25).

Finally in P2, contrary to what we observed in P1, there is no correlation between the seagrass meadow variables (shoots m^{-2}, LAI, percentage of dead matte) and sediment parameters (percentage of sand, silt). The correlation between these values is low and not significant. The only significant relationship occurs between dead matte and the percentage of clay (Fig. 3.4).

Discussion

This study on *Posidonia* meadows started out from one clear fact: both the meadows considered are suffering a strong regressive trend, of which there are three characteristics: (1) the lower limit of most meadows has become more shallow and is *c.* 20 m in depth; (2) shoot density is low, often <50 m^{-2}, which is considered the limit for a real meadow (Giraud, 1977); (3) large amounts of visible dead matte, a clear sign of recent regression, as dead matte tends to disappear with the activity of currents or becomes covered by sediment.

At area P1, the seagrass bed is very heterogeneous, and shows different health conditions, coverage and sediment texture. Of note is the different sediment composition at the same depth (Fig. 3.3, 20 m). In fact, it is thanks to this variety of conditions that a number of descriptors could be selected to identify the regressive characteristics: sediment grain-size classes and seagrass meadow parameters were all

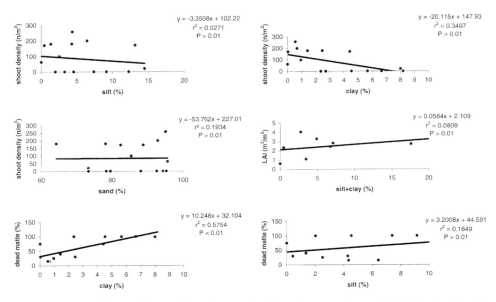

Fig. 3.4 Regression analyses of meadow and sediment parameters at site P2, showing the relationship between shoot density and percentage of silt, sand and clay in the sediment for all depths. Variation in LAI and percentage dead matte with percentage silt and clay is also shown

significantly correlated, and showed clear trends of meadow regression as the fine fraction of sediment increased. Upper limits for these parameters at which *Posidonia* cannot survive were also identifiable (around 10% for silt and clay and 20% for the whole pelitic component). This trend was strongly confirmed by the significant results obtained when the variables were considered independently of the effects of depth, thus eliminating the physiological reduction linked to lower photosynthetic efficiency.

The regression of the seagrass meadow in P1 is caused without any doubt by changes in sediment composition, as this area is untrawlable because of rocks on the seabed. This result was very important in evaluating what we observed in P2. In fact, in this case, active regression was also observed with an increase in the lower limit, low shoot density and a high percentage coverage of dead matte, but the sediment characteristics were quite different from those we observed in P1. Sandy sediment was dominant in most of the samples, and this type of sediment was always related to the best health condition in the P1 meadows. In P2, few of the seagrass meadow parameters considered were significantly correlated to the sediment parameters (Fig. 3.4). The only exception was the relation between dead matte and percentage of clay. Differences among the two areas in the relationship between shoot density and sediment composition are also emphasised when comparing the two sets of data together (Fig. 3.5).

Nevertheless, the regressive trend in P2 is revealing, because it is extreme on regular bottoms and absent where rocks or irregular bottoms are present. Therefore,

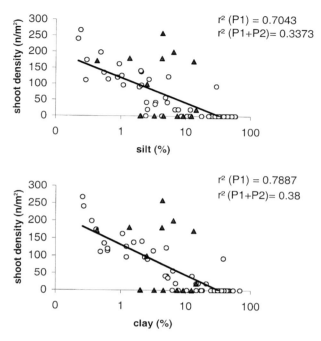

Fig. 3.5 Regression analysis of meadow parameters (shoots density) and sediment parameters (% silt and % clay) at sites P1 (○) and P2 (▲) displayed in the same figure for comparative purposes.

while sediment conditions at P2 are good for the development of seagrass beds, and no other factors for regression can be found, it seems evident that the cause must be trawling, as suggested elsewhere (Ardizzone & Migliuolo, 1982; Ardizzone & Pelusi, 1984).

The widespread regression of many *Posidonia* meadows in the Mediterranean Sea is a cause for concern, because of their ecological importance for littoral communities. The relationship between the level of mud and the health of *Posidonia* has been discussed many times and has been quantitatively observed in this paper. While it is clear that there are many factors that can increase the quantity of fine sediments suspended in the water or that settle on the seabed, it is difficult to prove that the continuous human changes made to the shoreline have caused the observed sediment alteration. Consequently, it is difficult to reverse this kind of regression. Hence, it is even more unacceptable that a further cause of regression, such as illegal trawling, can persist uncontrolled. A correct diagnosis of the source of seagrass alteration is important for the implementation of management measures. Such measures include the use of artificial reefs to protect the seabed from trawling, which have already proved to be an effective conservation measure adopted in some parts of the Mediterranean (Bombace, 1989; Relini & Moretti, 1986; Guillen *et al.*, 1994).

References

Ardizzone, G.D. & Belluscio, A. (1995) Le praterie di *Posidonia oceanica* delle coste laziali. In: *Il mare del Lazio*, pp. 194–217. Università degli Studi di Roma 'La Sapienza', Rome.

Ardizzone, G.D. & Migliuolo, A. (1982) Modificazione di una prateria di *Posidonia oceanica* del Medio Tirreno sottoposta ad attività di pesca a strascico. *Naturalista Siciliano*, **6** (Suppl.), 509–15.

Ardizzone, G.D. & Pelusi, P. (1984) Yield and damage evaluation of bottom trawling on *Posidonia* meadows. *International Workshop on Posidonia Oceanica Beds* (eds C.F. Boudouresque, A. Jeudy de Grissac, & J. Olivier), pp. 63–72. GIS Posidonie, Marseille.

Augier, H. & Boudouresque, C.F. (1970) Vegetation marine de l'ile de Port Cros. La baie de Port Man et le probleme de la regression de l'herbier de Posidonie. *Bulletin du Museum d'Histoire Naturelle de Marseille*, **30**, 145–64.

Bethoux, J.P. & Copin-Montegut, G. (1986) Biological fixation of atmospheric nitrogen in the Mediterranean Sea. *Limnology and Oceanography*, **31**, 1353–8.

Blanc, J.J., Clereifond, P., Froget, C.H., Jeudy de Grissac, A., Onoratini, G. & Orsolini, P. (1980) Facteurs de la sedimentation marine actuelle. In: *Application a l'amenagement de la marge continentale superieure de la Provence*, pp. 1–120. Centre National pour l'Exploitation des Oceans.

Bombace, G. (1989) Artificial reefs in the Mediterranean Sea. *Bulletin of Marine Science*, **44**, 1023–32.

Boudouresque, C.F. & Jeudy de Grissac, A. (1983) L'herbier *Posidonia oceanica*: les interactions entre la plante et le sediment. *Journal de recherche in Oceanographie*, **8**, 99–122.

Giraud, G. (1977) Contribution a la description et la phenologie quantitative des herbiers de *Posidonia oceanica* (L.) Delile. Doctorat de specialité en oceanologie, Université Aix-Marseille II, Faculté des Sciences de Luminy, Marseille.

Guillen, J.E., Ramos, A.A., Martinez, L. & Sanchez Lizaso, J.L. (1994) Antitrawling reefs and protection of *Posidonia oceanica* (L.) Delile meadows in the Western Mediterranean Sea: demand and aims. *Bulletin of Marine Science*, **55**, 645–50.

Jones, J.B. (1992) Environmental impact of trawling on seabed: a review. *New Zealand Journal of Marine and Freshwater Research*, **26**, 59–67.

Peres, J.M. (1984) La regression des herbiers a *Posidonia oceanica*. In: *International Workshop on Posidonia Oceanica Beds* (eds C.F. Boudouresque, A. Jeudy de Grissac & J. Olivier), pp. 445–54. GIS Posidonie, Marseille.

Peres, J.M. & Picard, J. (1975) Causes de la rarefaction et de la disparition des herbiers de *Posidonia oceanica* sur le cotes francaises de la Mediterranèe. *Aquatic Botany*, **1**, 133–9.

Relini, G. & Moretti, S. (1986) Artificial reef and *Posidonia* bed protection of Loano (Western Ligurian Riviera). *Fisheries and Agriculture Organisation Fisheries Report*, **357**, 104–9.

Short, F.T. & Wyllie-Echeverria, S. (1996) Natural and human-induced disturbance of seagrasses. *Environmental Conservation*, **23**, 17–28.

Part 2
Effects of fishing on benthic fauna and habitats

The nest-building bivalve *Limaria hians* lives in association with maerl beds and is extremely vulnerable to bottom-fishing disturbance. Scale bar = 1 cm. (Reproduced with the permission of Jason Hall-Spencer.)

Chapter 4
Fishing mortality of populations of megafauna in sandy sediments

M.J.N. BERGMAN and J.W. VAN SANTBRINK

Netherlands Institute for Sea Research, Department of Marine Ecology, PO Box 59, 1790 AB Den Burg, Texel, The Netherlands

Summary

1. For a number of invertebrate species (gastropods, starfish, crustaceans and annelids) direct mortality due to the single passage of a trawl ranged from about 5% to 40% of the initial densities in the trawl track and varied from 20% to 65% for bivalve species.
2. The direct mortality of all the species studied was largely attributed to the mortality of animals that died in the trawl track, either as a direct result of physical damage inflicted by the passage of the trawl or indirectly owing to disturbance, exposure and subsequent predation. Mortality of animals caught in the net was of minor importance.
3. The annual fishing mortality of megafaunal populations (animals > 1 cm) in the Dutch sector of the North Sea ranged from 5% to 39% and the mortality of half of the species was > 20%. The 12-m beam trawl fishery caused greater annual fishing mortality than the combined action of the other fisheries acting in the same area. Differences in fishing mortality due to the 12-m and 4-m beam trawl fleets were less pronounced in coastal areas, whereas the 4-m beam trawl fleet might cause higher mortalities for some species that occur only within the 12-mile zone.
4. Generally, fragile infaunal and epifaunal species that live in reach of the groundrope and tickler chains suffer significant direct mortalities due to trawling. The long-term impact of fishing mortality on population structure and spatial distribution of faunal species, depends on their life-cycle characteristics (e.g. dispersal of eggs, survival of larvae and subadults, age of maturation and natural mortality).
5. Owing to trawling activities over the recent decades, several benthic species have decreased in abundance and some have disappeared in certain regions in the southern North Sea. To achieve an integrated approach to fisheries and ecosystem management, the following measures have to be considered: a significant reduction of trawling effort, development of gears less damaging for habitats and fauna, and designation of areas closed to fisheries for species and habitats that cannot be protected otherwise.
Keywords: bottom-trawling impacts, fishing mortality, megafaunal populations, sustainable fisheries.

Introduction

The distribution of most macrofaunal species in the North Sea is closely related to sediment grain-size composition, which is determined by water depth, current speed and organic content of the sediment (Creutzberg *et al.*, 1984; Duineveld *et al.*, 1991). Bottom trawling causes mechanical perturbation of the seabed, and thus interferes with the physical and chemical properties of the habitat and leads to direct mortality in benthic fauna (Krost, 1990). In addition to marketable fish, invertebrate species (e.g. coelenterates, annelids, molluscs, echinoderms and crustaceans) are caught in the nets of commercial trawlers, and a certain percentage will not survive after being discarded into the sea (Graham, 1955; Bridger, 1970; Houghton *et al.*, 1971; Margetts & Bridger, 1971; de Groot, 1973; Fonds, 1994; Bergman *et al.*, 1998). However, a significant fraction of the animals that occur in the path of bottom fishing gear are not captured but are killed as a result of direct contact with the gear or exposure to predators (Holme, 1983; Brown, 1989; Rees & Eleftheriou, 1989; Hall *et al.*, 1990; Krost, 1990; Langton & Robinson, 1990; Bergman & Hup, 1992; Brylinski *et al.*, 1994; Kaiser & Spencer, 1996). Quantitative estimates of the direct mortality of megafaunal species (those animals >1 cm in size) in the trawl track, although scarce, demonstrate a significant change in the composition of the community immediately after trawling (Kaiser *et al.*, 1998).

In this chapter, we present the results of field studies that aimed to calculate the direct total mortality of invertebrate infaunal and epifaunal species. These calculations included the mortality of those animals caught in the net and of those that are damaged within the trawl track by the passage of a commercial trawl. We estimated the direct mortality caused by the single passage of a commercial otter trawl or a beam trawl, the latter being the most common demersal gear in the southeastern North Sea after the 1960s (Polet *et al.*, 1998). The annual fishing mortality for megafaunal invertebrate populations in the Dutch sector of the North Sea was estimated with respect to three parameters; the direct mortality due to the single passage of a trawl, the spatial distribution of the megafauna and the trawling frequency of the different fleets (based on fishing effort data for 1994).

Methods

Direct mortality of invertebrate species

The direct mortality of invertebrates was determined for three types of commercial beam trawls (12-m wide and 4-m wide with tickler chains, and 4-m wide with chain matrices) and for otter trawls (Table 4.1). The trawls were rigged for sole fishing and were representative of commercial trawling that occurred in the study area. A number of replicated field studies, in which the different gears were studied in parallel transects, were carried out in shallow coastal sandy areas and in deeper offshore areas with a silty sediment in the Dutch and German sector in spring and late summer of 1992–1995 (Fig. 4.1). The direct mortality associated with a particular type of trawl incorporated mortality of animals caught in the net and mortality of those damaged in the trawl track. Mortality was determined by measuring the difference in densities before (t_0-sampling) and 24–48 h after (t_1-sampling) trawling a well-defined strip. It was assumed that all fatally damaged or exposed animals in the strip would be consumed by predators in the 24–48 h interval

Table 4.1   Characteristics of the commercial trawls used in the field experiments

	Width (m)	Weight (kg × 10³)	Tickler chains		Number of net ticklers	Roller dia. (cm)	Mesh size (stretched)		Towing speed (nm h⁻¹)
			Number	Weight (kg × 10³)			front (mm)	codend (mm)	
12-m Beam ticklers	12[a]	5.9–7.8	9–10	1.1–2.2	8–10	25	260	80	5–7
4-m Beam ticklers	4[a]	1.4–1.5	5	0.1–0.3	5–6	15	170	80	3.5–5.5
4-m Beam matrices	4[a]	2.7	–	0.95[b]	–	25	120	80	3.5
Otter:		1	–	–	–	20	120	80,100	3.5–4
net (+ bridles)	20[a], (32)[a]								
between wings	15[a]–20[a]								
between doors	35–55								

[a]In contact with seabed, [b]chain matrices (see Fonteyne, Chapter 2).

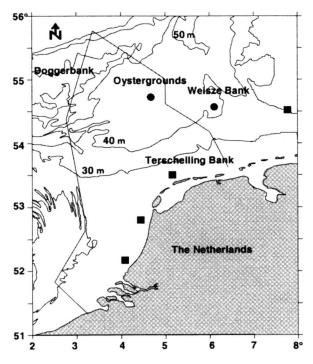

Fig. 4.1   Locations of the field experiments that examined the direct mortality caused by different types of trawls in the south-eastern North Sea: ■, sandy coastal areas (median grain size 200–370 μm; silt content 1–5%); ●, silty offshore areas (median grain size 150–170 μm; silt content 3–10%).

between trawling and t_1-sampling period. The proportion of a particular species that survived capture and discarding by the trawler was subtracted from the difference between the initial and residual densities in the trawled strip of seabed. Direct mortality (M_{dir}) was expressed as a percentage of the initial density in the trawl track:

$$M_{dir}(\%) = 100 \times \frac{D_{t_0} - [D_{t_1} + C \times (1 - 0.01 \times M_{dis})]}{D_{t_0}}$$

where D_{t_0} is the density (n m^{-2}) in the strip before trawling (t_0-sampling), D_{t_1} is this density after trawling (t_1-sampling), C is the number caught by the trawl (n m^{-2} swept area), M_{dis} is the mortality of animals caught in the trawl during the fishing of the the strip of seabed (% of the numbers caught; Fonds, 1994; Bergman *et al.*, 1998).

The area of a trawled strip of seabed was *c.* 2000 m long × 60 m wide, and the distance between parallel strips was *c.* 300 m. Megafauna were sampled (10 samples per strip of seabed) using the Triple-D benthos dredge (sampling depth 10 cm, stretched net mesh size 1.4 cm, sample size *c.* 25 m^2), which was designed to sample large-sized, relatively low-abundance in- and epifaunal species (Bergman & van Santbrink, 1994). In some studies, additional samples of the macrofauna (>1 mm) were collected using a Reineck boxcorer ($n = 20$) or Van Veen grab ($n = 12$–18). Since it was impossible to create a homogeneously trawled strip of seabed, we aimed for a trawling intensity of 150%, i.e. the surface of a strip was trawled on average 1.5 times. It is not uncommon for commercial trawlers to re-fish the same tracks within a few hours or days. In some studies, an extra heavily trawled strip (on average 300%) was created using 12-m beam trawls. Our own observations and those of our colleagues suggest that this is not unrealistic, as we have seen trawlers literally fishing in a line one behind the other. Side-scan sonar recordings of the seabed were made immediately after each sampling period and each trawling series to check the positions of the tracks of the Triple-D dredge relative to those of commercial trawls.

Direct mortality was calculated for sedentary and relatively immobile species, while excluding mobile epibenthic species that could migrate to or from the trawled strip within the 24–48-h interval after trawling occurred. Mortality for a particular species was calculated when the initial density was at least 5 per 100 m^2 (Triple-D) or 10 per m^2 (boxcore). The differences between the geometric mean initial and remaining densities were statistically tested (paired *t*-test on log-transformed data; one-sided for sedentary species and two-sided for mobile species). For each species, direct mortality was calculated for the different types of trawls in coastal sandy sediments and in offshore silty sediments. The direct mortality calculated due to a trawl intensity of on average 150%, was converted to that which would have been caused by the single passage of a trawl, i.e. 100% disturbance of the seabed. For the heart urchin *Echinocardium cordatum*, the measured mortality was corrected also for the proportion of the population that was actually in reach of the Triple-D, which was estimated at *c.* 25% in sandy areas and 60% in silty areas (Bergman & Hup,

Table 4.2 Direct mortality (% of the initial density in the trawl track) of macrofaunal species sampled with a Reineck boxcorer ($n = 20$) due to the single (100%) passage of a 12-m beam trawl in the silty Oystergrounds and with 4-m beam trawls in the sandy coastal zone south of Terschelling Bank

	Size (mm)	Type of trawl	Mortality (%)
Bivalves	length		
Arctica islandica	2–3	12 m	20
Corbula gibba	1–11	12 m	9
Donax vittatus	20–35	4 m	10
Mysella bidentata	2–3	12 m	4
Nucula nitidosa	2–10	12 m	4
Spisula sp. juv.	1–6	4 m	20
Tellimya ferruginosa	2–7	4 m	19*
Gastropods	height		
Cylichna cylindracea	3–8	12 m	14
Turritella communis	5–15	12 m	20*
Echinoderms	diameter		
Amphiura sp.	2–6	12 m	9
Crustaceans	length	12 m	
Callinassa subterranea	5–40	12 m	4
Cumacea	3–7	12 m	22*
Gammaridea	2–11	12 m	28
Annelids	length		
Pectinaria koreni	4–20	12 m	31*
Magelona papillicornis		12 m	30*
Scoloplos armiger		12 m	18
24 annelid spp. (excl. *Pectinaria*)		12 m	<0.5

*Indicates significance at $P<0.05$, using a paired t-test on log-data.

1992; unpublished data). This is an underestimate in the reproductive season when the animals move closer to the surface of the sediment.

To determine the relative vulnerability of species, species have been ranked according to their direct mortality caused by each trawl type in each study. Species were included only when direct mortality was determined for at least five different strips of seabed (i.e. five replicates).

Annual fishing mortality in megafaunal populations

The annual fishing mortality in a number of megafaunal populations in the Dutch continental shelf was calculated using three variables: the spatial distribution of

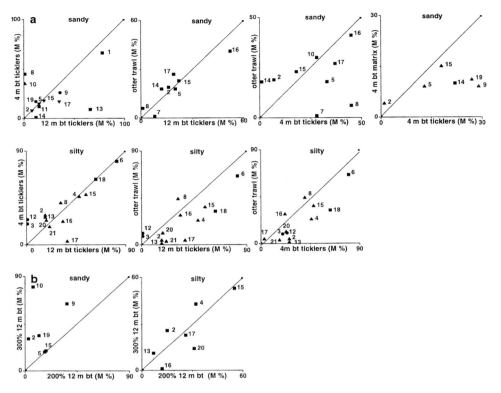

Fig. 4.2 Comparison of the direct mortality (M_{dir}), expressed as a percentage of the initial density, caused by (a) different commercial trawls and (b) two different trawling intensities (200% and 300%), in sandy and silty areas of the south-eastern North Sea (bt = beam trawl). Symbols denote the number of studies from which the results are obtained: ■, 1 study; ●, 2 studies; ▲, 3 studies; ▼, 4 studies. No difference in direct mortality was caused by different gears when the species fell on the line of equitability. Species codes: 1. *Angulus fabulus*; 2. *Chamelea gallina*; 3. *Corbula gibba*; 4. *Dosinia lupinus*; 5. *Ensis* spp; 6. *Gari fervensis*; 7. *Mactra corallina*; 8. *Phaxas pellucidus*; 9. *Spisula solida*; 10. *Spisula subtruncata*; 11. *Lunatia catena*; 12. *Turritella communis*; 13. *Astropecten irregularis*; 14. *Ophiura texturata*; 15. *Echinocardium cordatum*; 16. *Corystes cassivelaunus* (male); 17. *C. cassivelaunus* (female); 18. *C. cassivelaunus* (juv.); 19. *Thia polita*; 20. *Aphrodita aculeata*; 21. *Pelonaia corrugata*.

species in 1996 (Bergman & van Santbrink, 1998); the fishing frequency of the different trawl fleets in 1994; and the direct mortality due to a single passage of a particular commercial trawl. It was assumed that the distribution patterns of species in 1996 were roughly similar to those in 1994 and that the distribution of a species within an ICES statistical rectangle was homogeneous. The mean trawling frequencies per ICES rectangle by the Dutch, Belgian, German and British fleets in 1994 were calculated from the surface area of the ICES rectangles and the numbers of fishing hours (Polet *et al.*, 1998) of the bottom trawling fleets (as above). To simulate the clustered distribution of fishing effort that exists in the North Sea (Rijnsdorp *et al.*, 1998), each ICES rectangle was divided in nine subrectangles, over which the total trawling effort of a fleet in that ICES rectangle was redistributed. The direct mortality of megafauna due to a single passage of a commercial trawl was calculated in the first section of this chapter (Table 4.3). As the direct mortality of species due to 12-m and 4-m beam trawls that are rigged with tickler chains did not differ greatly (Fig. 4.3), the mean mortality for these beam trawls was used in the calculations of annual fishing mortality. In the calculations of the fishing mortality, recruitment and growth were not included. More detailed information on the calculation of the fishing mortality is given in Bergman & van Santbrink (1999).

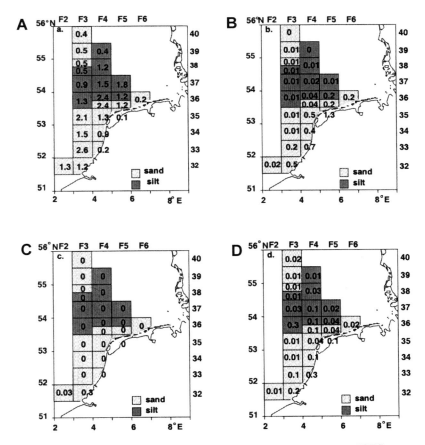

Fig. 4.3 Distribution of trawling effort (mean trawling frequency per ICES rectangle, i.e. trawled area/total area) for different types of trawl fisheries in the Dutch continental shelf in 1994 (Polet *et al.*, 1998): (A) 12-m beam trawl fishery rigged with tickler chains; (B) 4-m beam trawl fishery rigged with tickler chains; (C) 4-m beam trawl fishery rigged with chain matrices; (D) otter trawl fishery (flatfish and roundfish).

Table 4.3 Mean direct mortality (% of the initial density in the trawl track) of megafaunal species sampled with the Triple-D ($n = 10$) due the single passage of a trawl with different trawls in silty and sandy areas of the southern North Sea (see Fig. 4.1)

	12-m Beam trawl ticklers Silty		4-m Beam trawl ticklers Silty		Otter trawl Silty		12-m Beam trawl ticklers Sandy		4-m Beam trawl ticklers Sandy		4-m Beam trawl matrices Sandy		Otter trawl Sandy	
Size class (cm)	No. of strips	Mort. (%)	No. of strips	Mort. (%)	No. of strips	Mort. (%)	No. of strips	Mort. (%)	No. of strips	Mort. (%)	No. of strips	Mort. (%)	No. of strips	Mort. (%)
Bivalves														
Abra alba 1–1.5	4*	18	3**	38	3	<0.5	–	–	–	–	–	–	–	–
Angulus fabulus 1–1.5	–	–	–	–	–	–	1*	64	1*	52	–	–	–	–
Angulus tenuis 2–3	–	–	–	–	–	–	–	–	2	15	–	–	–	–
Arctica islandica 8–11	3*	5	2*	22	2	8	–	–	–	–	–	–	–	–
Chamelea gallina 1–3	4*	12	3**	18	3	3	4	5	7*	4	3*	2	1	13
Corbula gibba 1	2	<0.5	2*	14	2	6	–	–	–	–	–	–	–	–
Dosinia lupinus 1–4	4**	31	3**	33	3*	16	–	–	–	–	–	–	–	–
Ensis spp. 10–20	3**	<0.5	3**	7	3**	<0.5	4*	10	7**	9	3*	6	1	12
Gari fervensis 4–6	1*	68	1*	66	1*	52	–	–	–	–	–	–	–	–
Mactra corallina 3–5	2	16	2	25	2*	22	1	5	2	8	–	–	1	<0.5
Mysia undata 1.5–3	1*	35	–	–	–	–	–	–	–	–	–	–	–	–
Phaxas pellucidus 1.5–3	4**	27	3*	29	3*	32	1	<0.5	1*	33	–	–	1	4
Spisula solida 2–5	–	–	–	–	–	–	2*	26	3***	20	3*	6	–	–
Spisula subtruncata (l) 1.5–3	–	–	–	–	–	–	3	<0.5	3*	23	2*	29	1	21
Spisula subtruncata (h) 1.5–3	–	–	–	–	–	–	–	–	1*	47	–	–	–	–

Table 4.3 *continued*

	Size class (cm)	12-m Beam trawl ticklers Silty		4-m Beam trawl ticklers Silty		Otter trawl Silty		12-m Beam trawl ticklers Sandy		4-m Beam trawl ticklers Sandy		4-m Beam trawl matrices Sandy		Otter trawl Sandy	
		No. of strips	Mort. (%)	No. of strips	Mort. (%)	No. of strips	Mort. (%)	No. of strips	Mort. (%)	No. of strips	Mort. (%)	No. of strips	Mort. (%)	No. of strips	Mort. (%)
Gastropods	*height*														
Lunatia catena	1–3	–	–	–	–	–	–	1	10	1	10	–	–	–	–
Turritella communis	3–6	3*	5	2	17	2	7	–	–	–	–	–	–	–	–
Echinoderms	*diam.*														
Astropecten irregularis	3–6	4*	12	3**	18	3	0	1*	52	1	6	–	–	–	–
Echinocardium cordatum	3.5–5	4****	40	3***	35	3**	26	4****	14	7******	12	3***	10	1*	16
Ophiura texturata	5–11	–	–	–	–	–	–	1	8	3	2	1	7	1	12
Crustaceans	*width*														
Corystes cassivelaunus m.	2–3	4*	22	3	15	3*	20	2**	39	2*	31	–	–	1	30
Corystes cassivelaunus f.	2	4**	28	3	2	3	3	4*	25	5	7	3	7	1	19
Corystes cassivelaunus j.	<1.5	1*	48	1*	49	1	23	–	–	–	–	–	–	–	–
Thia scutellata	1–1.5	–	–	–	–	–	–	2	8	3*	19	3*	7	–	–
Other groups	*length*														
Aphrodita aculeata	3–14	4*	8	3	16	3	7	–	–	–	–	–	–	–	–
Golfingia spec.	3–7	2	20	2	5	2*	33	–	–	–	–	–	–	–	–
Pelonia corrugata	3–7	3*	14	3	11	3	2	–	–	–	–	–	–	–	–

Total number of strips studied is given. Mean trawling intensity in each strip was 150%. Significant differences in the % mortality that results from trawling ($P<0.05$) are indicated for the number of experimental strips in which significant differences occurred (number of *) using paired *t*-test on log-transformed data. Replicate studies were averaged after weighing based on the 95% confidence intervals. *Angulus fabulus* was sampled with the Van Veen grab ($n = 12$–18). *S. subtruncata* (l) indicates a low density (0.1 m^{-2}), (h) indicates a high density (24 m^{-2}).

The survival rate S of a species x in a subrectangle r after trawling that subrectangle with a particular gear g is described by the following power function:

$$S_{x,r,g} = (1 - Md_{g,x}/100)^{f(r,g)}$$

where $Md_{g,x}$ is the direct mortality estimate of species x for a gear g expressed as a percentage of the initial density in sandy or silty sediment, and $f(r,g)$ is the trawling frequency in subrectangle r ($r = 1, ..., 9$) with gear g.

The fishing mortality of a species x in a subrectangle r due to a gear g is then given by $1 - S_{x,r,g}$. The fishing mortality F in an ICES rectangle R is the average of the fishing mortalities in the nine subrectangles. Fishing mortality is multiplied by 100 to express it as a percentage of the initial population density:

$$F_{x,R,g}(\%) = 100 \times \left(\sum_r (1 - S_{x,r,g})\right)/9.$$

Based on the calculation of $F_{x,R,g}$, the fishing mortality for a species in the entire Dutch sector ($F_{x,g}$) was calculated for the different types of fisheries by dividing the sum of the specimens killed as a result of a particular fishery in each ICES rectangle by the sum of the actual numbers of specimens present in those rectangles:

$$F_{x,g}(\%) = \sum_R (n_{x,R} \times F_{x,R,g}) / \sum_R n_{x,R}$$

where $n_{x,R}$ is the initial numbers of specimens of species x in ICES rectangle R.

In this calculation, the decline in numbers of specimens calculated separately for each type of fishery occurs more slowly than the actual decline due to the combined action of the different types of trawl fleets. Therefore, the fishing mortality due to individual fisheries is slightly overestimated. Hence, to avoid this overestimate, the overall fishing mortality $F_{overall}$ for a species (i.e. due to the four types of fisheries considered) was calculated using the overall survival rate, instead of using the sum of individual fishing mortalities. This overall survival rate was calculated as the product of the survival rates within each type of fishery. For example, if the survival rate due to a certain fishery is 0.4 and due to another 0.5, the total survival rate is $0.4 \times 0.5 = 0.2$. The overall fishing mortality, when expressed as a percentage of the initial density, is given by:

$$F_{overall,x}(\%) = 100(1 - \Pi_g(1 - F_{x,g}/100)).$$

Results

Direct mortality of invertebrate species

After a single tow of a 4- or 12-m beam trawl in sandy and silty areas, some small-sized bivalves and crustaceans and some annelid worms showed direct mortalities up to 22% or 31%, respectively, of the initial densities in the trawl track (Table 4.2). The mortality of many other small annelids was negligible. After a single trawl event, the direct mortality for larger-sized species in different taxonomic groups was related to the type of trawl used and sediment-type (Table 4.3). Mortalities up to 38% (in a few cases up to 68%) of the initial densities occurred for large-sized bivalve species. For gastropods, various starfish species and the sea mouse *Aphrodita aculeata* direct mortalities were generally <18%, and for the sea urchin *Echinocardium cordatum* between 10 and 40%. Up to 39% of the adult population, and up to 49% of the juveniles, of the crab species studied died directly as a result of trawling.

Direct mortality was related to sex, density or size for some megafaunal species (Table 4.3). Female *Corystes cassivelaunus* generally suffered lower mortality than males (2–28% vs. 15–39%). When the bivalve *Spisula subtruncata* occurred in low densities (about 0.1 m^{-2}), direct mortality was about 23%, but when it occurred in high densities (>24 m^{-2}), mortality was as high as 47%. Direct mortality due to beam trawling was negligible for large (>1 cm) specimens of the gastropod *Lunatia catena*, whereas smaller specimens suffered a mortality of up to 49%. The opposite was the case for the bivalve *Chamelea gallina*: larger specimens (>2 cm) were more vulnerable (26% mortality) compared with small specimens (7%). A similar trend was also observed for the polychaete *Aphrodita aculeata*, with mortalities of up to 31% for large specimens (>7 cm), whereas no mortality was observed for smaller animals. Mortality was higher for the bivalves *Chamelea gallina*, *Mactra corallina* and the urchin *Echinocardium cordatum*, in silty compared with sandy areas of the seabed.

The effects of gear and trawling intensities

Figure 4.2 presents the direct mortality of invertebrates due to different types of trawls that were tested in the same area simultaneously. For the majority of benthic species, differences in direct mortality due to trawling with 4-m or 12-m beam trawls rigged with tickler chains were not obvious either in sandy coastal or in silty offshore areas. In the hard-sandy coastal zone, mortality due to 4-m beam trawls with tickler chains did not differ consistently from that caused by 4-m beams with chain matrices, although higher mortalities were found for at least three infaunal species (*Spisula solida*, *Ophiura ophiura*, *Thia scutellata*). Otter trawling, although tested at only one sandy location, caused direct mortalities in most benthic species in the same order of magnitude as beam trawls. In silty areas (three different locations), however, otter trawling clearly caused less mortality than beam trawling for a number of species (e.g. *Chamelea gallina*, *Dosinia lupinus*, juvenile *Corystes cassivelaunus*).

In a sandy seabed, direct mortalities in the 300%-trawled strip appeared to be higher than in a parallel 200%-trawled strip for a number of bivalve species (e.g. *Chamelea gallina*, *Spisula solida* and *S. subtruncata*) and the crab *Thia scutellata*. In a similar study in a silty area, however, no clear differences in mortalities due to different trawl intensities were found.

Relative vulnerability of invertebrate species

The relative vulnerability of megafaunal species (Table 4.4) reflects their mean ranking with respect to direct mortality in different sediments. In general, *Echinocardium cordatum*, *Corystes cassavilaunus* (male), and bivalves such as *Phaxas pellucidus*, *Dosinia lupinus*, *Mactra corallina*, *Abra abra*, *Spisula solida* and *S. subtruncata* appeared to be the most vulnerable species. Bivalves such as *Ensis* spp., *Corbula gibba* and *Chamelea gallina*, and starfish (e.g. *Astropecten irregularis*, *Ophiura texturata*) were relatively resistant to bottom trawling.

Mean trawling effort per ICES rectangle in 1994

The distribution of the mean effort of different trawler fleets across the ICES rectangles of the Dutch sector in 1994 (Fig. 4.3) demonstrated the dominance of the 12-m beam trawl fishery, which occurred at an average frequency of 1.23. This contrasts with the mean frequencies for the 4-m beam trawl fishery with ticklers (0.13), that with chain matrices (0.01) and the otter trawl fishery (0.06). Obviously,

Table 4.4. Ranking of megafaunal species according to their mean relative vulnerability to trawling on silty and sandy areas. The species with the highest mean rank number is the most vulnerable. The number of study strips on which the ranking was based is indicated

Silty areas	No. of strips	Mean rank	Sandy areas	No. of strips	Mean rank
Echinocardium cordatum	10	8.0	*Corystes cassivelaunus* (male)	5	9.1
Phaxas pellucidus	10	7.3	*Spisula subtruncata*	10	6.9
Dosinia lupinus	10	6.5	*Spisula solida*	8	6.6
Mactra corallina	6	6.4	*Echinocardium cordatum*	15	5.6
Golfingia sp.	6	5.9	*Corystes cassivelaunus* (female)	13	5.5
Corystes cassivelaunus (male)	10	4.9	*Thia scutellata*	8	4.3
Abra alba	10	4.9	*Ophiura texturata*	6	3.6
Turritella communis	7	4.4	*Ensis* spp.	15	3.1
Arctica islandica	7	4.4	*Chamelea gallina*	15	3.0
Corystes cassivelaunus (female)	10	3.9			
Aphrodita aculeata	10	3.9			
Pelonia corrugata	9	3.7			
Chamelea gallina	10	3.7			
Corbula gibba	6	3.3			
Astropecten irregularis	10	3.1			
Ensis spp.	9	0.8			

the different types of fisheries were not distributed homogeneously over the sector. The 12-m beam trawl fishery occurred predominantly offshore. The 4-m beam trawls rigged with tickler chains were mainly deployed in the coastal zone. The 4-m beam trawls rigged with chain matrices were used exclusively in the mobile, medium-grained sandy areas in the two southernmost rectangles. Otter trawls rigged for flatfish and roundfish were used throughout the sector. Although the otter trawl fishery for roundfish is currently widely practised, the use of these trawls for flatfish has declined strongly during recent decades (Polet *et al.*, 1998).

Annual fishing mortality in megafaunal populations

The fishing mortality in 1994 of invertebrate megafaunal populations in the Dutch continental shelf varied from 5% to 39% (Table 4.5). For all species considered, the 12-m beam trawl fishery caused the highest annual fishing mortality. For species that are restricted to sandy areas, where 4-m beam trawl fisheries are concentrated mainly, the difference in fishing mortality between both fleets was less pronounced (e.g. *Spisula solida*, *S. subtruncata*, *Ensis* spp.). This also applies to *Ensis* spp. dominated by the strictly coastal species *Ensis americanus*. Otter trawl fisheries caused annual fishing mortalities that were generally similar to the 4-m beam trawl fleet. The fishing mortality caused by the 4-m beam trawl fleet rigged with chain matrices was less than 0.5% for all species studied.

Discussion

Direct mortality of invertebrate species

The single passage of a 4-m or a 12-m wide beam trawl or an otter trawl caused direct mortalities in a number of invertebrate species ranging from 5% to 50% and up to 68% for some bivalve species (Tables 4.2 and 4.3). Direct mortalities in crustacean and starfish species might be underestimated, as some of the 'survivors' in the trawled strip during the t_1-sampling may, in fact, be predatory immigrants. This might explain why no mortality was apparent for the larger faster moving specimens of the mobile carnivore *Lunatia catena*, in contrast to the considerable mortality observed in smaller specimens. Mortality of the fragile heart urchin *Echinocardium cordatum* was estimated at 10–40% of their initial density, after a correction for the animals' depth distribution. Assuming that most animals in reach of the trawl would be killed, mortality might increase up to about 90% during the relatively short reproductive season in summer (Buchanan, 1966) when animals migrate to the surface of the sediment. Some sedentary bivalves (*Lutraria lutraria*, *Mya truncata*, *Nucula nitidosa*) and anemone species apparently increased in density after trawling. This suggests that a larger fraction of the population came into the reach of the Triple-D (sampling depth 10 cm), possibly because of resuspension of the top layers

Table 4.5 Fishing mortality (%) in the populations of invertebrate megafaunal species in the Dutch sector due to bottom trawl fisheries in 1994

		Fishing mortality (%) in the Dutch sector in 1994				
	Length class (cm)	12-m Beam trawl ticklers fishery	4-m Beam trawl ticklers fishery	4-m Beam trawl matrices fishery	Otter trawl fishery	All trawl fisheries
All sediments						
Chamelea gallina	<2	5	<0.5	<0.5	<0.5	5
Chamelea gallina	>2	22	1	<0.5	1	24
Corystes cassivelaunus f.	>1.5[a]	17	1	<0.5	<0.5	18
Corystes cassivelaunus m.	>1.5[a]	26	1		2	28
Echinocardium cordatum	>3	20	2	<0.5	3	24
Ensis spp.	>10	7	3	<0.5	1	11
Mactra corallina	>1	13	1	–	1	15
Phaxas pellucidus	>1.5	16	<0.5		1	17
Sandy sediment						
Lunatia catena	<1.5	27	17	–	–	39
Ophiura texturata	>0.5[b]	5	1	<0.5	1	7
Spisula solida	>1	17	8	<0.5	–	24
Spisula subtruncata	>1	12	6	<0.5	2	19
Thia scutellata	>0.5[a]	17	3	<0.5	–	19
Silty sediment						
Abra alba	>0.5	24	1		<0.5	25
Aphrodita aculeata	>7	20	<0.5		1	21
Arctica islandica	>8	11	<0.5		<0.5	11
Astropecten irregularis	>2.5[b]	14	<0.5		0	14
Corystes cassivelaunus j.	<1.5[a]	29	1		1	30
Dosinia lupinus	>0.5	24	1		1	26
Gari fervensis	>2.5	33	<0.5		3	35
Pelonaia corrugata	>1	14	<0.5		<0.5	14
Turritella communis	>1.5	12	1		<0.5	13

–, no direct mortality estimate measured for this species. blank cells, no overlap in trawling and species distribution. When size is not based on length, this is indicated by: [a]carapace width; [b]diameter oral disk.

of the sediment (Kaiser & Spencer, 1996; Kaiser *et al.*, 1998). It can be assumed that these species were only slightly affected by trawling, as they usually live at depths greater than the penetration depth of the trawls (*c.* 6 cm) (Laban & Lindeboom, 1991).

The direct mortality calculated in the studies incorporates mortality of animals caught in the net and of those damaged in the trawl track. Commercial trawls generally have a low catch efficiency for invertebrates: for most species studied less than 10%, and for almost half the number of species less than 5% (Craeymeersch *et al.*, 1998). Therefore, despite the fact that the mortality measured in invertebrate discards is high, ranging from 26% to 88% for bivalve species, from 25% to 67% for crustaceans and from 11% to 21% for starfish (Fonds, 1994; Bergman *et al.*, 1998), the mortality of these discarded animals is still only a few per cent of their initial density in the trawl track. Hence, discard mortality plays only a minor role in the direct mortality that is presented in this chapter. It can therefore be concluded that most direct mortality takes place among animals in the trawl track, either as a direct result of physical damage inflicted by the passage of the trawl or indirectly due to disturbance, exposure and subsequent predation.

Variation between different trawls and areas

The lower mortality of invertebrates associated with otter trawling suggests that the groundrope plus bridles of the otter trawls do not disturb silty sediments as deeply as beam trawls. Mortality due to the doors could not be measured, but as their width accounts for less than 10% of the total gear width, total mortality is probably only slightly underestimated. In the comparison of 4-m and 12-m beam trawls, none of the trawl types caused higher mortalities for the majority of megafauna. In these studies the direct mortality of several infaunal species was higher in silty areas than in sandy areas (Table 4.3). This suggests that beam trawls penetrate more deeply into the softer silty seabed, leading to higher mortalities.

Relative vulnerability of invertebrate species

Most of the larger-sized species that appeared to be relatively resistant to trawling (Table 4.4) were rather robust (e.g. *Astropecten irregularis*, *Chamelea gallina*), or burrow deeply into the sediment (e.g. *Ensis* spp.). Robust bivalve species such as *Corbula* and *Astarte* were able to survive contact with otter trawl doors (Rumohr & Krost, 1991). The most vulnerable species were those that were physically the most fragile (e.g. *Echinocardium cordatum*, *Phaxas pellucidus*), or lived in the uppermost layer of the sediment where they were within reach of the trawl (e.g. *Spisula* spp.). In studies near Northern Ireland, the fragile epibenthic bivalve *Modiolus modiolus* also showed high mortality due to otter trawling and scallop dredging (Brown, 1989). Although some species might appear to be capable of withstanding physical contact with fishing gear, subsequent mortality is often delayed. For example, 55% of the common whelks caught in 12-m beam trawls died after 6 weeks of maintenance in laboratory conditions (Mensink *et al.*, unpublished data). Additional mortality in developing embryos is likely when the egg capsules that are normally attached to the seabed are torn loose by the trawl.

In general, small-sized species (Table 4.2) tend to show lower direct mortalities, when compared with larger-sized species (Table 4.3) and smaller individuals of megafaunal species tend to show lower mortalities than larger-sized ones (e.g. *Chamelea gallina* and *Aphrodita aculeata*). Results of other studies show similar trends in bivalves in the Irish Sea (Bergman *et al*., 1998), and in *Echinocardium cordatum* (5–10 mm) in the North Sea (Bergman & Hup, 1992). This trend is probably caused by a size-related impact on bottom fauna. Trawling affects small-sized benthos mainly by disturbing the sediment, whereupon the animals are resuspended and possibly translocated by currents (which is comparable to natural disturbances such as storms), whereas it affects larger-sized benthos through direct physical contact with the gear (groundrope, tickler chains). Therefore, the low impact of trawling on benthos inhabiting mobile sediments suggested by Kaiser & Spencer (1996) might be correct for small-sized animals (<1 cm). Larger-sized species inhabiting the dynamic zone, like *Spisula subtruncata* (1.5–3.0 cm), suffer mortalities up to 47% (Table 4.3). The higher direct mortality found in the bivalve *Chamelea gallina* in stable silty areas when compared with those living in mobile sandy areas is probably related to the deeper penetration of beam trawls into softer seabeds.

Our studies indicate that direct mortality within a species may vary with sex and density. Females of *Corystes cassivelaunus*, which burrowed in the seabed, appeared relatively more protected against the impact of trawling than males living at the seabed. Trawling might induce changes in the sex structure of such populations. The density-related mortality in *Spisula subtruncata* observed after 4-m beam trawling might be related to differences in the texture of the upper sediment layers, leading to increased penetration in the high density areas.

Annual fishing mortality in megafaunal populations

The fishing mortality in 1994 of the invertebrate megafaunal populations in the Dutch sector ranged from 5% to 39%, for which half the number of species had mortalities $>20\%$ (Table 4.5). The fishing mortality due to the 12-m beam trawl fishery was much higher than that due to the 4-m beam trawl fishery, mainly owing to the low effort of the latter fishery in offshore areas. For species restricted to sandy coastal areas, where the effort of the 4-m beam trawl fisheries was concentrated, the difference in fishing mortality was less pronounced (e.g. *Spisula* spp). For those species exclusively restricted to the 12-mile zone (such as *Ensis americanus*), where 12-m beam trawling is forbidden, may the 4-m fleet account for the highest mortality even averaged over the entire Dutch sector. The low annual fishing mortality due to the fleet's use of 4-m beams rigged with chain matrices (for all species $<0.5\%$) is due to the very restricted spatial distribution of effort, i.e. the two southernmost rectangles in the Dutch sector, where none of the selected species occurred in high densities. The fishing mortality due to otter trawl fisheries tended to be slightly lower than due to 4-m beam trawl fisheries. The fishing mortalities due to otter trawling, however, might be overestimated, as the effort data included otter trawling for roundfish, whereas the direct mortality was based solely on otter trawling for flatfish (which is assumed to have the highest impact on the seabed and the benthic fauna).

Despite the annual fishing mortality of 5–39% demonstrated in a number of invertebrate megafaunal populations, apparently these species were able to maintain a certain population density. The long-term impact of fishing mortality on populations depends on life-history characteristics of species in question (e.g. dispersal of eggs, survival of larvae and subadults, migration behaviour, age of maturation and natural mortality rate). Analyses of historical datasets of the occurrence, abundance and catchability of invertebrate species indicate that species showing a high direct fishing mortality or life-history characteristics less suited to offset fishing mortality have become scarce or have disappeared in certain regions of the southern North Sea. Demersal fisheries are thought to be responsible for the decreased abundances of bivalve species such as *Ostrea edulis*, *Modiolus modiolus*, *Arctica islandica*, the edible crab *Cancer pagurus*, lobster *Homarus gammarus*, gastropods (e.g. *Buccinum undatum*), anthozoa (*Alcyonium* spp.), sponges (*Halichondria* spp.) and tube-building polychaetes (*Pectinaria* spp., *Sabellaria* spp.) (Bergman *et al.*, 1991; Vooys *et al.*, 1993; Cadée *et al.*, 1995; Lindeboom & de Groot, 1998; Rumohr *et al.*, 1998). Generally, these fragile infaunal and epifaunal species live within reach of the groundrope and tickler chains. Modelling results indicate that the observed decline in stocks of a number of crabs, lobsters, urchins and gastropods appeared to be related to changes in the type of bottom trawl used (otter trawl vs. beam trawl) and in fishing effort (Philippart, 1998). Evidence of long-term changes in benthic communities has been found in other parts of the world. In Loch Gareloch (west Scotland), fragile sessile anthozoan species were adversely affected by trawling, while, in general, opportunistic species increased in abundance, and measures of diversity and evenness were consistently lower (Tuck *et al.*, 1998). In Hauraki Gulf (New Zealand), decreases were observed in the densities of echinoderms, long-lived surface dwellers, large epifauna, the total number of species and individuals, and in the Shannon–Weiner diversity index. In addition, increased densities were observed in deposit-feeders and small opportunistic species (Thrush *et al.*, 1998).

Since fishery-related, long-term changes in benthic communities were observed in many fishing grounds, the reduction of this impact needs to play an essential role in discussions on the development of sustainable fisheries. In the framework of an integrated fisheries and ecosystem management, a sustainable North Sea ecosystem has to be a common objective. In addition to a reduction in fishing effort and development of more selective fishing gears, areas closed for fisheries are recommended. The outlines of closed areas (location, dimensions, duration of closure) have to be attuned to the life-cycle strategies of the species to be protected (Bergman & Lindeboom, 1999).

Acknowledgements

The authors thank the crew of RV *Mitra* (RWS/DNZ) for their assistance throughout the field experiments. Funding for this study was provided by a research grant (AIR2 CT1664) from the Commission of the European Communities within the frame of the EC programme in the Fisheries sector. This article is publication No. 3341 of the Netherlands Institute for Sea Research (NIOZ).

References

Bergman, M.J.N. & Hup, M. (1992) Direct effects of beam trawling on macrofauna in a sandy sediment in the southern North Sea. *ICES Journal of Marine Science*, **49**, 5–11.

Bergman, M.J.N. & Lindeboom, H.J. (1999) Natural variability and the effects of fisheries in the North Sea: towards an integrated fisheries and ecosystem management? In: *Biogeochemical Cycling and Sediment Ecology* (eds J.S. Gray, W. Ambrose Jr. & A. Szaniawska). *Proceedings of the Poland NATO Workshop*. Kluwer, Dordrecht, pp. 173–84.

Bergman, M.J.N. & van Santbrink, J.W. (1994) A new benthos dredge (TRIPLE-D) for quantitative sampling of infaunal species of low abundance. *Netherlands Journal of Sea Research*, **33**, 129–33.

Bergman, M.J.N. & van Santbrink, J.W. (1998) *The distribution of larger sized invertebrate species (megafauna) in the Dutch sector of the North sea*. BEON Report 1998-2, pp. 55–89. Netherlands Institute for Sea Research, Den Burg, Texel, The Netherlands.

Bergman, M.J.N. & van Santbrink, J.W. (in press) Fishing mortality in megafaunal benthic populations due to trawl fisheries in the Dutch continental shelf in the North Sea in 1994. *ICES Journal of Marine Science*.

Bergman, M.J.N., Lindeboom, H.J., Peet, G., Nelissen, P.H.M., Nijkamp, H. & Leopold, M.F. (1991) *Beschermde gebieden Noordzee, noodzaak en mogelijkheden*. NIOZ Report 1991-3, 195 pp. Netherlands Institute for Sea Research, Den Burg, Texel, The Netherlands.

Bergman, M.J.N., Ball, B., Bijleveld, C., Craeymeersch, J.A., Mundy, B.W., Rumohr, H. & van Santbrink, J.W. (1998) Direct mortality due to trawling. In: *The effects of different types of fisheries on the North Sea and Irish Sea benthic ecosystems* (eds H.J. Lindeboom & S.J. de Groot), pp. 167–185. *NIOZ Report* 1998-1/*RIVO-DLO Report* C003/98. Netherlands Institute for Sea Research, Den Burg, Texel, The Netherlands.

Bridger, J.P. (1970) Some effects of the passage of a trawl over the seabed. *International Council for the Exploration of the Sea*, C.M./B:10, pp. 10.

Brown, R.A. (1989) Bottom trawling in Strangford Lough: problems and politics. In: *Third North Sea Seminar 1989* (eds C.C. ten Hallers & A. Bijlsma), pp. 117–27.

Brylinski, M., Gibson, J. & Gordon, D.C. (1994) Impacts of flounder trawls on the intertidal habitat and community of the Minas Basin, Bay of Fundy. *Canadian Journal of Fisheries and Aquatic Science*, **51**, 650–61.

Buchanan, J.B. (1966) The biology of *Echinocardium cordatum* (Echinodermata: Spatangoidea) from different habitats. *Journal of the Marine Biological Association (UK)*, **46**, 97–114.

Cadée, G.C., Boon, J.P., Fischer, C.V., Mensink, B.P. & Hallers-Tjabbes, C.C. (1995) Why the whelk (*Buccinum undatum*) has become extinct in the Dutch Wadden Sea. *Netherlands Journal of Sea Research*, **34**, 337–9.

Craeymeersch, J.C., Ball, B., Bergman, M.J.N., Damm, U., Fonds, M., Munday, B.W. & van Santbrink, J.W. (1998) Catch efficiency of commercial trawls In *The effects of different types of fisheries on the North Sea and Irish Sea benthic ecosystems* (eds H.J. Lindeboom & S.J. de Groot), pp. 157–67. *NIOZ Report* 1998-1/*RIVO-DLO Report* C003/98. Netherlands Institute for Sea Research, Den Burg, Texel, The Netherlands.

Creutzberg, F., Wapenaar, P., Duineveld, G.C.A. & Lopez Lopez, N. (1984) Distribution and density of benthic fauna in the southern North Sea in relation to bottom characteristics and hydrographic conditions. *Rapports et Procès-verbaux des Réunions Conseil International pour l'Exploration de la Mer*, **183**, 101–10.

Duineveld, G.C.A., Kunitzer, A., Niermann, U., de Wilde, P.A.W.J. & Gray, J. (1991) The macrobenthos of the North Sea. *Netherlands Journal of Sea Research*, **28**, 53–65.

Fonds, M. (1994) Mortality of fish and invertebrates in beam trawl catches and the survival chances of discards. In: *Environmental impact of bottom gears on benthic fauna in relation to natural resources management and protection of the North Sea* (eds S.J. de Groot & H.J. Lindeboom), pp. 131–46. *NIOZ Report* 1994-11/*RIVO-DLO Report* CO 26/94. Netherlands Institute for Sea Research, Den Burg, Texel, The Netherlands.

Graham, M. (1955) Effect of trawling on animals of the seabed. *Deep Sea Research*, **3**, 1–16.

De Groot, S.J. (1973) De invloed van trawlen op de zeebodem. *Visserij*, **26**, 401–9.

Hall, S.J., Basford, D.J. & Robertson, M.R. (1990) The impact of hydraulic dredging for razor clams *Ensis* sp. on an infaunal community. *Netherlands Journal of Sea Research*, **27**, 119–25.

Holme, N.A. (1983) Fluctuations in the benthos of the western English Channel. *Oceanologica Acta, Proceedings of the 17th European Marine Biological Symposium*, Brest, France, pp. 121–4.

Houghton, R.G., Williams, T. & Blacker, R.W. (1971) Some effects of double beam trawling. *International Council for the Exploration of the Sea*, C.M./B:5, 16 pp.

Kaiser, M.J. & Spencer, B.E. (1996) The behavioural response of scavengers to beam-trawl disturbance. In: *Aquatic Predators and their Prey* (eds S.P.R. Greenstreet & M.L. Tasker), pp. 116–23. Blackwell Scientific, Oxford.

Kaiser, M.J., Edwards, D.B., Armstrong, P.J., Radford, K., Lough, N.E.L., Flatt, R.P. & Jones, H.D. (1998) Changes in megafaunal benthic communities in different habitats after trawling disturbance. *ICES Journal of Marine Science*, **55**, 353–61.

Krost, P. (1990) The impact of otter-trawl fishery on nutrient release from the sediment and macrofauna of Kieler Bucht (Western Baltic). *Berichte aus dem Institut für Meereskunde an der Christian-Albrechts-Universität, Kiel*, **200**, 157 pp. (in German, English summary).

Laban, C. & Lindeboom, H.J. (1991) Penetration depth of beam trawl gear. In: *Effects of beam trawl fishery on the bottom fauna in the North Sea: the 1990 studies BEON Report* 13, pp. 37–52. Netherlands Institute for Sea Research, Den Burg, Texel, The Netherlands.

Langton, R.W. & Robinson, W.E. (1990) Faunal associations on scallop grounds in the western Gulf of Maine. *Journal of Experimental Marine Biology and Ecology*, **144**, 157–71.

Lindeboom, H.J. & de Groot, S.J. (eds) (1998) *The effects of different types of fisheries on the North Sea and Irish Sea benthic ecosystems. NIOZ-Report* 1998-1/*RIVO-DLO Report* C003/98, 404 pp. Netherlands Institute for Sea Research, Den Burg, Texel, The Netherlands.

Margetts, A.R. & Bridger, J.P. (1971) The effect of a beam trawl on the sea bed. *International Council for the Exploration of the Sea*, C.M./B:8, 9 pp.

Philippart, C.J.M. (1998) Long-term impact of bottom fisheries on several by-catch species of demersal fish and benthic invertebrates in the south-eastern North Sea. *ICES Journal of Marine Science*, **55**, 342–52.

Polet, H., Ball, B., Blom, W., Ehrich, S., Ramsay, K. & Tuck, I. (1998) Fishing gears used by different fishing fleets In: *The effects of different types of fisheries on the North Sea and Irish Sea benthic ecosystems* (eds H.J. Lindeboom & S.J. de Groot), pp. 83–120. *NIOZ Report* 1998-1/ *RIVO-DLO Report* C003/98. Netherlands Institute for Sea Research, Den Burg, Texel, The Netherlands.

Rees, H.L. & Eleftheriou, A. (1989) North Sea benthos: a review of field investigations into the biological effects of man's activities. *Journal du Conseil International pour l'Exploration de la Mer*, **45**, 284–305.

Rijnsdorp, A.D., Buys, A.M., Storbeck, F. & Visser, E.G. (1998) Micro-scale distribution of beam trawl effort in the southern North Sea between 1993 and 1996 in relation to the trawling frequency of the sea bed and the impact on benthic organisms. *ICES Journal of Marine Science*, **55**, 403–19.

Rumohr, H. & Krost, P. (1991) Experimental evidence of damage to benthos by bottom trawling with special reference to *Arctica islandica*. *Meeresforschung*, **33**, 340–5.

Rumohr, H., Ehrich, S., Knust, R., Kujawski, T., Philippart, C.J.M. & Schroeder, A. (1998) Long term trends in demersal fish and benthic invertebrates. In: *The effects of different types of fisheries on the North Sea and Irish Sea benthic ecosystems* (eds H.J. Lindeboom & S.J. de Groot), pp. 280–53. *NIOZ Report* 1998-1/*RIVO-DLO Report* C003/98. Netherlands Institute for Sea Research, Den Burg, Texel, The Netherlands.

Thrush, S.F., Hewitt, J.E., Cummings, V.J., Dayton, P.K., Cryer, M., Turner, S.J., Funnell, G.A., Budd, R.G., Milburn, C.J. & Wilkinson, M.R. (1998) Disturbance of the marine benthic habitat by commercial fishing: impacts at the scale of the fishery. *Ecological Applications*, **8**, 866–79.

Tuck, I., Ball, B. & Schroeder, A. (1998) Comparison of undisturbed and disturbed areas. In: *The effects of different types of fisheries on the North Sea and Irish Sea benthic ecosystems* (eds H.J. Lindeboom & S.J. de Groot), pp. 245–80. *NIOZ Report* 1998-1/*RIVO-DLO Report* C003/98. Netherlands Institute for Sea Research, Den Burg, Texel, The Netherlands.

De Vooys, C.G.N., Witte, J.IJ., Dapper, R., van der Meer, J. & van der Veer, H.W. (1993) *Lange termijn veranderingen op het Nederlands continentaal plat van de Noordzee: trends in evertebraten van 1931–1990*. *NIOZ Report* 1993-17, 68 pp. Netherlands Institute for Sea Research, Den Burg, Texel, The Netherlands.

Chapter 5
Effects of otter trawling on the benthos and environment in muddy sediments

B. BALL[1], B. MUNDAY[1] and I. TUCK[2]

[1]*Martin Ryan Marine Science Institute, National University of Ireland, Galway, Galway, Ireland*
[2]*Fisheries Research Services, Marine Laboratory Aberdeen, PO Box 101, Victoria Road, Aberdeen, AB11 9DB, UK*

Summary

1. Undisturbed muddy sediments have a rich and diverse fauna that include large deep burrowing animals and erect epifauna.
2. Muddy sediments accumulate in high depositional areas where disturbance from currents and storms are uncommon. As such, they may act as sinks (accumulation areas) for toxic pollutants or biota (e.g. TBT, toxic algal spores) and are susceptible to eutrophication effects due to the depositional nature of sediments and associated high organic carbon content.
3. Such areas may be less capable of sustaining disturbance than more dynamic coarser sediments and accordingly have much longer recovery times.
4. The very stable nature of muddy sediment habitats makes them susceptible to disturbance from fishing in a number of ways, including the removal of target species and by-catch from the grounds, mortality of animals discarded, and those damaged by the gear but not retained in the trawl.
5. Otterboard trawling causes visible physical effects on the seabed that may still be discernible after 18 months, in sheltered areas.
6. Such physical disturbance also leads to community changes in the benthos. These include reduction in diversity, biomass and of individual organism size. These changes may persist for a long time (> 18 months) and may be severe where trawling intensity is very high, even leading to an impoverished community that is in an alternative stable state adapted to regular fishing disturbance.
7. Remedial action and good management are often hindered by a lack of knowledge on the details of deterioration and recovery rates in fished muddy sediments.

Keywords: muddy sediments, trawling, community change, habitat alteration, alternative stable state.

Introduction

Muddy sediments are perhaps the most widespread of all seabed types. They are found at depths ranging from a few metres (e.g. sheltered sea lochs) to thousands of metres on the deep ocean floor. Such areas support a wide range of commercial fisheries in Europe (Piñeiro *et al.*, 1998) and elsewhere (Cryer & Stotter, 1997). Although fisheries are, in general, concentrated on the continental shelf (Pauly & Christensen, 1995), they may extend to the edge of the continental shelf and beyond. Muddy sediments are therefore subject to regular fishing disturbance, and as technological developments allow deeper fishing, disturbance will ultimately extend into the deep sea, where the potential impacts are likely to be much more severe.

Muddy sediments form in high-depositional areas, where disturbance from currents and storms is less common. These sediments are often heavily bioturbated, with extensive burrow systems > 1 m deep into the seabed (Atkinson, 1986). Faunal communities of muddy seabeds are often rich and diverse, and may include large, long-lived, deep-burrowing animals and fragile emergent epifauna (Fig. 5.1). By their nature, such sediments are generally sheltered from severe forms of natural disturbance. As a result, the faunal and epifaunal communities associated with muddy sediments are also protected from natural disturbance. Such stable environments are often found in low-energy systems, and may be substantially affected by even low frequencies of disturbance (Macdonald *et al.*, 1996). In terms of species richness, the richest benthic communities often occur in stable sediments where long-lived species, which settle only occasionally, can survive and add cumulatively to the richness (Sanders, 1968). This contrasts with coarser sediments that are subject to a greater level of natural disturbance and whose faunal communities are thus more resilient to disturbance effects. Therefore, the communities in muddy sediment areas may be less capable of sustaining disturbance than those of more dynamic coarser sediments, and accordingly show greater effects of fishing disturbance and have much longer recovery times.

Fig. 5.1 Sediment profile image of a rich and diverse muddy sediment fauna.

The soft nature of muddy sediments makes them more susceptible to the physical impacts of trawl gear compared with harder and coarser sediments. These impacts include trenches left in the seabed by trawl doors, and a flattening out of the seabed over which the net passes. Trawl doors penetrate deeply into mud compared with other sediments, and this results in potentially greater effects on infaunal species. Muddy seabeds are often heavily bioturbated, with many 'volcano-like' mounds produced by the exhalent shafts of burrowing megafauna. Flattening of these mounds will smother adjacent epifaunal species, and also produce considerable resuspension of sediment, which may lead to a degree of smothering over a larger area. Natural rates of siltation are low in such stable habitats, and excessive sedimentation may lead to the death of some species. In addition, trawling tends to remove the large-bodied, long-lived macrobenthic species (Lindeboom & de Groot, 1998). Once these species are removed, recovery will naturally take a long time, and regular disturbance may prevent recovery. The removal of such species may, in some instances, lead to a diminished bioturbation zone. Such an effect might increase the susceptibility of such areas to eutrophication (Rumohr *et al.*, 1996), and also lengthen the time required to recover from the physical impact.

Methods

The specific results presented in this paper relate to two studies as part of a multinational effort (Lindeboom & de Groot, 1998). The first involved an investigation of a *Nephrops* otter trawl fishery at two study sites in the north-western Irish Sea. One site was located offshore (depth 75 m) and was very heavily trawled by commercial trawlers, while the other inshore site (depth 35 m) was fished less frequently, primarily at dawn and dusk. Sampling was by means of replicated quantitative grabs. Short-term impacts (direct mortality) were calculated as the difference between the initial density of benthos prior to trawling and the density of the remaining, surviving animals 24 h after experimental trawling, using a commercial pattern otter trawl. Medium-term impacts were estimated by comparing the fauna of these fished grounds, both before and after experimental trawling, with shipwreck sites (unfished 'pseudo'-controls). The first shipwreck, *Iron Man*, was located on a muddy, fine-sand substratum in approximately 35 m depth of water, and acted as a control for the inshore site. The second wreck site, *41 Fathom Fast*, lies on a sandy silt substratum approximately 75 m deep, in a heavily fished area, and acted as a control for the offshore site. While the exact date of sinking was not available for either wreck, anecdotal evidence suggests that both have been in place for more than 50 years and are avoided by all fishing trawlers. Full details of the methodologies used are given in Lindeboom & de Groot (1998) and Ball *et al.* (1999).

The second study involved an experimental investigation of the effects of trawling disturbance on the benthic community of a sheltered Scottish sea loch. This study was carried out in Loch Gareloch, in the upper Firth of Clyde, a sea loch which had been closed to fishing for over 25 years. Following an initial survey of reference and treatment sites, experimental fishing disturbance was performed at monthly intervals for 16 months, using a commercial pattern otter trawl. During this period, the effects of the extensive and repeated experimental trawl disturbance on the seabed and benthic community structure were examined by quantitative infaunal sampling, underwater television, acoustic surveys, and comparison of the reference and treatment sites. The subsequent patterns of recovery were then followed over a further 18 months. Full details of the methodologies used are given in Lindeboom & de Groot (1998) and Tuck *et al.* (1998).

Results

Irish Sea studies

In the Irish Sea, the lightly fished inshore site showed short-term effects of fishing disturbance. The number of species, species richness, diversity and biomass all decreased 24 h after trawling, whilst the number of individuals increased slightly. Many of the larger species of molluscs declined in number, as did many of the small crustacea. At the offshore station, however, most of the individuals collected during the quantitative grab surveys represent either species that have a small adult size or juveniles of other larger species. This paucity of fauna and the associated low biomass (mean biomass is *c.* 20 g m^{-2}) rendered any quantitative assessment of the short-term effects of trawling impossible, because of intersample variance.

At the offshore station, the number of species, number of individuals and biomass, all showed a decrease with increasing distance along transects away from the wreck. All of the parameters measured showed a significant decrease between the wreck sites and the fished ground, prior to experimental trawling. Within the fished area away from the wreck 24 h after experimental trawling, there was further decrease in most of these parameters (Table 5.1). The fauna of the wreck and fished area are dramatically different as shown by the distinctive groupings in MDS plots (Fig. 5.2).

Table 5.1 Offshore *Nephrops* fishing grounds

Parameters	Wreck	Pre-trawling	Post-trawling
Species/0.1 m^2	47 ± 2.9	14 ± 3.8	15 ± 2.6
Individuals/0.1 m^2	340 ± 67.0	72 ± 55.4	51 ± 18.6
Shannon's diversity	4.13 ± 0.20	2.80 ± 0.32	3.12 ± 0.27
Evenness	0.74 ± 0.03	0.72 ± 0.10	0.82 ± 0.06
Species richness	5.53 ± 0.35	2.20 ± 0.38	2.38 ± 0.35
Biomass g/0.1 m^2	7.8 ± 14.4	2.2 ± 1.4	1.8 ± 1.3

Mean (± SD) community parameters measured along transects sampled in the vicinity of the *41 Fathom Fast* wreck and from nearby *Nephrops* trawling grounds before (pre-trawling) and 24 h after (post-trawling) experimental trawling.

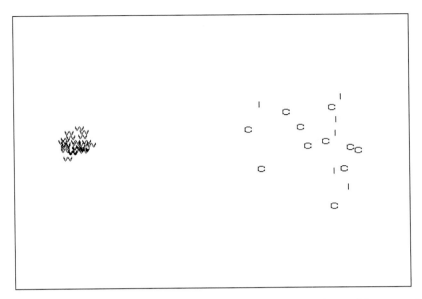

Fig. 5.2 MDS ordination plots of the *41 Fathom Fast* wreck stations with pre- and post-trawling offshore fishery stations. W = wreck, C = pre-trawled, I = post-trawled.

Sixty-nine of the species found at the offshore wreck site were not found at the offshore experimental fishing site. These include polychaetes, crustaceans, bivalves, gastropods and echinoderms. In particular, large specimens of some mollusc species (*Phaxas pellucidus, Cylichna cylindracea, Corbula gibba, Nuculoma tenuis*) and echinoderms (*Amphiura chiajei, Brissopsis lyrifera, Echinocardium cordatum*) were quite common along the offshore wreck transects. In contrast, while the juveniles of some of these species were occasionally sampled at the offshore trawling station, large specimens were never found. Only five polychaete species (large carnivores and small opportunists) were found to be most abundant at the offshore site, but not near the wreck.

The lightly fished inshore site, by contrast, was affected less severely in the medium term, indicated by the more scattered distribution of samples in the MDS plots (Fig. 5.3). The total number of individuals and biomass was substantially higher at the wreck sites compared with the fished ground prior to experimental trawling. Within the fishing grounds 24 h after experimental trawling, there was a further decrease in biomass and Shannon's diversity (Table 5.2). Fifty-eight of the species found at the wreck site were not found at the inshore site. These included worms from the families Phyllodocidae (active predators) and Ampharetidae (sedentary tube dwellers), as well as a number of bivalves and echinoderms. By comparison, a small number of polychaete species were most common at the inshore site compared with the wreck site. These included *Nephtys incisa* and *Glycera rouxi* (predators) and *Mediomastus fragilis* (typical of enriched/disturbed muds) and some spionids. In addition, large specimens of some molluscs (*Turritella communis, Dosinia lupinus, Azorinus chamasolen*) and echinoderms (*Amphiura chiajei, Brissopsis*

Stress=0.16

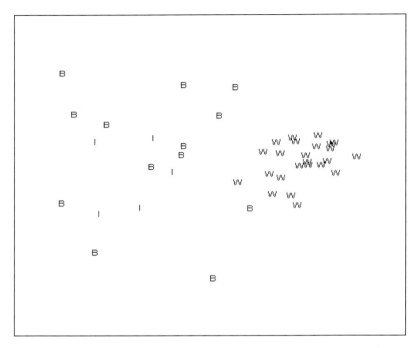

Fig. 5.3 MDS ordination plots of the *Iron Man* wreck stations with pre- and post-trawling inshore fishery stations. W = wreck, B = pre-trawled, I = post-trawled.

lyrifera, Echinocardium cordatum) were quite common along the transects sampled close to the inshore wreck. While large specimens of these molluscs and of *Amphiura chiajei* were also found at the inshore experimental fishing station in small numbers, the spatangid echinoids *Brissopsis lyrifera* and *Echinocardium cordatum* were never found.

Table 5.2 Inshore *Nephrops* fishing grounds

Parameters	Wreck	Pre-trawling	Post-trawling
Species/0.1 m^2	49 ± 6.7	39 ± 6.7	39 ± 5.1
Individuals/0.1 m^2	299 ± 97.9	161 ± 53.0	162 ± 49.1
Shannon's diversity	4.83 ± 0.16	4.45 ± 0.19	4.25 ± 0.36
Evenness	0.80 ± 0.03	0.84 ± 0.04	0.80 ± 0.05
Species richness	7.98 ± 0.75	5.24 ± 0.60	5.25 ± 0.78
Biomass g/0.1 m^2	16.4 ± 11.0	5.3 ± 3.1	3.9 ± 0.9

Mean ($\pm$ SD) community parameters measured along transects sampled in the vicinity of the *Iron Man* wreck and from nearby *Nephrops* trawling grounds before (pre-trawling) and 24 h after (post-trawling) experimental trawling.

Gareloch study

At the Gareloch site, the number of species and individuals increased throughout the period of trawling disturbance, although no significant changes occurred for biomass. At the same time, measures of diversity and evenness decreased in the experimentally trawled area relative to the reference site. These changes were mostly due to an increase in abundance of opportunistic species. The cirratulid polychaetes *Chaetozone setosa* and *Caullierella zetlandica* were found to be most resistant to disturbance with *Mediomastus fragilis* and *Pseudopolydora paucibranchiata* also increasing in abundance following disturbance. In contrast the bivalves *Nucula nitidosa* and *Corbula gibba*, and polychaetes *Scolopolos armiger*, *Nephtys cirrosa* and *Terebellides stroemi*, were identified as species sensitive to trawling disturbance (Table 5.3). Multivariate analysis and abundance biomass comparison plots

Table 5.3 Summary of two-way ANOVA of 20 commonest species from Loch Gareloch fishing disturbance experiment

Species	Density (1 m^{-2})	Phylum	Site	Date	Interaction	Change	% Change
Chaetozone setosa	79.06	Annelida	<0.0001	<0.0001	<0.05	+ve	61
Mediomastus fragilis	68.24	Annelida	<0.0001	<0.0001	<0.05	+ve	105
Caulleriella zetlandica	47.54	Annelida	<0.0001	<0.0001	<0.0001	+ve	61
Pseudopolydora paucibranchiata	39.45	Annelida	n.s.	<0.0001	<0.0001	+ve	117
Abra alba	34.62	Mollusca	n.s.	<0.0001	<0.05	?	50
Lagis koreni	23.11	Annelida	<0.0001	<0.0001	n.s.		73
Melinna palmata	17.96	Annelida	n.s.	<0.0001	<0.001	+ve	41
Thyasira flexuosa	15.36	Mollusca	n.s.	<0.0001	n.s.		63
Scalibregma inflatum	12.06	Annelida	n.s.	<0.0001	<0.05	?	85
Nucula nitidosa	11.66	Mollusca	<0.005	<0.001	<0.005	−ve	71
Scolopolos armiger	10.19	Annelida	<0.0001	<0.05	<0.01	−ve	56
Pholoe inornata	9.64	Annelida	n.s.	<0.0001	n.s.		70
Nephtys cirrosa	9.42	Annelida	<0.0001	<0.0001	<0.001	−ve	48
Terebellides stroemi	9.07	Annelida	<0.01	<0.0001	<0.005	−ve	52
Nuculoma tenuis	6.45	Mollusca	<0.005	<0.0001	n.s.		72
Corbula gibba	6.03	Mollusca	<0.0001	<0.0001	<0.001	−ve	45
Nemertea sp.	5.75	Nemertea	<0.0001	<0.0001	n.s.		75
Aphelochaeta marioni	5.36	Annelida	n.s.	<0.0001	<0.05	?	51
Abra nitida	5.18	Mollusca	n.s.	<0.01	n.s.		68
Goniada maculata	5.11	Annelida	<0.05	<0.001	<0.05	−ve	27

P values provided for site, date and interaction effects (n.s., not significant at 5% level). Densities provided are averages from all samples collected throughout the experiment. Where significant ($P < 0.05$) interactions were found, the change indicated represents change in abundance (relative to the reference site) associated with the disturbance of the treatment site (i.e. +ve, increase in abundance following disturbance; −ve, decrease in abundance following disturbance; ?, change variable and unclear).

confirmed that community changes occurred following disturbance, with some differences between treatment and reference sites still apparent, in box plots for the median values of the W statistic, after 18 months of recovery (Fig. 5.4). Epifauna was relatively scarce at the study site, and no effects were identified on the numbers of individuals or numbers of species, but some short-term effects were noted on individual species abundances. Physical effects on the seabed were dramatic, with side-scan sonar observations showing trawl door tracks criss-crossing the treatment area (Fig. 5.5). Physical effects on the seabed were also identified through changes in the RoxAnn roughness parameter (E1). Using both side-scan sonar and RoxAnn techniques, it was possible to identify the effects of disturbance after 5 months of fishing, but these effects were almost indistinguishable 18 months after fishing ceased. The trawling disturbance did not have a significant effect on median sediment particle size or organic carbon in the sediment.

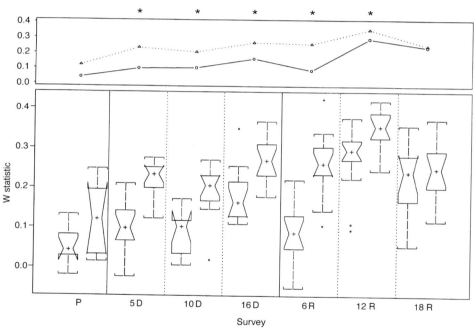

Fig. 5.4 Box plots (lower panels) and time series for the median values of the W statistic for each site through the experiment (upper panels). Box plots are arranged in pairs in time (survey) order, with the reference plot on the right for each pair. Surveys in which medians of two sites were significantly different are marked by an asterisk. Surveys labelled as follows: P – preliminary survey, 5D – 5 months of disturbance, 10D – 10 months of disturbance, 16D – 16 months of disturbance, 6R – 6 months of recovery, 12R – 12 months of recovery, 18R – 18 months of recovery. The notches in the boxes indicate 95% confidence intervals of the median. If the intervals around two population medians do not overlap, the population medians can be considered significantly different ($P < 0.05$). The dashed lines represent whiskers and extend to the largest observation that is less than or equal to the upper quartile plus 1.5 times the interquartile range (or the smallest observation that is greater than or equal to the lower quartile minus 1.5 times the interquartile range).

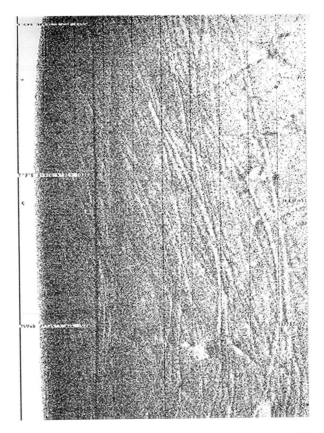

Fig. 5.5 Portion of the side-scan sonar record obtained following 10 months of otter trawling disturbance, showing disturbed seabed in the treatment area.

Discussion

Intensive demersal trawling over muddy seabeds leads to apparent long-term alteration of the seabed. In the Scottish sea loch study, trawl door tracks were still visible from side-scan sonar records 18 months after fishing. It is generally considered that the trawl tracks on harder sediment disappear sooner than this (beam trawling on sand, 52 h: Lindeboom & de Groot, 1998; suction dredge on sand, 40 days: Hall *et al.*, 1990; scallop dredging on sand, 6 months: Currie & Parry, 1996), although hydraulic dredge tracks in the Bering Sea have been reported to last from a few days to a few years (Kauling & Bakus, 1979). Rumohr recorded the penetration depths of commercial otter trawl doors in Kiel Bay of 8–17 cm in mud and 0–5 cm in sand, while bobbins on the groundrope left marks to a depth of 2–5 cm (Krost *et al.*, 1990). The depth of penetration of the trawl doors into a muddy seabed is generally greater than for other sediment types, but is of course dependent on the size, weight and towing angle of the doors, as well as on boat speed and warp length (Jones, 1992). In addition, alterations in the boat's course as well as changes in bottom topography may cause the otter boards to 'jump', resulting in deeper penetration of the seabed (Khandriche *et al.*, 1986). Penetration into the seabed by other gears (beam trawls on sand, 8 cm: Laban & Lindeboom, 1991; scallop dredges on sand, 6 cm: Currie & Parry, 1996) may not be as deep, but covers the full width of the fished tracks. Hydraulic and suction dredges penetrate deeper than this (up to 0.6 m deep; Hall *et al.*, 1990), but these gears specifically target burrowing infaunal bivalves. Clearly, the length of time required for fishing tracks to disappear will depend on the size of the tracks and the rate of infilling. In an area of strong tidal or wave action, tracks are likely to disappear rapidly. Also, in areas where levels of bioturbation are high, and a regular turnover of sediment produces large numbers of mounds on the seabed, tracks will be filled relatively quickly.

It is clear from both studies that fishing disturbance altered macrobenthic community structure. Throughout the manipulative fishing experiment in Loch Gareloch (16 months), disturbance increased numbers of species and numbers of individuals, but reduced diversity, due to a dramatic increase in opportunist polychaetes. Over the longer time scale (50 years), the Irish Sea studies suggest that fishing disturbance may ultimately lead to an altered, but stable, community comprising a reduced number of species and hence diversity, with a fauna that was comprised primarily of small polychaetes. Fragile echinoderms and large bivalves were rarely found at the heavily fished sites, where the fauna was dominated by carnivorous polychaetes and opportunist species. The Irish Sea study also identified differences in effects between areas of different fishing effort, suggesting that the degree of community change is related to fishing intensity. The severity of the impact will also depend on other environmental variables (e.g. depth, current regime, etc.) relating to the background level of natural disturbance. A diminished bioturbation zone in the upper layers of the substratum (caused by intensive trawling removing the large macro/mega fauna) could make such areas very susceptible to eutrophication (Rumohr, 1996), and also lengthen recovery from physical impact. The rate at which such areas can recover is not clear, but it appears that more than 12 months is required. However, the areal extent of disturbance was limited, and it is quite conceivable that recovery would take longer at the scale of a commercial fishing ground (Hall *et al.*, 1994; Thrush *et al.*, 1998). Such long recovery times suggest that even fishing during a restricted period of the year may be sufficient to maintain, in an altered state, communities that occupy fine muddy sediment habitats (Lindeboom & deGroot, 1998). In recent years, new concerns have arisen as fishing activity moves into deeper waters at the shelf edge and on the slope. Little or no information is available about the benthos of such regions or the likely impacts of trawling on such deep and stable environments. Such areas are likely to have slow recovery rates and impacts may be very long lasting (> 5 years) (Thiel & Schriever, 1990).

Experimental studies in coarser habitats have generally found that recovery of the infaunal community is more rapid (suction dredging on sand, 40 and 56 days: Hall *et al.*, 1990; Hall & Harding, 1997; beam trawling on sand, 6 months: Kaiser *et al.*, 1998; scallop dredging on sand, 14 months: Currie & Parry, 1996). The fauna in habitats that are subject to regular natural disturbance are often considered to be morphologically and behaviourally adapted to a dynamic environment, and may therefore be less affected by fishing disturbance. Such variability underlines the importance of identifying species sensitive to fishing disturbance.

For example, in muddy sediments, the Scottish sea loch study indicated a level of sensitivity to trawling by certain species. This has, in some cases, been supported by the longer-term data from the Irish Sea. However, an obstacle to such comparisons is variations in the composition of the fauna at the two locations. Both studies support the inclusion of the molluscan families (Nuculidae and Corbulidae) as sensitive indicator species. The Irish Sea study suggests the Veneridae and Turritellidae may also be sensitive. Amongst the polychaeta, the Orbiniidae and Trichobranchidae appear sensitive, while the Nephtyidae provide conflicting results. *Nephtys cirrosa* (Scotland), *N. kersivalensis* and *N. hombergii* (Irish Sea) all appear sensitive. By contrast, *Nephtys incisa* appears to favour trawled sites in the Irish Sea, which may reflect its large robust size and predatory lifestyle. This phenomenon seems to be supported by the co-occurrence of a similar large scavenger/predator, *Glycera rouxi*, in many of the fished samples. Other polychaete families resistant to trawling disturbance include the cirratulids, spionids and paraonids, all of which have short life spans. Echinoderms appear particularly sensitive to trawling disturbance, as indicated by the complete absence of large Amphiuridae at the heavily fished offshore Irish Sea location, and their comparatively higher abundance at the wreck sites. The irregular sea urchins showed apparent high mortality, but, because of their low initial densities, it proved difficult to show any statistically significant impact.

Epifauna are generally scarce in muddy sediment habitats, and detection of fishing effects on such species has therefore been limited. Fishing effects on epifauna are likely to be greater on harder substrata, where more diverse epifaunal communities are present and the impact of fishing gear is likely to alter the physical structure of the habitat, such as in maerl beds (Biomaerl Team, 1998) and epifaunal reefs (Cranfield *et al.*, 1999). The significance of bottom micro-topography has been highlighted in recent studies from America and Australia (Auster *et al.*, 1996; Sainsbury *et al.*, 1997). These studies have shown that when target fish species are under pressure and stocks are low, the removal of epifauna and microhabitats can push species beyond their population threshold. Detailed studies of such processes will require manipulative experiments in areas free from commercial fishing activities.

Otter trawls are in more common usage than beam trawls in muddy sediments. In direct comparative trials, otter trawls generally catch less benthos than beam trawlers, and disturb the seabed to a lesser depth over the whole fished track (Houghton *et al.*, 1971; Kaiser *et al.*, 1996), although trawl-door penetration may be greater than any part of a beam trawl. The present studies have, however, shown that the cumulative effect of intense otter trawling is as important as gear weight and design in impacting on the benthos.

Conclusions

The most obvious physical effects of fishing on muddy sediments are the tracks left by the trawl doors. These may last up to 18 months, although they would disappear more rapidly in heavily bioturbated areas. Although limited in its area of impact, penetration into the seabed by trawl doors on muddy sediment is likely to be greater than in other habitats.

The effects of otter trawling on the infauna include a reduction in the abundance of large-bodied fragile organisms, and an increase in abundance of opportunists, and may ultimately lead to an altered but stable community comprising a reduced number of species and faunal diversity, and with the fauna comprising primarily small polychaetes. Subsequent recovery of infaunal communities in muddy habitats following experimental disturbance appears to take longer than in other habitats. Communities in more dynamic environments may be adapted to natural disturbance, and thus are less affected by fishing disturbance.

Epifauna are generally scarce in muddy habitats, and as a result, the direct effects of fishing are likely to be limited, and less than for harder sediment habitat types. However, the removal of epifauna and alteration of microhabitats may affect ecosystem productivity.

Both studies described here identified effects of fishing on benthic communities, but neither may show the full picture. The protection to the seabed offered by wrecks may be confounded with 'wreck effects' (Hall *et al.*, 1993), and the effects of seabed disturbance on a limited scale may not match the effects at the scale of the fishery (Hall *et al.*, 1994; Thrush *et al.*, 1998). The only way to test fully the effects of fishing on benthic communities is through large-scale experiments in areas closed to fishing.

Acknowledgement

This study was partly funded by the EC as part of the IMPACT II project, contract number AIR 94 1664.

References

Atkinson, R.J.A. (1986) Mud-burrowing megafauna of the Clyde Sea Area. *Proceedings of the Royal Society of Edinburgh*, **90B**, 351–61.

Auster, P.J., Malalesta, R.J., Langton, R.W., Watling, L., Valentine, P.C., Donaldson, C.L.S., Langton, E.W., Shepard, A.N. & Babb, I.G. (1996) The impact of mobile fishing gear on seafloor habitats in the Gulf of Maine (Northwest Atlantic). Implications for conservation of fish populations. *Reviews in Fisheries Science*, **4**, 185–202.

Ball, B., Fox, G. & Munday, B. (1999) Long and short term consequences of a *Nephrops* otter trawl fishery on the benthos and environment of the Irish Sea. *ICES Journal of Marine Science*.

Biomearl Team (1998) Mearl grounds: habitats of high biodiversity in European Seas. In: *3rd European Marine Science and Technology Conference*, Lisbon. 23–27 May 1998, Vol. 1, pp. 170–8.

Cranfield, H.J., Michael, K.P. & Doonan, I.J. (1999) Changes in the distribution of epifaunal reefs and oysters during during 130 years of dredging for oysters in Foveaux Strait, southern New Zealand. *Aquatic Conservation.*

Cryer, M. & Stotter, D. (1997) *Trawling and tagging of scampi off the Alderman Islands, western Bay of Plenty, September 1995. New Zealand Fisheries Data Report*, No. 84.

Currie, D.R. & Parry, G.D. (1996) Effects of scallop dredging on a soft sediment community: a large-scale experimental study. *Marine Ecology Progress Series*, **134**, 131–50.

Fox, G.M., Ball, B.J., Munday, B.W. & Pfeiffer, N. (1996) The IMPACT II study: preliminary observations on the effect of bottom trawling on the ecosystem of the *Nephrops* grounds in the N.W. Irish Sea. In: *Irish Marine Science 1995* (eds B.F. Keegan and R. O'Connor), pp. 337–54. Galway University Press Ltd, Galway.

De Groot, S.J. (1984) The impact of bottom trawling on benthic fauna of the North Sea. *Ocean Management*, **9**, 177–90.

Hall, S.J. & Harding, M.J.C. (1997) Physical disturbance and marine benthic communities: the effects of mechanical harvesting of cockles on non-target benthic infauna. *Journal of Applied Ecology*, **34**, 497–517.

Hall, S.J., Basford, D.J. & Robertson, M.R. (1990) The impact of hydraulic dredging for Razor clams *Ensis* sp. on an infaunal community. *Netherlands Journal of Sea Research*, **27**, 119–25.

Hall, S.J., Robertson, M.R., Basford, D.J. & Heaney, S.D. (1993) The possible effects of fishing disturbance in the Northern North Sea: an analysis of spatial patterns in community structure around a wreck. *Netherlands Journal of Sea Research*, **31**, 201–8.

Hall, S.J., Raffaelli, D. & Thrush., S.F. (1994) Patchiness and disturbance in shallow water benthic assemblages. In: *Aquatic Ecology, Scale, Pattern and Process* (eds P.S. Giller, A.G. Hildrew and D.G. Raffaelli), pp. 333–75. Blackwell Scientific, Oxford.

Hill, A.E., Brown, J. & Fernand, L. (1996) The western Irish Sea gyre: a retention system for Norway lobster (*Nephrops norvegicus*). *Oceanologica Acta*, **19**, 357–68.

Houghton, R.G., Williams, T. & Blacker, R.W. (1971) Some effects of double beam trawling. *ICES, Gear and Behaviour Committee*, C.M. 1971/B:5, 12 pp.

Jones, J.S. (1992) Environmental impact of trawling on the seabed: a review. *New Zealand Journal of Marine and Freshwater Research*, **26**, 59–67.

Kaiser, M.J., Hill, A.S., Ramsay, K., Spencer, B.E., Brand, A.R., Veale, L.O., Prudden, K., Rees, E.I.S., Munday, B.W., Ball, B. and Hawkins, S.J. (1996) Benthic disturbance by fishing gear in the Irish Sea: a comparison of beam trawling and scallop dredging. *Aquatic Conservation: Marine and Freshwater Ecosystems*, **6**, 269–85.

Kaiser, M.J., Edwards, D.B., Armstrong, P.J., Radford, K., Lough, N.E.L., Flatt, R.P. & Jones, H.D. (1998) Changes in megafaunal benthic communities in different habitats after trawling disturbance. *ICES Journal of Marine Science*, **55**, 353–61.

Kauling, T.J. & Bakus, G.J. (1979) *Effects of hydraulic clam harvesting in the Bering Sea. Tetra Tech. Rep.* TC3324. Prepared for: North Pacific Fishery Management Council, 183 pp.

Khandriche, A., Werner, F. & Erlenkeuser, H. (1986) Auswirkungen der Oststürme vom Winter 1978/79 auf die Sedimentation im Schlickbereich der Eckernförder Bucht (westliche Ostsee). *Meyniana*, **38**, 125–52.

Krost P., Bernhard, M., Werner, F. & Hukriede, W. (1990) Otter trawl tracks in Kiel Bay (Western Baltic) mapped by side-scan sonar. *Meeresforschung*, **32**, 344–54.

Laban, C. & Lindeboom, H.J. (1991) Penetration depth of beam trawl gear. In: *Effects of beam trawl fishery on the bottom fauna in the North Sea: the 1990 studies. BEON Report* 13, pp. 37–52. Netherlands Institute for Sea Research, Den Burg, Texel, The Netherlands.

Lindeboom, H.J. & de Groot, S.J. (1998) *IMPACT-II: the effects of different types of fisheries on the North Sea and Irish Sea benthic ecosystems. NIOZ Report* 1998-1/*RIVO-DLO Report* C003/98, 404 pp. Netherlands Institute for Sea Research, Den Burg, Texel, The Netherlands.

MacDonald, D.S., Little, M., Eno, N.C. & Hiscock, K. (1996) Disturbance of benthic species by fishing activities: a sensitivity index. *Aquatic Conservation: Marine and Freshwater ecosystems,* **6**, 257–68.

Pauly, D. & Christensen, V. (1995) Primary production required to sustain global fisheries. *Nature,* **374**, 255–7.

Massy, A.L. (1912) Report on a survey of trawling grounds on the coasts of Counties Down, Louth, Meath and Dublin. Part III, Invertebrate fauna. *Scientific Investigations of the Fisheries Branch of Ireland,* 1911, **1**, 1–225.

Piñeiro, C.G., Casas, M. & Bañón, R. (1998) Current situation of the Deep Water Fisheries exploited by Spanish fleets in the North and Northeast Atlantic: a review. *International Council for the Exploration of the Sea,* C.M. 1998/O:10.

Rumohr, H., Bonsdorff, E. & Pearson, T.H. (1996) Zoobenthic succession in Baltic sedimentary habitats. *Archives of Fisheries and Marine Research,* **44**, 179–214.

Sainsbury, K.J., Campbell, R.A., Lindholm, R. & Whitelaw, A.W. (1997) Experimental management of an Australian multispecies fishery: examining the possability of trawl-induced habitat modification. In: *Global Trends: Fisheries Management* (eds E.K. Pikitch, D.D. Huppert and M.P. Sissenwine), pp. 107–12. American Fisheries Society, Symposium 20, Bethesda, MD.

Sanders, H.L. (1968) Marine benthic diversity: a comparative study. *American Naturalist,* **102**, 245–82.

Snelgrove, P.V.R. & Butman, C.A. (1994) Animal–sediment relationships revisited: cause versus effect. *Oceanography and Marine Biology: An Annual Review,* **32**, 111–77.

Thiel, H. & Schriever, G. (1990) Deep-sea mining, environmental impact and the DISCOL project. *Ambio,* **19**, 245–50.

Thrush, S.F., Hewitt, J.E., Cummings, V.J., Dayton, P.K., Cryer, M., Turner, S.J., Funnell, G.A., Budd, R.G., Milburn, C.J. & Wilkinson, M.R. (1998) Disturbance of the marine benthic habitat by commercial fishing: impacts at the scale of the fishery. *Ecological Applications,* **8**, 866–79.

Tuck, I.D., Hall, S.J., Robertson, M.R., Armstrong, E. & Basford, D.J. (1998) Effects of physical trawling disturbance in a previously unfished sheltered Scottish sea loch. *Marine Ecology Progress Series,* **162**, 227–42.

Chapter 6
The effects of scallop dredging on gravelly seabed communities

C. BRADSHAW, L.O. VEALE, A.S. HILL and A.R. BRAND
Port Erin Marine Laboratory, University of Liverpool, Port Erin, Isle of Man, IM9 6JA, UK

Summary

1. Gravelly seabed communities around the Isle of Man, Irish Sea, are very heterogeneous in terms of both epi- and infauna, and vary over a wide range of spatial scales. This paper reviews the results of a large study which investigated the ecological effects of disturbance by scallop dredging at both a large (fishing grounds) and a small scale (experimental plots).
2. Commercial dredging for scallops and queen scallops is a significant factor in the structuring of benthic communities on gravelly substrata.
3. Community composition was related to the intensity of commercial dredging effort; this was also confirmed by dredging experiments undertaken in an area closed to commercial fishing.
4. The effect of scallop-dredge disturbance on a gravelly seabed may differ from that of bottom fishing on other soft sediments, owing to the extreme patchiness of animal distribution, sediment stability, greater abundance of epifauna and to the combined effect of the heavy, toothed scallop gear and stones caught in the dredges.

Keywords: scallop dredging, benthic community disturbance, long-term effects, community change.

Introduction

Gravel sediments encompass a wide range of seabed sediment types with varying proportions of stones, shell material, sand and mud. The characteristic feature that sets gravel sediments apart from other soft sediment types is the presence of coarse particles, such as stones and shells, at the surface. The wide range of sediment grain sizes provides a heterogeneous habitat, and growth and settlement substrata for sessile epifauna such as hydroids, bryozoans and tunicates. Larger stones and their attached epifauna also provide shelter for invertebrates such as crabs, nudibranchs and gastropods, and for small bottom-dwelling fish such as gobies and dragonets. More uniform, finer areas of gravel are inhabited by burrowing anemones, bivalves and worms, and mobile epifauna such as edible crabs, brittlestars and starfish.

In the northern Irish Sea around the Isle of Man, gravel sediments are found down to water depths of about 70 m and vary from extremely stony to fine gravel substrata. Changes from one sediment type to another may be sudden and very localised. This small-scale patchiness increases habitat heterogeneity and the range of available niches for a wide range of animals.

The gravelly seabed around the Isle of Man supports important fisheries for great scallops (*Pecten maximus*), referred to in this paper simply as scallops, and queen scallops (*Aequipecten opercularis*), or queens. The annual scallop fishing season lasts from 1 November to 31 May inclusive; queens may be fished all year round. Scallops live on, or partly buried in, the surface sediments and are fished with toothed, Newhaven-type dredges. Queens are more epifaunal than scallops but are also fished mainly with Newhaven-type dredges modified to retain this smaller species. Toothed dredges scrape through the top 10 cm or so of the seabed, disturbing the associated benthic communities, damaging some animals and removing others as by-catch.

The direct, immediate effects of bottom fishing on the seabed and its benthic communities are intuitively obvious and have been well documented (e.g. see Jones, 1992 for a review), as have injury and mortality rates of many benthic species in the dredge tracks (e.g. Eleftheriou & Robertson, 1992; Bergman *et al.*, 1994). Suspension of fine sediments also occurs (Churchill, 1989; Jones, 1992; Black & Parry, 1994; Hill *et al.*, 1997). Most studies have been carried out on less coarse substrata such as sand (e.g. Eleftheriou & Robertson, 1992; Bergman & van Santbrink, 1994; Bergman *et al.*, 1994; Kaiser & Spencer, 1996b) or mud (e.g. Margetts & Bridger, 1971; Lopez & Mejuto, 1988) and only a few recent studies have examined the effect of bottom fishing on coarser gravelly, shelly or maerl bottoms (Kaiser & Spencer, 1994; Auster *et al.*, 1996; Kaiser *et al.*, 1996; Hall-Spencer & Moore, this volume, Chapter 7).

Other naturally occurring and interrelated factors also influence the community structure of the benthos. Physical factors, which in turn affect sediment particle size, include water depth, wave or current action and storms. There is seasonal variation in species abundance and diversity due to the differing life cycles of different species and as a result of variable recruitment from year to year. The separation of the effects of these non-anthropogenic factors from those of scallop fishing is central to the study of dredge disturbance.

This chapter gives an overview of some of the studies carried out during the first phase of a MAFF-funded project to study the effects of scallop dredging carried out at Port Erin Marine Laboratory over the last 4 years (Hill *et al.*, 1996, 1997). Here, we consider the effects of scallop dredging that have occurred over a time period of less than 10 years. The longer-term effects (*c.* 60 years) are currently under investigation but are outside the scope of this paper.

Methods

A variety of different approaches have been used to study the effects of dredge disturbance around the Isle of Man. These include:

1. Sampling of the benthos of fishing grounds subjected to varying commercial fishing intensities. Fine mesh dredges (a short-toothed dredge with a 2-cm mesh liner), beam trawls and queen scallop dredges were used for epibenthic sampling, and 0.1-m^2 Day grabs for infaunal sampling.

2. Sediment description of samples from separate Day grabs. Sediment particle size distribution and organic carbon content were determined using methods described by Holme & McIntyre (1984). RoxAnn data were also obtained during epibenthic tows to give acoustic measures of seabed roughness and hardness.

3. Studies of the damage sustained by by-catch species in scallop and queen dredges, determined by subjective visual assessment and corroborated by mortality experiments in the laboratory.

4. Experimental disturbance with a commercial scallop dredge of plots in an area closed to commercial dredging. Two experimental plots (1 and 3) were dredged 10 times every 2 months from January 1995 onwards and two adjacent plots (2 and 4) were left undredged as controls (Fig. 6.1, inset). Three plots outside the closed area and subject to commercial dredging (W, N1 and N2) were also sampled. All seven plots were sampled three times during the year preceding the first experimental dredging. The benthos was sampled every 6 months by fine-meshed dredge and 0.1-m^2 Day grabs.

5. Diver surveys of the epibenthos of the closed area. Counts of the major epibenthic species were made by pairs of divers over 50×2 m transects at random locations throughout the closed area. Counts by the two divers were pooled to give numbers per 200 m^2 of seabed.

6. Fixed and towed video to examine scavenger aggregations immediately after experimental dredge disturbance.

For the benthic community analyses, the samples were sieved (5-mm sieve for epifauna, 0.5-mm sieve for infauna) and all animals identified to species or family, counted. A variety of uni- and multivariate analyses were used to distinguish differences between communities. Univariate measures of species abundance, evenness and diversity were compared using ANOVA, Kruskal–Wallis non-parametric ANOVA or Mann-Whitney U tests. Multivariate cluster analysis, non-metric multidimensional scaling (MDS) and subsequent analytical procedures (e.g. ANOSIM, SIMPER, BIOENV) were carried out with the PRIMER software package (Clarke & Warwick, 1994). Full details of all the methods used can be found in Hill *et al.* (1997).

Two aspects of these studies are of particular interest in this chapter: the area closed to fishing and the relatively high resolution fishing effort dataset for the northern Irish Sea scallop fishery.

Closed area

An exclusion zone of nearly 2 km^2 off the south-west of the Isle of Man has been closed to commercial fishing by towed gear (although not to static gear) since 1989 (Brand & Prudden, 1997). The area had been heavily dredged for 50 years prior to closure, and the surrounding area, known as the Bradda Inshore fishing ground, continues to be one of the most intensively fished grounds in the British Isles. The

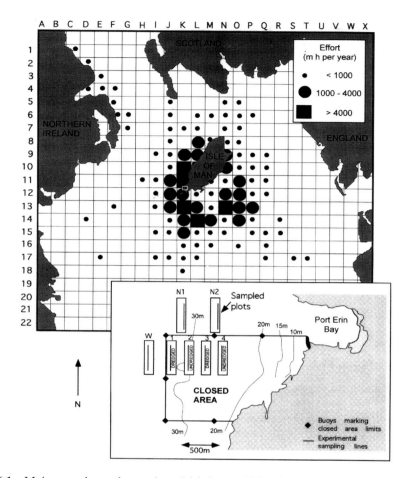

Fig. 6.1 Main map shows the northern Irish Sea and Isle of Man overlaying the 5-n mile grid for which effort data are collected. Mean annual dredging effort (queen and scallop boats combined) is shown for the period 1982–1995. The inset shows the position of the closed area and the arrangement of the experimental dredged and undredged plots.

presence of this undredged area in an otherwise very heavily fished zone, allows direct comparisons and provides a good control to study the effects of dredging and the recovery of dredged areas.

Commercial dredging effort

Since 1981, a proportion (*c.* 30%) of the Manx fishing fleet has recorded details of their daily fishing activities in voluntarily completed logbooks. Information recorded includes details of the type of boat, daily records of the area fished (on a 5 × 5-n mile grid), hours fished, catch size and gear type (including gear width). This information allows the calculation of an unusually precise measure of fishing effort in metre hours (metres dredge width × hours fished) for 5 × 5-n mile squares covering the entire northern Irish Sea (Fig. 6.1). This dataset is much more detailed, both temporally and spatially, than that available for many other European fisheries which are usually based on International Council for Exploration of the Sea (ICES) rectangles (0.5° latitude × 1° longitude).

Although the 5 × 5-n mile grid provides a relatively high resolution effort dataset, fishermen tend to dredge along known tows and avoid areas hazardous for the gear. As a result, there will inevitably be patches of low effort within high-effort grid squares and patches of high effort within low-effort grid squares. To increase further the resolution and accuracy of the effort data, a number of fishermen were asked to pinpoint areas of high or low effort within 5 × 5-n mile squares, based on the frequency of their fishing over the year. Pairs of geographically close sites (one high and one low effort) were then sampled using beam trawls and Day grabs.

Results

Heterogeneity of benthic communities

Benthic sampling around the Isle of Man between 1994 and 1998 has shown that the distribution of the fauna is extremely patchy, especially with respect to epifauna. Some species are ubiquitous and common, such as the starfish *Asterias rubens*, whereas others are found with more restricted geographical distributions. For example, *Echinus esculentus*, *Neptunea antiqua*, *Buccinum undatum* and *Pagurus* spp. are all more abundant in the north, east and offshore, *Cancer pagurus* and *Porania pulvillus* mainly in the west and *Spatangus purpureus* in only two localised areas in the east (Fig. 6.2).

Diversity and equitability of epifauna differed significantly between four widely separated sites, identified by the fishermen as undredged but adjacent to major fishing grounds (Fig. 6.3; Targets, Chickens, Port St Mary and Douglas). Each site is dominated by a different suite of species, and although considerable overlap exists in the species composition, multivariate cluster analysis showed significant differences between grounds (Fig. 6.4: ANOSIM analysis of between sites difference, global $R = 0.961$, $P < 0.001$).

At a smaller scale, high levels of variability have been found within the 2 km^2 of the closed area. Diver surveys and RoxAnn data have shown the seabed of the southern half of the closed area to be much coarser (i.e. with more dead shell and stones) than the north, shallow areas to be coarser than deep areas, patches of fine gravel megaripples in amongst coarser areas and a ridge of stony substratum running SW–NE in the southern half of the area. Faunal communities also show variability, which may be attributable to factors other than substratum, such as local water currents. For example, epifaunal assemblages recorded from diver surveys in the NE

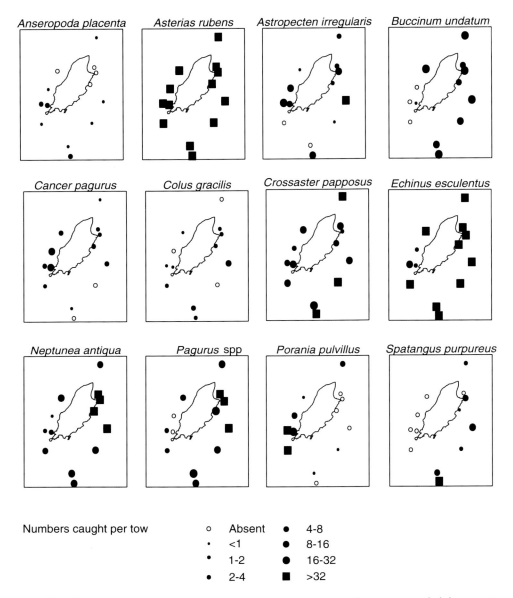

Numbers caught per tow

o	Absent	●	4-8
·	<1	●	8-16
*	1-2	●	16-32
●	2-4	■	>32

Fig. 6.2 Numbers of major by-catch species per 2-nautical mile tow, sampled by queen scallop fishing gear on grounds around the Isle of Man, during the 1994–1995 fishing season. Some species are ubiquitous and common, whilst others have localised distributions.

and the SW quadrants of the closed area were found to be significantly different at the 5% level (ANOSIM Global $R = 0.263$).

Moving to a progressively smaller scale, high variability has been found between replicate samples of the infaunal and epifaunal communities within single experimental plots (each covering an area of $c.$ 500 × 100 m). Table 6.1 is an

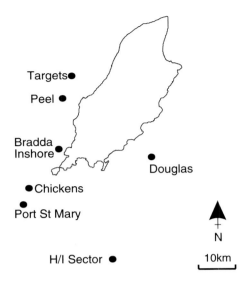

Fig. 6.3 Map of the Isle of Man showing the locations of sampling sites: Bradda Inshore, Chickens, Douglas, H/I Sector, Peel, Port St Mary and Targets.

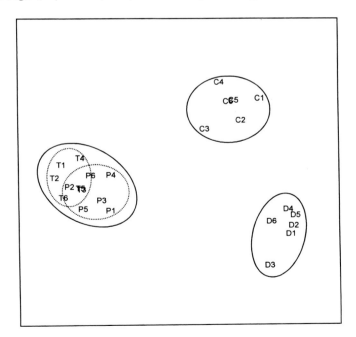

Fig. 6.4 MDS ordination plots based on group average cluster analysis of Bray–Curtis similarities of epifauna biomass. Six replicate beam trawl samples were collected from each of four different sites: D, Douglas; T, Targets; C, Chickens; and P, Port St Mary (see Fig. 6.3), all of which have low commercial fishing effort. The four sites cluster separately (solid lines indicate 43% similarity, dotted lines 56% similarity), suggesting different epifaunal assemblages at each site.

Table 6.1 Variances from one-way ANOVA tests to compare the abundances of three infaunal species between and within four experimental plots in the closed area

Source of variance	Infaunal species		
	Polycarpa spp.	*Moerella donacina*	*Chone fauveli*
Between plot	17.8	20.9	67.1
Within plot	85.2	60.9	150.1
Significance of ANOVA	not significant	inconclusive	inconclusive

Abundances were square root transformed to provide normality and the data are from grab samples taken before any experimental dredging began. Inconclusive results are due to low power. The results illustrate that variability between the 12 replicates within plots is higher than that between the four plots.

example showing the variability between the abundance of three infaunal species from replicate grabs on single sampling plots; the variability between replicates within plots is higher than that between the four plots.

Species abundance related to experimental fishing effort

From grab and fine-meshed dredge sampling of dredged and undredged experimental plots within the closed area, some species were found to be consistently more abundant on the undredged lines. Among the infauna, these included various polychaetes (e.g. *Eurysyllis tuberculata*, *Notomastus latericeus* and *Pista cristata*) and the amphipod *Urothoe marina*, and among the epifauna, the crabs *Hyas coarctatus* and *Monodaeus couchii*. In contrast, the brittlestar *Ophiura albida*, the bivalve *Dosinia exoleta* and the polychaete *Owenia fusiformis* were consistently more abundant on the dredged plots (Fig. 6.5). Physical characteristics of the substratum such as organic content and particle size explained few of the infaunal differences (BIOENV, Fig. 6.6), suggesting that it was the dredging activity that was structuring the community. Surprisingly, species diversity and dominance of epifaunal assemblages were not significantly different between dredged and undredged lines.

Diver surveys of the epifauna of undredged parts of the closed area have been carried out since 1989 to examine the community changes on an area of seabed recovering from commercial scallop dredging. Analysis of changes in mean animal numbers between 1989 and 1998 showed consistent significant increases in many species, including *Pecten maximus* and *Luidia ciliaris*, and upward trends in numbers of *Pagurus* spp., spider crabs and brittlestars (Fig. 6.7). Conversely, *Asterias rubens* appears to be decreasing in abundance.

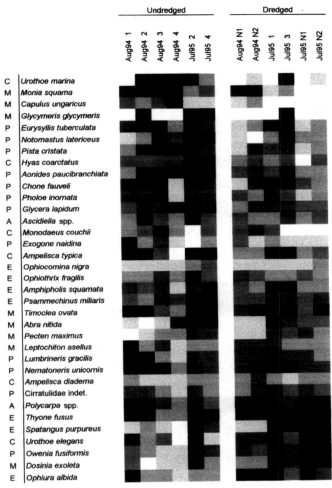

Fig. 6.5 Shade diagram of selected infaunal and epifaunal species from undredged and dredged plots from two sampling dates. Plots 1–4 are inside the closed area, N1 and N2 are outside. The first experimental dredging took place in January 1995 on plots 1 and 3. Abundances of species are based on counts from 12 grab samples and three fine-meshed dredge samples from each plot. Shade intensity indicates relative abundance by species (darkest = highest, lightest = lowest). A, Ascidiacea; C, Crustacea; E, Echinodermata; M, Mollusca; P, Polychaeta.

Dredge disturbance experiments in the closed area

Benthic communities were compared from experimentally disturbed lines and undisturbed lines within the closed area, and with areas outside the closed area that are subject to high levels of commercial dredging. Multivariate cluster analysis showed that since experimental dredging began in the closed area the infaunal communities of the dredged plots have become more similar to those of the

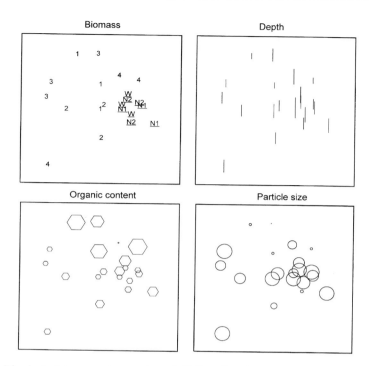

Fig. 6.6 Physical data superimposed on MDS ordinations of the infaunal data from the closed area in April 1994. Sites 1–4 are within the closed area and are undredged; W, N1 and N2 are outside and are subject to commercial dredging. No clear relationship can be seen between these physical factors and the community composition.

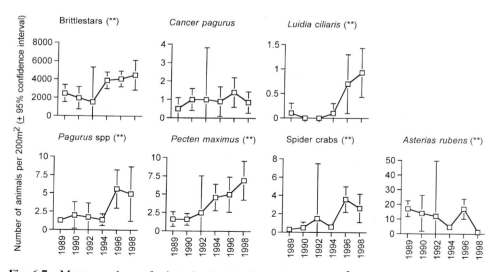

Fig. 6.7 Mean numbers of selected animals observed per 200 m^2 of paired diver surveys in the closed area between 1989 and 1998. ** indicates significant differences ($P < 0.01$) between years (Kruskal–Wallis one-way ANOVA on ranks).

commercially dredged ground than to undredged closed area lines, which have now been undredged since 1989 (Fig. 6.8). This is some of the strongest evidence in this study to indicate the direct longer-term effect of dredge disturbance on benthic community structure.

Closed areas may increase heterogeneity

There is evidence that communities outside the closed area, where heavy dredging takes place, are more homogeneous than those inside the closed area (Fig. 6.8) and that species numbers are slightly greater inside the closed area. The lack of any good relationship between the physical nature of the sediments or water depth and

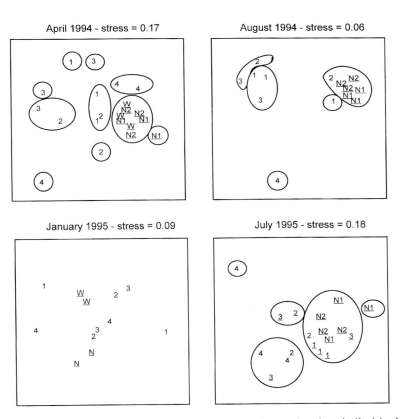

Fig. 6.8 MDS ordinations of infaunal data from the closed area showing similarities between sites within dredged and undredged plots. Dredged sites are underlined. Sites 1–4 are inside the closed area, and W, N1 and N2 outside. Sites 1 and 3 were experimentally disturbed every 2 months, commencing after the January 1995 sampling. Rings represent groupings of sites identified as similar by cluster analysis. Significant differences between dredged and undredged sites were not detected in January 1995, perhaps as a result of inadequate replication due to bad weather.

associated species assemblages suggests that these observations are a real effect of dredging and not due to greater variation in the physical habitat within the closed area.

Infaunal communities were compared before and after the annual closed season for scallop fishing on a number of fishing grounds. A decrease in Pielou's Evenness on some grounds after closure suggests that species present were more uniform in abundance before the closed season than after. Species that increased after closure on the Bradda Inshore ground included tube-forming, burrowing and upright sessile species such as *Pectinaria koreni*, *Tapes rhomboides*, *Cellaria* spp. and *Polycarpa* spp. The rapid proliferation of such growth forms may lead to an increase in microhabitat heterogeneity as other species settle on or seek shelter beneath them.

The changes described were most marked on the heavily dredged Bradda Inshore ground, suggesting that closure can allow the development of more heterogeneous communities; however, the effects of seasonality cannot be ruled out.

Damage and mortality of by-catch

On coarse seabeds, it might be expected that by-catch fauna would suffer badly from being retained in the dredge with shell material and stones (tows of 1-h duration are quite normal for commercial scallop boats), in addition to coming into contact with the heavy gear itself. The damage and mortality sustained by various by-catch animals were compared with the volume of stones in the dredge, but were not found to be significantly related.

The extent of immediate damage depends on the fragility of the species concerned. Brittle or fragile animals such as the urchins *Spatangus purpureus* and *Echinus esculentus*, the brittlestar *Ophiocomina nigra*, starfish *Anseropoda placenta* and edible crab *Cancer pagurus* all suffer badly from the impact of a passing dredge, whereas animals with more robust bodies (e.g. the sea star *Porania pulvillus*) or thick shells (e.g. hermit crabs and the gastropod *Colus gracilis*) have a lower sensitivity (Table 6.2). Some soft-bodied species are also more tolerant of high levels of damage than others (e.g. *Asterias rubens* and *Crossaster papposus*; in the case of these two starfish, this may be due to the ability to regenerate arms). Immediate high damage levels were usually reflected in subsequent death over the next 72 h in laboratory mortality experiments.

Queen gear catches and kills a greater number of individuals, species and biomass of by-catch animals than scallop gear. For example, queen gear killed 27.3–57.0 animals per metre hour of dredging compared with 4.6–8.9 animals per metre hour for scallop dredges. This is mainly due to the more closely spaced teeth collecting more animals, the smaller belly rings retaining more debris and more smaller animals (e.g. spider crabs, small holothurians, bivalves and coelenterates). In addition, many animals are damaged or killed by contact with the two gear types and are left on the seabed in the dredge tracks. The numbers of these animals are not yet known and are much more difficult to investigate.

Table 6.2 Mean damage grades incurred by a selection of by-catch species caught by scallop and queen scallop gear from June 1994 to October 1995. Subjectively assessed damage levels increase from 1 to 3, using set criteria appropriate for each species

	Mean damage grade	
Species	Scallop gear	Queen gear
Spatangus purpureus	3.0	2.8
Anseropoda placenta	2.4	2.2
Echinus esculentus	2.2	2.2
Cancer pagurus	2.1	2.3
Ophiura spp.	2.0	2.0
Ophiothrix fragilis	1.9	2.0
Buccinum undatum	1.9	1.8
Inachus dorsettensis	1.8	1.7
Modiolus modiolus	1.8	1.8
Neptunea antiqua	1.8	1.6
Psammechinus miliaris	1.7	2.0
Astropecten irregularis	1.7	1.6
Aequipecten opercularis	1.7	1.7
Crossaster papposus	1.4	1.3
Asterias rubens	1.4	1.5
Pagurus spp.	1.3	1.6
Porania pulvillus	1.0	1.0

The intensity of fishing effort is known with some degree of accuracy from the logbook scheme, and independent population estimates for many species on some grounds are available from other sampling programmes, so minimum estimates of the total numbers killed by the fishery and the annual incidental mortality due to fishing can be made for these species in some 5×5-nautical mile (n mile) squares, based on by-catch mortality. Annual incidental mortality was not more than 1% of the population for most benthic invertebrates, but was up to 14% for the edible crab, *Cancer pagurus*. Combined with the numbers removed by pot fishermen, this could potentially have an impact on the population of this species (Table 6.3).

Although the volume of stones in the dredge was not found to be significantly related to the extent of damage, a combination of the stones, the dredge teeth and the steel belly rings has the potential to be much more destructive to the benthos than, for instance, the same gear on a sandy substratum.

Table 6.3   Estimates of incidental fishing mortality caused by scallop gear for four grounds surveyed in June and October 1994

Fishing ground	Species	No. killed by commercial fleet per 5 × 5-n mile square	Estimated population per 5 × 5-n mile square	Mortality (No. killed/ population)
June H/I Sector	*Asterias rubens*	147	1 736 217	0.00
	Aequipecten opercularis	600	3 914 206	0.00
	Crossaster papposus	0	45 703	0.00
Bradda Inshore	*Asterias rubens*	7485	6 699 203	0.00
	Cancer pagurus	1663	369 058	0.01
Targets	*Echinus esculentus*	38 079	3 481 696	0.01
	Asterias rubens	9489	2 165 813	0.00
	Aequipecten opercularis	0	83 518	0.00
	Buccinum undatum	0	150 659	0.00
Peel	*Asterias rubens*	7186	1 129 296	0.01
	Cancer pagurus	4669	33 184	**0.14**
October H/I Sector	*Asterias rubens*	62	2 743 237	0.00
	Aequipecten opercularis	600	6 959 189	0.00
	Pagurus spp.	0	2 907 444	0.00
	Crossaster papposus	64	106 327	0.00
Bradda Inshore	*Asterias rubens*	9148	4 873 808	0.00
	Cancer pagurus	9980	93 036	**0.11**
	Anseropoda placenta	5838	359 454	0.02
	Ophiothrix fragilis	13 307	5 957 315	0.00
	Astropecten irregularis	1663	371 716	0.00
	Crossaster papposus	0	165 236	0.00
	Porania pulvillus	0	402 842	0.00
Peel	*Asterias rubens*	11 346	840 584	0.01
	Cancer pagurus	13 243	93 808	**0.14**
	Crossaster papposus	1868	20 751	0.09

Population estimates are based on data obtained from separate studies of commercial grounds (see Fig. 6.3 for site locations).

Scavenger and predator responses

There is some evidence from towed video and diver surveys of experimentally dredged plots that numbers of *Neptunea antiqua*, *Aporrhais pespelecani*, spider crabs, and *Crossaster papposus* increase immediately after dredging (although variability is high). Time-lapse video from by-catch-baited cameras showed that the abundance of

Table 6.4 Summary of ANOVA table comparing predator/scavenger counts before and after two baiting exercises. All counts increased after baiting

Species	Baiting 1	Baiting 2
Cancer pagurus	$F_{(1,4)} = 22.47, P < 0.05$	$F_{(1,6)} = 0.09, P = 0.77$
Liocarcinus spp.	$F_{(1,4)} = 1.86, P = 0.24$	$F_{(1,6)} = 1.10, P = 0.34$
Pagurus spp.	$F_{(1,4)} = 6.16, P = 0.068$	$F_{(1,6)} = 0.11, P = 0.75$
Galatheidae	$F_{(1,4)} = 1.0, P = 0.37$	–
Inachus dorsettensis	$F_{(1,4)} = 1.0, P = 0.37$	–
Asterias rubens	$F_{(1,4)} = 18.61, P < 0.05$	$F_{(1,6)} = 8.51 \ P < 0.03$
Astropecten irregularis	$F_{(1,4)} = 8.24, P < 0.05$	$F_{(1,6)} = 0.71, P = 0.43$
Ophiura spp.	$F_{(1,4)} = 0.19, P = 0.69$	–
Nudibranch indet 1	$F_{(1,4)} = 4.58, P = 0.10$	$F_{(1,6)} = 1.0, P = 0.36$
Nudibranch indet 2	$F_{(1,4)} = 3.17, P = 0.15$	$F_{(1,6)} = 2.63, P = 0.16$
Pecten maximus	$F_{(1,4)} = 1.0, P = 0.37$	–
Callionymus lyra	$F_{(1,4)} = 4.30, P = 0.11$	$F_{(1,6)} = 1.10, P = 0.33$
Flatfish	$F_{(1,4)} = 1.81, P = 0.25$	$F_{(1,6)} = 1.46, P = 0.27$

certain dominant species (*Asterias rubens, Astropecten irregularis, Cancer pagurus*; Table 6.4) increased significantly (ANOVA $P < 0.05$) for up to 60 h after the food source was placed. Several other scavengers, such as plaice *Pleuronectes platessa*, dab *Limanda limanda*, dragonets *Callionymus lyra*, *Pagurus* spp. and *Liocarcinus* spp., were also commonly present at the food source.

Localised aggregations of predator and scavenger numbers around the dredge spoil immediately after it has settled on the seabed will add to the immediate mortality resulting from the dredge impact, thus increasing the mortality of discards. Mobile epifauna, such as the species listed above, are abundant on these coarse substrata and may have more of an effect on dredge-impacted animals immediately after dredging than on more homogeneous fine sediments where mobile epifauna are sparser.

The active scavengers *Neptunea antiqua*, *Buccinum undatum* and *Asterias rubens* were abundant in the queen dredge by-catch. These species are known to be rather robust to encounters with fishing gear (Kaiser & Spencer, 1995), but there is no evidence in our study that they have benefited at the population level; in fact, their relative production was less in areas of high fishing effort. Scavenging species such as *Asterias rubens* and *Cancer pagurus* form the greatest proportion of the catch at high-fishing-effort sites, but their absolute production is generally equal to, or less than, that on the low- and medium-effort sites.

Links between community composition and commercial fishing effort

The relationship between community composition and commercial fishing effort. is not clear cut; few species are confined to either heavily or lightly fished grounds.

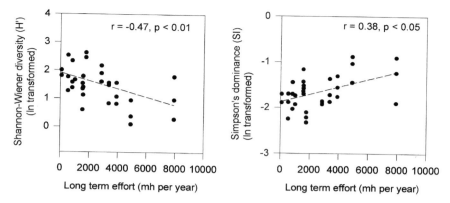

Fig. 6.9 Scatter plots of Shannon–Wiener diversity (H') and Simpson's dominance index (SI) against long-term fishing effort (mh per year, averaged over the years 1981–1996 from fishermen's logbook data) with results of the Spearman's rank correlation calculations shown. The dashed lines indicate trends in the data. H' and SI are calculated for the epifauna collected in each of three replicate queen scallop dredge samples at 13 sites around the Isle of Man.

Robust species such as *Asterias rubens* are found at all sites; however, many smaller crustaceans (e.g. spider crabs and squat lobsters) and molluscs (*Clausinella fasciata*, *Circomphalus casina* and nudibranchs) were absent in the by-catch of heavily dredged grounds. From multivariate comparisons of three benthic communities from around the island, water depth was the most important physical parameter examined in explaining species composition (BIOENV; Spearman weighted correlation $p_w = 0.492$ for epifauna, $p_w = 0.50$ for infauna). Sediment type (particle size, organic content) did not seem to be a major contributing factor, although it was significantly different between sites.

A comparison of queen by-catch composition on 13 sites grouped by mean annual commercial dredging effort over the last 15 years was statistically significant (ANOSIM global $R = 0.31$, $P = 0.03$). Increased fishing effort was also related to increased multivariate variability, as indicated by the index of multivariate dispersion, within a group. Univariate measures also bore out these differences: species richness, number of species, number of individuals and Shannon–Wiener diversity were all less in areas of high mean annual fishing effort, and Simpson's dominance was higher in areas of high mean annual fishing effort (unpublished observations and Fig. 6.9).

Discussion

Patchiness and scale

The way in which scallop dredging affects benthic communities is evidently complex. At its simplest, some animals are injured or killed, or their habitat or food source disrupted in some way so as to be detrimental to their survival. Other animals benefit from the dredging activity, either directly, through improved habitat or increases in food, or indirectly through reduced competition from other animals whose numbers have decreased. Effects may act at the level of the individual, the species, the functional group (e.g. trophic group, r/K specialists, etc.) or the whole community.

Heterogeneity occurs at all scales; within a square metre of seabed, dead shells and stones provide substrata for sessile epifauna and shelter for small invertebrates and fish (Jean & Hily, 1994), while the gravel sediment between them may be home to burrowing anemones and worms. On a scale of, say, 100 m^2 there may be patches of coarser and finer sediment, ripples or waves, or aggregations of brittlestars. On a larger scale still, that of fishing grounds of many square nautical miles, there are localised concentrations of certain species and community types, their distribution perhaps determined by local water currents, sediment type or fishing activity.

The underlying patchiness of the gravelly bottom benthos necessitates the study of dredging impact on many spatial scales. Changes to a benthic community may occur at the small or large scale, and studying one scale at the expense of another may lead to the inability to detect effects at all (Hewitt *et al.*, 1998). In our study, a common problem has been the difficulty in interpreting data due to high variability between replicates. The heterogeneity of gravel communities means sufficient replication to eliminate within-site variability can be logistically difficult. Experiments and sampling regimes must be designed to cater for this variability and must be sufficiently powerful to detect changes if they occur and to reduce the risk of type II errors. For this reason, replication has been increased in the continuations of some of the studies described here.

Broad-scale effects

Although it is useful and interesting to examine overall faunal changes, investigation of the reasons behind the changes can give more insight into the processes occurring as a result of dredge disturbance. Which animals are responsible for changes in species number or species abundance? Is dredging affecting different components of the system in different ways; for instance, are particular trophic groups or animals with particular life history strategies benefiting or being harmed? If particular species or functional groups are found to be the underlying cause of community change, is there an explanation for their increased or decreased success?

Animal morphology

Species with brittle, hard tests, such as the echinoids and crustaceans, are known to be sensitive to contact with scallop dredges and beam trawls (Eleftheriou & Robertson, 1992; Kaiser & Spencer, 1995). It is therefore not surprising that relative production of this group was found to be lower on heavily dredged sites. In contrast, animals with tough, protective shells such as gastropods and hermit crabs are better able to survive capture (Kaiser & Spencer, 1995).

There is some evidence from other studies that animals may suffer higher mortality after the passage of bottom fishing gear, even if they are not damaged. Ramsay & Kaiser (1998) found that undamaged, but disorientated, whelks were more susceptible to predation.

Sessile epifauna

The presence of a rich epifauna is one of the main differences between gravelly substrata and those of finer sediments. The epifauna is also a component of the benthic community that might be expected to be especially sensitive to dredge disturbance.

Alcyonium digitatum is a slow-growing, sedentary species and might be expected to suffer from the effects of dredging. However, it was present in great abundance on some high-effort grounds. *Alcyonium* has been shown to be tolerant of high levels of suspended sediment in laboratory experiments (Hill *et al.*, 1997), sloughing off settled particles with large amounts of mucus. This mechanism may help it to survive in heavily disturbed areas.

The total biomass of sessile epifauna (comprising mainly hydroids, bryozoans and tunicates) was compared between high- and low-effort fishing grounds. Although there were obvious differences between grounds, these could not be attributed to fishing effort. More detailed analysis using SIMPER suggested that different species were affected in different ways by dredging. For example, the hydroids *Nemertesia* spp., *Sertularia cupressina* and *Sertularella gayi* and the bryozoan *Cellaria* spp. were all more abundant on low-effort grounds, whereas the hydroids *Hydrallmania falcata* and *Abietinaria abietina*, dead men's fingers (*Alcyonium digitatum*), cup corals (*Caryophyllia smithii*) and the bryozoan *Flustra foliacea* were all more abundant on high-effort grounds. This difference in distribution is presumably related to the different biology of the species concerned, especially their ability to recolonise and grow rapidly. Some of these seemingly fragile species are evidently more resistant to disturbance than might be assumed.

Scavengers and predators

It might be expected that fishing disturbance would provide a substantial supply of injured and dead animals, which could benefit scavenger populations as a food source (e.g. Kaiser & Spencer, 1996a). Our study supports this hypothesis in the short term at the individual level, as animals aggregate around the dredged areas, but no longer-term (population scale) increases have been demonstrated.

It is unlikely that long-term increases in the populations of these animals will be seen on high-effort grounds. The frequency of dredging (i.e. food supply), even in the highest effort areas, is unlikely to provide a reliable or continuous enough food supply to sustain elevated populations of benthic scavengers or predators. The commercial scallop fishing fleet generally exploits the grounds around the Isle of Man sequentially and will only spend, at most, a few weeks each year fishing on any one ground (Brand & Prudden, 1997). The degree of damage sustained by these opportunistic species may also outweigh the benefits of an increased food source. In this inability to benefit at the population level from increased food supplies caused by fishing, benthic scavengers contrast markedly with more mobile animals such as seabirds, which benefit greatly from fisheries discards (Camphuysen *et al.*, 1995).

Disturbance hypotheses

Various ecological models of disturbance effects have been suggested over the last 20 years. Several authors have supported the hypothesis that species diversity reaches a maximum at intermediate levels of disturbance (e.g. Connell & Keough, 1978; Caswell & Cohen, 1991; Dial & Roughgarden, 1998) but also agree that the ultimate effect depends on the size and frequency of disturbance events and the temporal and spatial scales on which they occur, relative to the life histories of the species involved and interactions among the animals in the community (Connell and Keough, 1978; Petraitis *et al.*, 1989; Caswell *et al.*, 1991).

It is interesting to consider part of this study in relation to these theories: species diversity (as measured in the by-catch assemblage from queen gear) was found to be significantly inversely related to mean annual fishing effort (Fig. 6.9a). For this finding to concur with intermediate disturbance hypotheses, it must be assumed that some factor, be it natural disturbance or past fishing history, has already shifted the low fishing effort areas up to an intermediate level of disturbance. This is not unlikely since these relatively shallow, gravelly seabeds are subjected to appreciable natural disturbance from storms, water currents and bioturbation.

A common problem in studies of this type is to find a benthic community that has not been disturbed by human activities. Commercial scallop fishing has been operating since the late 1930s in the north Irish Sea, and may have already altered the community structure sufficiently that a return to its pre-dredging state is impossible, possibly owing to permanent changes in the substratum (Margetts & Bridger, 1971; Caddy, 1973; Auster *et al.*, 1996). Areas that are unfished today are unfished for good reason. The reason is either that they are different in some way and do not have good scallop stocks or are unsuitable for the fishing gear; this makes them poor control areas. Even the closed area is not a perfect control, as it had been heavily dredged from 1937 to 1989 and its benthic community is therefore in a 'recovery' phase. However, changes in the substratum and the surrounding faunal assemblages may mean that it will never return to its pre-dredging state.

There is evidence from our study that dredge disturbance makes benthic communities more uniform (Fig. 6.8). At the scale of this dredging experiment, this can be explained by dredging activity homogenising the sediment and its associated patchy fauna into a more uniform community (Caddy, 1973; Schwin-ghamer *et al.*, 1996). In contrast, another part of our work suggested that by-catch assemblages on grounds that had experienced high commercial fishing effort had higher multivariate variability compared with grounds that had medium and low fishing effort. At this scale, even grounds with the highest dredging intensity probably contain many small islands of undredged seabed between the dredge tracks, creating a heterogeneous habitat. These latter results agree with ecological models of community change and succession proposed by Caswell & Cohen (1991), who found that, as levels of perturbation increase (up to a point), spatial heterogeneity increases.

In summary, the short-term effects of scallop and queen scallop dredging on gravelly seabeds of the north Irish Sea are measurable but complex. This complexity is due to dredging: (a) impacting different components of the benthos in different ways; and (b) affecting the heterogeneous benthos at a variety of spatial scales. Additionally, the size of the sampling unit relative to the size of the 'patch' will affect subsequent interpretation of the results.

Acknowledgements

Many thanks to the crew of the RV *Roagan* and to all the volunteers who have helped with the sorting of numerous samples over the last 4 years. The study was funded by the UK Ministry of Agriculture, Fisheries and Food as part of contract numbers CSA 2332 and CSA 4142 and drew also on the resources and data provided by the Isle of Man Department of Agriculture, Fisheries and Forestry scallop research project. The manuscript was improved by suggestions from Prof. P.G. Moore and Dr J.M. Hall-Spencer.

References

Auster, P.J., Malatesta, R.J., Langton, R.W., Watling, L., Valentine, P.C., Donaldson, C.L.S., Langton, E.W., Shepard, A.N. & Babb, I.G. (1996) The impacts of mobile fishing gear on seafloor habitats in the Gulf of Maine (Northwest Atlantic): implications for conservation of fish populations. *Reviews in Fisheries Science*, **4**, 185–202.

Bergman, M.J.N. & van Santbrink, J.W. (1994) Direct effects of beam trawling on macrofauna in sandy areas off the Dutch coast. In: *Environmental impact of bottom gears on benthic fauna in relation to natural resources management and protection of the North Sea* (eds S.J. de Groot & H.J. Lindeboom), pp. 179–208. *NIOZ Report* 1994-11/*RIVO-DLO Report* C026/94. Netherlands Institute for Sea Research, Den Burg, Texel, The Netherlands.

Bergman, M.J.N., van Santbrink, J.W., Craeymeersch, J.A. & Fonds, M. (1994) Direct mortality of invertebrate macrobenthos due to trawling with commercial beam and otter trawls. In: *Environmental impact of bottom gears on benthic fauna in relation to natural resources management and protection of the North Sea* (eds S.J. de Groot & H.J. Lindeboom). *NIOZ Report* 1994-11/*RIVO-DLO Report* C026/94. Netherlands Institute for Sea Research, Den Burg, Texel, The Netherlands.

Black, K.P. & Parry, G.D. (1994) Sediment transport rates and sediment disturbance due to scallop dredging in Port Phillip Bay. *Memoirs of the Queensland Museum*, **36**, 327–41.

Brand, A.R. & Prudden, K.L. (1997) *The Isle of Man scallop and queen fisheries: past, present and future. Report to the Manx Government*. University of Liverpool, Liverpool.

Caddy, J.F. (1973) Underwater observations on the tracks of dredges and trawls and some effects of dredging on a scallop ground. *Journal of the Fisheries Research Board of Canada*, **30**, 173–80.

Camphuysen, C.J., Calvo, B., Durink, J., Ensor, K., Follestad, A., Furness, R.W., Garthe, S., Leaper, G., Skov, H., Tasker, M.L. & Winter, C.J.N. (1995) *Consumption of discards by seabirds in the North Sea. Final Report EC DG XIV research contract BIOECO/93/10. NIOZ Report*, Chapter 795-5. Netherlands Institute for Sea Research, Den Burg, Texel, The Netherlands.

Caswell, H. & Cohen, J.E. (1991) Communities in patchy environments: a model of disturbance, competition and heterogeneity. In: *Ecological Heterogeneity* (eds J. Kolasa & S.T.A. Pickett), pp. 97–122. Springer, New York.

Churchill, J.H. (1989) The effect of commercial trawling on sediment resuspension and transport over the Middle Atlantic Bight continental shelf. *Continental Shelf Research*, **9**, 841–64.

Clarke, K.R. & Warwick, R.M. (1994) *Change in Marine Communities: An Approach to Statistical Analysis and Interpretation*, 144 pp. Natural Environment Research Council, UK.

Connell, J.H. & Keough, M.J. (1978) Disturbance and patch dynamics of subtidal marine animals on hard substrata. In: *The Ecology of Natural Disturbances and Patch Dynamics* (eds S.T.A. Pickett & P.S. White), pp. 3–13. Academic Press, New York.

Dial, R. & Roughgarden, J. (1998) Theory of marine communities: the intermediate disturbance hypothesis. *Ecology*, **79**, 1412–24.

Eleftheriou, A. & Robertson, M.R. (1992) The effects of experimental scallop dredging on the fauna and physical environment of a shallow sandy community. *Netherlands Journal of Sea Research*, **30**, 289–99.

Hewitt, J.E., Thrush, S.F., Cummings, V.J. & Turner, S.J. (1998) The effect of changing sampling scales on our ability to detect effects of large-scale processes on communities. *Journal of Experimental Marine Biology and Ecology*, **227**, 251–64.

Hill, A.S., Brand, A.R., Wilson, V.A.W., Veale, L.O. & Hawkins, S.J. (1996) Estimation of by-catch composition and the numbers of by-catch animals killed annually on Manx scallop fishing grounds. *Aquatic Predators and their Prey* (eds S.P.R. Greenstreet & M.L. Tasker). Blackwell Science, Oxford.

Hill, A.S., Brand, A.R., Veale, L.O. & Hawkins, S.J. (1997) *Assessment of the effects of scallop dredging on benthic communities. Final Report to MAFF*. Contract CSA 2332. University of Liverpool, Liverpool.

Holme, N.A. & McIntyre, A.D. (1984) *Methods for the Study of Marine Benthos*. Blackwell Science, Oxford.

Jean, F. & Hily, C. (1994) Quantitative sampling of soft-bottom macroepifauna for assessing the benthic system in the Bay of Brest (France). *Oceanologica Acta*, **17**, 319–30.

Jones, J.B. (1992) Environmental impact of trawling on the seabed: a review. *New Zealand Journal of Marine and Freshwater Research*, **26**, 59–67.

Kaiser, M.J. & Spencer, B.E. (1994) A preliminary assessment of the effect of beam trawling on a benthic community in the Irish Sea. In: *Environmental impact of bottom gears on benthic fauna in relation to natural resources management and protection of the North Sea* (eds S.J. de Groot & H.J. Lindeboom), pp. 87–94. *NIOZ Report 1994-11/RIVO-DLO Report* C026/94. Netherlands Institute for Sea Research, Den Burg, Texel, The Netherlands.

Kaiser, M.J. & Spencer, B.E. (1995) Survival of by-catch from a beam trawl. *Marine Ecology Progress Series*, **126**, 31–8.

Kaiser, M.J. & Spencer, B.E. (1996a) Behavioural responses of scavengers to beam trawl disturbance. In: *Aquatic Predators and their Prey* (eds S.P.R. Greenstreet & M.L. Tasker), pp. 116–23. Blackwell Science, Oxford.

Kaiser, M.J. & Spencer, B.E. (1996b) The effects of beam trawl disturbance on infaunal communities in different habitats. *Journal of Animal Ecology*, **65**, 348–58.

Kaiser, M.J., Hill, A.S., Ramsay, K., Spencer, B.E., Brand, A.R.B., Veale, L.O., Prudden, K., Rees, E.I.S., Munday, B.W., Ball, B. & Hawkins, S.J. (1996) Benthic disturbance by fishing gear in the Irish Sea: a comparison of beam trawling and scallop dredging. *Aquatic Conservation: Marine and Freshwater Ecosystems*, **6**, 269–85.

Lopez, J.E. & Mejuto, J. (1988) Infaunal benthic recolonization after dredging operations in La Coruna Bay, NW Spain. *Cahiers de Biologie Marine*, **29**, 37–49.

Margetts, A.R. & Bridger, J.P. (1971) The effect of a beam-trawl on the sea bed. *Gear and Behaviour Committee, ICES*, C.M. 1971/B:8.

Petraitis, P.S., Latham, R.E. & Niesenbaum, R.A. (1989) The maintenance of species diversity by disturbance. *The Quarterly Review of Biology*, **64**, 393–418.

Ramsay, K. & Kaiser, M.J. (1998) Demersal fishing disturbance increases predation risk for whelks (*Buccinum undatum* L.). *Journal of Sea Research*, **39**, 299–304.

Schwinghamer, P., Guigne, J.Y. & Siu, W.C. (1996) Quantifying the impact of trawling on benthic habitat using high resolution acoustics and chaos theory. *Canadian Journal of Fisheries and Aquatic Sciences*, **53**, 288–96.

Chapter 7
Impact of scallop dredging on maerl grounds

J.M. HALL-SPENCER[1,2] and P.G. MOORE[1]

[1]*University Marine Biological Station, Millport, Isle of Cumbrae, KA28 0EG, UK*
[2]*Division of Environmental and Evolutionary Biology, Institute of Biomedical & Life Sciences, University of Glasgow, Glasgow, UK*

Summary

1. The single passage of Newhaven scallop dredges can bury and kill 70% of the living maerl in their path and extract *c.* 85% of the scallops present.
2. On a dredge track, most of the flora and megafauna to a depth of 10 cm beneath the maerl sediment surface is damaged. Only small, strong-shelled animals are resistant to damage within that stratum.
3. For every 1 kg of scallops caught, 8–15 kg of other organisms are captured from maerl habitats.
4. Dredge tracks remain visible for up to 2.5 years in maerl habitats.
5. Scallop dredging has indirect effects through sediment redistribution, altered habitat structure and modified predator/prey relationships.
6. Maerl is a 'living sediment'; it is slow to recover from disturbance by towed gear due to infrequent recruitment and extremely slow growth rates.
7. Maerl has an associated deep-burrowing megafauna that is resistant to towed gear impact.
8. Pristine maerl communities are highly susceptible to scallop dredging with long-term (> 4 year) reductions in the population densities of epibenthic species and decadal consequences for the maerl itself.
9. Previously impacted maerl beds support modified benthic communities that recover more quickly from scallop dredging (1–2 years).
Keywords: maerl, benthos, scallop dredging, long-term impacts, Scotland.

Introduction

Maerl beds are unusual coastal benthic habitats derived from living calcareous rhodophytes with structural properties that are intermediate between solid and soft sedimentary environments. They typically support a highly diverse flora and fauna (Grall & Glémarec, 1998; Hall-Spencer, 1998; Birkett *et al.*, 1999) but are under immediate threat from human impacts, notably demersal trawling in the Mediterranean (Biomaerl team, 1998; Borg *et al.*, submitted) and scallop dredges in NE Atlantic coastal waters (Hily, Potin & Floc'h, 1992; Hall-Spencer, 1995; Hall-Spencer & Moore, in press).

Large maerl thalli are among the oldest marine macrophytes in the North Atlantic and contribute to deposits that take hundreds to thousands of years to accumulate since even optimal growth rates are extremely slow (Potin *et al.*, 1990; Birkett *et al.*, 1999). For this reason, management of the exploitation of two of the main maerlforming species in Europe, *Lithothamnion corallioides* and *Phymatolithon calcareum*, is obligatory under the EC Directive on the Conservation of Natural Habitats and Wild Fauna and Flora (1992).

Many maerl beds are productive scallop fishing grounds, but concern has been expressed over their vulnerability to dredging disturbance (MacDonald *et al.*, 1996). Of 242 maerl beds surveyed around the UK in the past 10 years, 29% were found to support the great scallop, *Pecten maximus* (Marine Nature Conservation Review database, 1998). In the UK, scallops are fished using Newhaven dredges, which are amongst the most robust types of fishing gear used in Europe as they are designed to withstand collision with boulders (cf. Hall-Spencer *et al.*, 1999). Short-term effects of this gear have been studied on sand and gravel habitats (Chapman *et al.*, 1977; Eleftheriou & Robertson, 1992; Kaiser *et al.*, 1996; Hill *et al.*, 1996; Bradshaw *et al.* this book, Chapter 6), but long-term effects and the influence on maerl fauna were hitherto unknown.

We recently showed that scallop dredges have profound and lasting effects on living maerl, since up to 70% of thalli on dredged tracks are killed through burial (Hall-Spencer & Moore, in press). Here, we present results obtained by filming dredges in use, and monitor recovery of maerl benthos in previously dredged and undredged areas.

Methods

Study sites

Scallop-dredging experiments were undertaken on two maerl beds in the Clyde Sea area, Scotland. A previously undredged area was located at Creag Gobhainn (56°00.66′N, 5°22.32′W) and an area that had been commercially dredged for scallops over the past 40 years was located in Stravanan Bay (55°45.32′N, 5°04.27′W). These sites were surveyed in detail from 1994 to 1998, using a combination of Sprint Remote Operated Vehicle (Perry Tritech Ltd), RoxAnn, Van Veen grab sampling (at least six grabs per site per quarter) and >200 h field observations using SCUBA. Details relating to the environmental characteristics, scallop-dredging history and macrobenthic ecology of these grounds have been presented elsewhere (Hall-Spencer, 1995, 1998; Hall-Spencer & Atkinson, 1999; Hall-Spencer *et al.*, 1999; Hall-Spencer & Moore, in press).

Fishing gear

A gang of three Newhaven dredges (dredge mouth width 77 cm, with spring-loaded teeth 10 cm long × 0.8 cm wide mounted 8 cm apart) was towed by RV *Aora* (260 hp). The dredge mouths were maintained perpendicular to the seabed by a horizontal metal bar, towed on warps, and held off the seabed by a rubber roller at each end. A 92-cm long mat of linked 7-cm diameter steel rings extended behind each tooth bar to withstand abrasion. This chain mail formed a robust belly to the bag and each dredge weighed *c.* 85 kg on land. A 6.0-mm lens colour UWTV camera (Simrad Osprey model OE1362) and compact lamp (Simrad Osprey model OE1132) were mounted on a purpose-built adjustable bracket. This bracket was armoured to protect the camera, cable and lights from collisions with boulders. A submersible TV cable with kevlar braid reinforcement and 2.5-mm thick polyurethane sheath (Hydrocable Systems Ltd, Aberdeen) was used to eliminate problems of cable stretch. Images of the dredge mouths were recorded on videotape (VHS; Ferguson Ltd, UK) with a time/date overlay provided by a video timer (VTG-88; FOR-A Co. Ltd, Japan).

Experimental protocol

At both sites, three transects approximately 100 m long were dredged using single tows of the gear described above. These transects were dredged during the summer of 1994 at water depths between -10 and -15 m Chart Datum. Immediate effects of the scallop dredges were noted on each transect and one transect at each site was monitored by divers for 4 years. Each transect was marked with buoys laid *c.* 10 m apart to delimit the width of a dredging corridor. The long-term transects were surveyed the day before dredging by divers who deployed 1 m^2 quadrats haphazardly along two parallel strips of seabed, one situated between the marker buoys (test plot) the other situated 30 m away, outside the buoys (control plot).

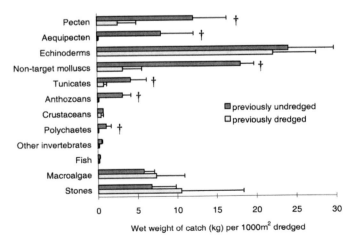

Fig. 7.1 Mean and maximum weights (indicated by histograms and range bars, respectively) of scallops, other organisms and stones (cobbles and pebbles) caught in scallop dredges on transects on previously dredged and undredged areas. † denotes categories in which mean catches from a previously undredged site were more than four times higher than on the previously dredged ground.

Visible organisms, megafaunal burrows/tubes and nests of *Limaria hians* were enumerated in 40 replicate quadrats on each plot. *Limaria hians* nests were counted as the communal byssus nests of this thin-shelled bivalve were an important structural feature of the previously undredged site (Hall-Spencer, 1998).

On each transect, the scallop dredges were towed at 2–3.5 km h^{-1}. On retrieval, *Pecten maximus* were counted, wet weighed and measured. Other elements of catches were sorted into categories (queen scallop *Aequipecten opercularis*, echinoderms, non-target molluscs, tunicates, polychaetes, crustaceans, other invertebrates, fish, macroalgae and stones) and wet weighed. Megafauna were identified, counted and inspected for external signs of damage (cracked shells, missing limbs, etc.). Within an hour of dredging, divers measured the length of dredged transects, collected *P. maximus* remaining on dredge paths and recorded the degree of damage to uncaught elements of the affected benthos. To estimate the efficiency of the dredges, the numbers and size of scallops that remained within the dredge paths immediately after fishing were compared with the numbers and sizes of scallops retained in the dredge bags.

The day after dredging, and two to four times a year over the following 4 years, the benthos of marked test and control plots was monitored by divers deploying 20–40 replicate 1-m^2 quadrats. The number of replicates depended on dive time available. Here, we present data on variation with time and between treatments in the numbers of the burrowing anemone *Cerianthus lloydii*, burrows of the crustacean *Upogebia deltaura* and byssus nests of the bivalve *Limaria hians* as examples of contrasting benthic lifestyles. Analyses using ANOVA were performed on log-transformed data to test for differences in the abundances of these members of the benthos between treatment and control transects before and at time intervals after fishing.

Results

General observations

Video recordings showed that the gear behaved similarly on both sites. The rollers and chain rings of each dredge were in contact with the seabed while the dredge teeth projected fully into the maerl substratum (10 cm) and harrowed the seabed. Maerl was flicked upwards as the dredge teeth jerked back and forward on spring-loaded bars, creating a cloud of suspended sediment. Boulders <1 m^3 were overturned and cobbles often became wedged between the teeth and were dragged through the sediment or were flicked along ahead of the gear. Large, fragile organisms were usually broken on impact (e.g. the sea urchin *Echinus esculentus* and the starfish *Luidia ciliaris*), whereas strong-shelled organisms usually passed into the dredges intact (e.g. *Pecten maximus* and the gastropod *Buccinum undatum*). Other animals (e.g. *Aequipecten opercularis*, the swimming crab *Liocarcinus depurator* and most fish) swam as the gear approached and often escaped capture.

Direct observations on dredged transects showed profound immediate impacts to the benthos. Each transect consisted of a *c.* 2.5-m wide track along which natural bottom features (e.g. crab pits and burrow mounds) were erased. Sand and silt was brought to the sediment surface and living maerl was buried. Dislodged rocks lay overturned, exposing crevice-dwelling fauna and burying surface epiliths. Thick-shelled molluscs that were small enough to pass between the dredge teeth remained in the dredge tracks intact (e.g. *Gibbula magus, Clausinella fasciata*). However, most megafauna on or within the top 10 cm of maerl were either caught in the dredges or left damaged on the dredge tracks (e.g. the crab *Cancer pagurus*, the bivalves *Ensis arcuatus, Laevicardium crassum*, and the sea urchins *Spatangus purpureus, Echinocardium pennatifidum* and *Echinus esculentus*). Close examination of dredge tracks also revealed a littering of animal fragments derived from smaller members of the maerl community (e.g. *Cerianthus lloydii* and the polychaete *Terebellides stroemi*).

On each ground, video and direct observations by divers revealed that several species rapidly aggregated to feed along the dredge tracks including whelks (*Buccinum undatum*), crabs (*Liocarcinus depurator, Necora puber, Pagurus bernhardus*), brittlestars (*Ophiocomina nigra*) and fish (*Pleuronectes platessa, Scyliorhinus canicula, Pomatoschistus pictus, Gadus morhua, Callionymus lyra*). Not only were damaged animals consumed but intact members of the infauna were taken before they could reburrow. Over a 24-h period, less mobile or nocturnal scavengers also moved onto dredge transects (e.g. the isopod *Natatolana borealis* and the starfish *Asterias rubens*) and swarms of lyssianassid amphipods were observed when dead crustaceans were checked by divers. After 3 days the shells of damaged animals were picked clean and scavengers had begun to disperse. Further details of these general observations are given by Hall-Spencer (1995).

Catch composition

The dredges were highly size-selective and none of the *P. maximus* caught was <7 cm in length. Video recordings showed that the brittlestar *Ophiocomina nigra* entered the dredges in their thousands but were not included in the catch as they passed through the 7-cm rings of the dredge bags and littered the dredge tracks immediately after dredging. From an area of 780 m² at the previously undredged site; 42 *P. maximus* were caught and seven were left on the dredge tracks immediately after fishing. From an area of 800 m² at the previously dredged site; nine scallops were caught and two were left on the dredge paths. Thus, catches were much higher at the previously undredged site and, overall, 86% of the marketable size scallops lying in the paths of the gear were captured. Thus, capture rates for scallops in maerl are much higher than for other substrata (Caddy, 1973; Dare *et al.*, 1993).

On each dredge tow, 8–15 kg of by-catch organisms were caught per 1 kg of *P. maximus*. By-catch quantity and composition varied greatly between sites (Fig. 7.1). In each faunal category, more organisms were caught at the previously undredged site which reflected higher population densities of macrobiota than in the previously dredged area. A major difference was that byssus nests of *Limaria hians* and associated organisms were abundant in catches from the previously undredged area but were absent from the dredged site. Per unit area, more than four times the amount of non-target molluscs, tunicates, anthozoans and polychaetes were dredged from the previously undredged site than from the previously dredged site.

The survivorship of discarded by-catch was not investigated. However, animals protected by thick shells (e.g. *P. maximus, B. undatum, Pagurus bernhardus*) were mostly unbroken in both areas. Overall though, > 50% of by-catch organisms were badly damaged owing to either direct impact with the dredges or maceration by cobbles and pebbles that were churned in the chain-mail bags as the gear was towed (Hall-Spencer, 1995).

Temporal dynamics

While no differences were observed between test and control plots prior to experimental scallop dredging, substantial changes were recorded in the months thereafter (Figs 7.2–7.4). Immediately after dredging, control plots were unchanged but most sessile macrofauna and flora was removed or damaged on test plots, and an ephemeral increase in abundance of motile scavengers followed (see above). Dredge tracks remained visible for 0.5–2.5 years depending on depth and exposure to wave action. In a shallow, exposed area (Stravanan Bay, −10 m), dredge tracks were erased by sediment redistribution during storms but in deeper and more sheltered areas, tracks remained clearly visible after storms but were gradually erased through bioturbation.

Monitoring over a period of 4 years revealed different rates of recovery for species associated with maerl habitats. After 1 month, no significant differences were found in the population densities and species diversity of mobile epibenthos (e.g. crabs, gastropods, starfish) between test and control plots. After 6 months, population densities and diversity of fleshy macroalgae were not significantly different on test and control plots. In contrast, slow-growing and/or infrequently recruiting sessile organisms (see below) remained depleted on test plots for 4 years after dredging occurred. The temporal dynamics of three species that exemplified contrasting life styles and responses are given in Figs 7.2–7.4.

In our study, the thalassinidean shrimp *Upogebia deltaura* was the deepest burrowing animal (to 68 cm) and the most abundant large crustacean found within the maerl substratum at both sites. Monitoring on control plots showed that the density of burrow openings changed little over the 4-year period with only slightly fewer openings at the previously undredged site (3–6 m^{-2}) than at the dredged site (4–10 m^{-2}). An investigation into *U. deltaura* burrow structure at these sites showed that each burrow had a single occupant and *c.* 70% of burrows had two openings while the remainder had three (Hall-Spencer & Atkinson, 1999). Immediately after dredging, nearly all burrow openings were erased from test plots (Fig. 7.2). However,

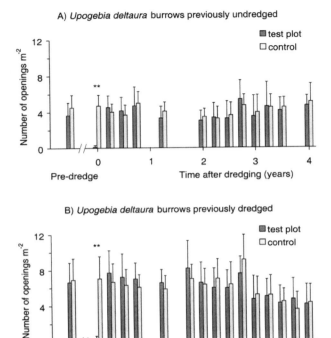

Fig. 7.2 Numbers of openings of deep burrows of the crustacean *Upogebia deltaura* recorded within test and control plots prior to and over 4 years after experimental scallop dredging at (A) previously undredged ground and (B) previously dredged ground. Error bars are ±SE (*n* = 20), significant differences (ANOVA on log-transformed data) are indicated (**P < 0.01, *P < 0.05).

the shrimps rebuilt the upper sections of their burrows within a week and no significant long-term differences were found between burrow densities on control and test plots at either site (Fig. 7.2). At each site, other deep-burrowing megafaunal species (e.g. the bivalve *Mya truncata* and the holothurian *Neopentadactyla mixta*) were also resistant to scallop-dredging disturbance, with no long-term changes to their population density.

Effects on sessile organisms living within the surface layer of maerl were more pronounced, as illustrated by the anemone *Cerianthus lloydii*. At both sites, this tube-dwelling anemone was one of the most abundant animals visible to divers. Monitoring of control plots revealed that they were slightly more abundant during the summer at the undredged site (25–35 m^{-2}) than in the dredged area (18–24 m^{-2}), and their observed densities at both sites decreased to 5–10 m^{-2} in winter. These anemones were able to retract deeply (to 44 cm) within the maerl when disturbed and immediately after dredging very few were seen on test plots, although their torn tubes and polyps littered the maerl surface. Repeat surveys 1 month after fishing revealed that *c.* 25% of the anemones had survived at each site (Fig. 7.3). On both the test and control plots of each ground, *C. lloydii* were seen reproducing through longitudinal fission in the spring months of 1995–98. Significant differences in the abundance of *C. lloydii* persisted between plots for 14 months at both sites after which the population densities became similar (Fig. 7.3). The numbers of several species followed this pattern (e.g. the polychaetes *Eupolymnia nebulosa*, *Lanice conchilega* and the ascidian *Ascidiella aspersa*) with a dramatic decrease in population densities immediately after dredging followed by a return to pre-dredge and control plot levels over the following 1–2 years (unpublished data). After 2 years, no significant differences were found in the population densities of organisms visible to divers between test and control plots at the previously dredged site.

In contrast, at the previously undredged site there was a group of sessile, surface-dwelling species that exhibited consistently lower abundance on test plots for 4 years (e.g. *Phymatolithon calcareum*, sponges, the anemone *Metridium senile*, the bivalves

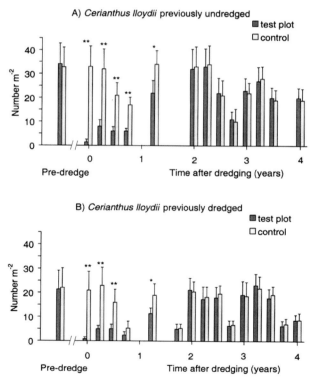

Fig. 7.3 Numbers of the tube-dwelling sea anemone *Cerianthus lloydii* recorded within test and control plots prior to and over 4 years after experimental scallop dredging at (A) previously undredged ground and (B) previously dredged ground. Error bars are ±SE (*n* = 20), significant differences (ANOVA on log transformed data) are indicated (**P < 0.01, * P < 0.05).

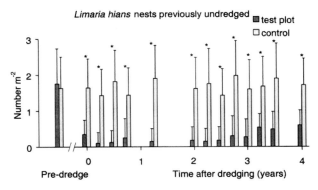

Fig. 7.4 Numbers of byssus nests of the bivalve *Limaria hians* recorded within test and control plots prior to and over 4 years after experimental scallop dredging on the previously undredged ground. Error bars are $\pm SE$ ($n = 20$), significant differences (ANOVA on log-transformed data) are indicated ($*P < 0.05$).

Modiolus modiolus and *Limaria hians*). *Limaria hians*, for example, was absent from the dredged site, although their dead shells were abundant – as noted on three other previously scallop-dredged maerl beds in the Clyde Sea area (Hall-Spencer, 1995). At the previously undredged site, these bivalves built roughly circular byssus nests on the maerl surface that were usually 10–30 cm in diameter. These were covered in a variety of associated fauna (mainly hydroids and tunicates). *Limaria hians* proved to be particularly susceptible to scallop dredging as most of its byssus nests were removed and those that remained were torn, exposing the inhabitants to opportunistic scavengers. The abundance of *L. hians* nests remained significantly depleted post-impact compared with the control plot for 4 years (Fig. 7.4), as did those organisms that were commonly found on (or in) its nests (e.g. the hydroid *Thuiaria articulata* and the ascidian *Diplosoma listerianum*).

Discussion

The immediate effects of scallop dredging over maerl grounds were similar to those recorded for other shallow sublittoral biotopes (Caddy, 1973; Chapman *et al.*, 1977; Eleftheriou & Robertson, 1992), although extensive redistribution of coarse sediment left tracks that lasted up to 2.5 years. Such persistent towed-gear tracks reflect the sheltered conditions of sea lochs (Tuck *et al.*, 1998). Collateral damage to sessile/slow-moving organisms was high and attracted aggregations of benthic predators and scavengers; a typical response to towed gear impact (Kaiser & Spencer, 1994; Ramsay *et al.*, 1998; Bradshaw *et al.*, this volume, Chapter 6). Video records of dredge behaviour and catch analysis revealed interesting differences compared with findings for other sediments. Thus, the ratio of marketable *Pecten maximus* caught to those left on dredge tracks showed that Newhaven scallop dredges were *c.* 85% efficient on maerl, compared with estimates of 14–27% efficiency on coarse, stony sediments (Gruffydd, 1972; Chapman, Mason & Kinnear, 1977) up to a maximum of 41% on smooth muddy gravel (Dare *et al.*, 1993). The maximum penetration of the dredge teeth (10 cm cf. 3 cm recorded by Chapman *et al.*, 1977) and the presence of only a few boulders to cause lifting of the gear are thought to explain the high capture efficiency recorded on maerl grounds. Paradoxically, selectivity of the gear was poor. By-catch organisms constituted > 80% of the biomass of dredge contents, cf. 25% recorded for the same gear on gravel grounds in the Irish Sea (Kaiser *et al.*, 1996). The proportion of by-catch was highest on a previously undredged ground where sessile, epibenthic organisms were more abundant. As in similar studies undertaken in the Irish Sea (Hill *et al.*, 1996), much of the by-catch was fatally injured either by impact with the gear or with stones within the dredge bags.

The high porosity of maerl beds allows a suite of species to occur deep within the sediment (Keegan & Könnecker, 1973; Hall-Spencer & Atkinson, 1999). We found that deep-burrowing species escaped dredge damage and were relatively unaffected in the long-term whereas species that occurred within more superficial maerl strata or on the substratum surface were progressively more susceptible to damage. Monitoring of surface dwellers showed patterns predicted by MacDonald *et al.* (1996). That is, species with regular recruitment and rapid growth (e.g. annual macroalgae) recovered quickly whereas numbers of irregularly recruiting, slow-growing species (e.g. *Modiolus modiolus*) remained significantly depleted. Another set of species also remained reduced in numbers owing to a lack of suitable substrata provided by *Limaria hians*. A similar situation was reported in Strangford Lough (Service & Magorrian, 1997) where *M. modiolus* beds were sensitive to trawling but supported diverse sessile communities in undredged areas.

Our surveys showed that the benthic community present on a previously dredged maerl bed was heavily modified with no long-lived/fragile organisms on or in the surface layer of sediment. Live *Modiolus modiolus* and *Limaria hians* were absent, for example, although their dead shells were abundant. This suggests that these bivalve species had thrived on this area of maerl in the past, since local benthic water currents (up to 11 cm s^{-1} recorded by Hall-Spencer & Atkinson 1999) were not strong enough to transport their shells onto the maerl from elsewhere in the bay and surrounding habitats lacked these species or their shells. The previously dredged macrobenthic community proved resistant to further scallop-dredge disturbance since it returned to pre-experimental status within 2 years. In contrast, scallop-dredging disturbance had a longer-lasting effect on a previously undredged community with changes still discernible after 4 years. As with other types of benthic habitat (Lindeboom, 1995; Tuck *et al.*, 1998), care is needed when interpreting investigations into the ecosystem effects of fishing, since observed impacts vary depending on the past exploitation of the ground.

The contrasting degrees of long-term effect recorded in the present study are particularly noteworthy given the small scale of our experimental impacts (single tows of three dredges with *c.* 230 m^2 ground contact per tow). A typical UK scallop boat, towing 16 or more dredges per side, has been estimated to impact 6.6 km^2 of seabed per 100 h fishing (Kaiser *et al.*, 1996). Our experiments were deliberately restrained, since manipulations at the scale of a fishing fleet (e.g. Currie & Parry,

1996) would have been unnecessarily destructive. That up to 70% of the living maerl can be buried and killed through light deprivation by the single passage of a scallop dredge gang is of serious concern, as this limits the ability of disturbed beds to regenerate (Hall-Spencer & Moore, in press). We found that many maerl grounds in the Clyde Sea area had been scallop-dredged sometime in the recent past, and only one site was located that remained unaffected (trawling being prohibited due to the presence of an underwater communication cable). It is clear from our studies that maerl beds are particularly vulnerable to scallop-dredging disturbance but the present extent of fishing disturbance to such beds is unknown. Not all maerl beds support *Pecten maximus* (Marine Nature Conservation Review database, 1998) and so the risk from scallop dredging in such areas is lower, while others are protected by natural obstructions to trawling (J.H.-S., personal observation). A detailed review of the extent, distribution and status of European maerl beds is needed.

It is vital to the lasting integrity of maerl beds, which represent an essentially non-renewable resource, that they be protected both from direct exploitation (Farrow, 1983; Biomaerl Team, 1998) and from the impacts of towed demersal gear (Hall-Spencer & Moore, in press; present study). Even permanent anchorages over protected maerl beds should be restricted since repeated anchoring can create disturbances analogous to towed fishing gear impacts (Hall-Spencer, 1995).

Acknowledgements

This research was funded by the European Commission under contracts PEM/93/08 (Evaluation of direct impact of fishing gears on substratum and benthos) and MAS3-CT95–0020 (Biomaerl project, Marine Science and Technology Programme (MAST III)). Local fishermen kindly kept clear of experimental plots, and we thank all staff at the University Marine Biological Station Millport for their assistance, especially Ken Cameron, John Chamberlain and Hugh Brown for diving support, Mick Parker and the crew of RV *Aora* and the voluntary students Marion Dolezel, Harry Goudge and Antje Kaffenberger.

References

Biomaerl Team (1998) Maerl grounds: habitats of high biodiversity in European seas. *Third European Marine Science and Technology Conference*, Lisbon 23–27 May 1998, *Project Synopses*, **1**, 170–8.

Birkett, D.A., Maggs, C. & Dring, M.J. (1999) 'Maerl': an overview of dynamic and sensitivity characteristics for conservation management of marine SACs, 90pp. Prepared for the Scottish Association of Marine Science UK Marine Special Areas of Conservation Project.

Borg, J.A., Lanfranco, E., Mifsud, J.R., Rizzo, M. & Schembri, P.J. (submitted) Does fishing have an impact on Maltese maerl grounds? *ICES Journal of Marine Science*.

Caddy, J.F. (1973) Underwater observations on tracks of dredges and trawls and some effects of dredging on a scallop ground. *Journal of the Fisheries Research Board of Canada*, **30**, 173–80.

Chapman, C.J., Mason, J. & Kinnear, J.A.M. (1977) Diving observations on the efficiency of dredges used in the Scottish fishery for the scallop, *Pecten maximus* (L.). *Scottish Fisheries Research Report*, **10**, 1–16.

Council Directive 92/43/EEC (1992) Conservation of natural habitats and of wild flora and fauna. *International Journal of the European Communities*, **L206**, 7–49.

Currie, D.R. & Parry, G.D. (1996) Effects of scallop dredging on a soft sediment community: a large-scale experimental study. *Marine Ecology Progress Series*, **134**, 131–50.

Dare, P.J., Key, D., Darby, C.D. & Connor, P.M. (1993) The efficiency of spring-loaded dredges used in the western English Channel fishing industry for scallops, *Pecten maximus* (L.). *International Council for the Exploration of the Sea*, C.M. 1993/B.15

Eleftheriou, A. & Robertson, M.R. (1992) The effects of experimental scallop dredging on the fauna and physical environment of a shallow sandy community. *Netherlands Journal of Sea Research*, **30**, 289–99.

Farrow, G.E. (1983) Recent sediments and sedimentation in the inner Hebrides. *Proceedings of the Royal Society of Edinburgh*, **83B**, 91–105.

Grall, J. & Glémarec, M. (1998) Biodiversité des fonds de maerl en Bretagne: approche fonctionelle et impacts anthropiques. *Vie Milieu*, **47**, 339–49.

Gruffydd, Ll.D. (1972) Mortality of scallops on a Manx scallop bed due to fishing. *Journal of the Marine Biological Association of the United Kingdom*, **52**, 449–55.

Hall-Spencer, J.M. (1995) *Evaluation of the direct impact of fishing gears on the substratum and on the benthos. EC Report PEM/93/08*, 120 pp. European Commission, Brussels.

Hall-Spencer, J.M. (1998) Conservation issues relating to maerl beds as habitats for molluscs. *Journal of Conchology Special Publication*, **2**, 271–86.

Hall-Spencer, J.M. & Atkinson, R.J.A. (1999) *Upogebia deltaura* (Leach) (Crustacea, Thalassinidea) on maerl bed habitats in the Firth of Clyde. *Journal of the Marine Biological Association of the United Kingdom*, **79**, 871–80.

Hall-Spencer, J.M., Frolia, C., Atkinson, R.J.A. & Moore, P.G. (1999) The impact of Rapido trawling for scallops, *Pecten jacobaeus* L., in the Gulf of Venice. *ICES Journal of Marine Science*, **56**, 111–24.

Hall-Spencer, J.M. & Moore, P.G. (in press) Scallop dredging has profound long-term impacts on maerl habitats. *ICES Journal of Marine Science*.

Hall-Spencer, J.M., Moore, P.G. & Sneddon, L. (1999) Observations on the striking anterior coloration pattern in *Galathea intermedia* (Crustacea: Decapoda: Anomura) and its possible function. *Journal of the Marine Biological Association of the United Kingdom*, **79**, 371–2.

Hill, A.S., Brand, A.R., Wilson, U.A.W., Veale, L.O. & Hawkins, S.J. (1996) Estimation of by-catch composition and the numbers of by-catch animals killed annually on Manx scallop fishing grounds. In: *Aquatic Predators and their Prey* (eds S.P.R. Greenstreet & M.L. Tasker), pp. 111–15. Blackwell Science, Oxford.

Hily, C., Potin, P. & Floc'h, J.-Y. (1992) Structure of subtidal algal assemblages on soft-bottom sediments: fauna/flora interactions of role of disturbances in the Bay of Brest, France. *Marine Ecology Progress Series*, **85**, 115–30.

Kaiser, M.J. & Spencer, B.E. (1994) Fish scavenging behaviour in recently trawled areas. *Marine Ecology Progress Series*, **126**, 41–9.

Kaiser, M.J., Hill, A.S., Ramsay, K., Spencer, B.E., Brand, A.R., Veale, L.O., Prudden, K., Rees, E.I.S., Munday, B.W., Ball, B. & Hawkins, S.J. (1996) Benthic disturbance by fishing gear in the Irish Sea: a comparison of beam trawling and scallop dredging. *Aquatic Conservation: Marine and Freshwater Ecosystems*, **6**, 269–85.

Keegan, B.F. & Könnecker, G. (1973) *In situ* quantitative sampling of benthic organisms. *Helgoländer Meeresuntersuchungen*, **24**, 256–63.

Lindeboom, H.J. (1995) Protected areas in the North Sea: an absolute need for future marine research. *Helgoländer Meeresuntersuchungen*, **49**, 591–602.

MacDonald, D.S., Little, M., Eno, N.C. & Hiscock, K. (1996) Disturbance of benthic species by fishing activities: a sensitivity index. *Aquatic Conservation: Marine and Freshwater Ecosystems*, **6**, 257–68.

Marine Nature Conservation Review database (1998) Joint Nature Conservation Committee, Peterborough, UK.

Potin, P., Floc'h, J.Y., Augris, C. & Cabioch, J. (1990) Annual growth rate of the calcareous red alga *Lithothamnion corallioides* (Corallinales, Rhodophyta) in the Bay of Brest, France. *Hydrobiologia*, **204**, 263–7.

Ramsay, K, Kaiser, M.J. & Hughes, R.N. (1998) Responses of benthic scavengers to fishing disturbance by towed gears in different habitats. *Journal of Experimental Marine Biology and Ecology*, **224**, 73–89.

Service, M. & Magorrian, B.H. (1997) The extent and temporal variation of disturbance to epilithic communities in Strangford Lough, Northern Ireland. *Journal of the Marine Biological Association of the United Kingdom*, **77**, 1151–64.

Tuck, I.D., Hall, S.J., Robertson, M.R., Armstrong, E. & Basford, D.J. (1998) Effects of physical trawling disturbance in a previously undredged Scottish sea loch. *Marine Ecology Progress Series*, **162**, 227–42.

Part 3
Fishing as a source of energy subsidies

The common starfish *Asterias rubens*. Fishing activities generate energy subsidies for marine scavengers such as starfish and whelks that are able to take advantage of this food source. Nevertheless, any increases in the populations on invertebrate scavengers may be offset by the effects of fishing mortality.

Chapter 8
The behavioural response of benthic scavengers to otter-trawling disturbance in the Mediterranean

M. DEMESTRE[1], P. SÁNCHEZ[1] and M.J. KAISER[2]
[1]*Instituto de Ciencias del Mar – CSIC, Paseo Joan de Borbó, s/n 08039, Barcelona, Spain*
[2]*School of Ocean Sciences, University of Wales – Bangor, Menai Bridge, Gwynedd, LL59 5EY, UK*

Summary

1. The behaviour of scavengers and predators was studied in response to otter-trawling disturbance in muddy sediments in the north-west Mediterranean.
2. Repeated trawling with a commercial fishing gear over the same plotted coordinates depleted the abundance of commercially important species such as hake. However, smaller scavenging and predatory species increased in abundance significantly with time.
3. As in previous studies, the aggregative response of scavengers was short-lived and lasted no more than several days, which indicated that additional food resources made available by the trawling activities were rapidly consumed.
Key words: fishing disturbance, muddy sediment, otter trawling, scavenging behaviour.

Introduction

Mediterranean bottom-fishing otter trawls are used in multispecies fisheries and produce a variety of fisheries discards that vary considerably in composition, depending upon the target species and the season in which the fishery occurs (Carbonell *et al.*, 1997). In general, these discards are composed of the non-commercial species or undersized animals that are returned to the sea either dead or alive. The survival of discards once returned to the sea depends on a number of factors, such as the time they spend in the codend of the net, the time spent on board ship exposed to the air and the composition of the by-catch (e.g. a large proportion of debris will increase mortality rates). In addition to those animals killed by discarding, trawling disturbance of the seabed also damages and kills a proportion of benthic fauna. Both of these activities provide an additional source of potential food that can be utilised by scavengers either at the sea surface, in midwater or on the seabed. These predictions are reflected in a number of recent studies that have demonstrated that scavenging species aggregate in trawled areas to consume prey items generated by this disturbance (Kaiser & Spencer, 1994, 1996a; Kaiser & Ramsay, 1997; Ramsay *et al.*, 1997, 1998). Some scavengers move considerable distances to consume food in the form of carrion, which indicates its importance in marine systems (Dayton & Hessler, 1972). The effects of fishing disturbance will vary

121

according to the physical nature of the seabed under study and the local natural disturbance regime (Currie & Parry, 1996; Kaiser & Spencer, 1996a; Sánchez *et al.*, 1998; Tuck *et al.*, 1998). This paper reports the findings of an experiment conducted in the north-western Mediterranean to quantify the short-term behavioural responses of potential benthic scavengers to otter trawling disturbance. The effects of trawling with a commercial bottom-fishing otter trawl on a muddy seabed were examined analysing both changes in the biomass and abundance of demersal and epibenthic species.

Methods

This study was carried out at a site with a muddy seabed off the Catalan coast (NW Mediterranean) situated close to Barcelona harbour (41°10'N, 2°04'E; 41°17'N, 1°56'E). In 1996, a preliminary research survey was conducted to establish whether the chosen area was suitable for the purposes of the study. This area is closed to trawl fisheries, as Spanish fishing regulations do not permit fishing with demersal otter trawls at depths <50 m in their zone of the Mediterranean. A commercial trawler was chartered and four hauls of different duration were performed (15–47 min) in order to establish the minimum haul duration required to obtain a catch similar to that caught by vessels fishing on commercial grounds. The total yield obtained was between 30 and 45 kg h^{-1}, which is comparable with the usual range obtained by the commercial fleet (20–75 kg h^{-1}). In addition, a total of nine infaunal samples were obtained with a Van Veen grab to assess the heterogeneity of the study site. The percentage content of organic matter was fairly uniform across the site and was typical for muddy sediments (3–4%) except for one station (5.2%).

Two vessels were used simultaneously during this study. One vessel, the RV *Garcia del Cid*, was used to collect biological samples. A commercial demersal otter trawler was used to create the fishing disturbance. The fishing gear used was typical for the Catalan Coast and is called a 'bou'. The net was fished with a warp length three times the water depth. The oval otter doors were made from iron, and each weighed 300 kg in air. The codend had a stretched mesh size of 38 mm.

Experimental disturbance was created along two waylines (a wayline describes the route of passage between two plotted positions) located at a depth of either 30 m or 40 m. A Sercel differential global satellite positioning system (DGPS) was installed on both the commercial trawler and the RV *Garcia del Cid*. The coordinates of two experimental waylines were transcribed onto the DGPS system, such that trawling and sample collection was possible with a high degree of precision (Fig. 8.1). The accuracy of trawling was checked by surveying each of the waylines using a Hydroscan 100 kHz side-scan sonar system (Fig. 8.2). Two intensities of disturbance were created by repeatedly trawling along each wayline either seven or 14 times. This protocol effectively ensured that the seabed was entirely disturbed once or twice by the trawl gear, and created a trawled corridor wide enough to ensure accurate sampling (Kaiser & Spencer, 1996a). The trawler fished the first wayline (single

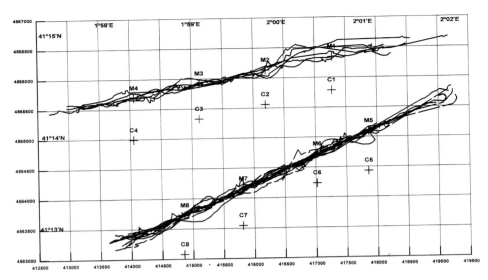

Fig. 8.1 Plotted DGPS positions of the commercial fishing vessel and the positions (M and C) of the towed dredge samples collected during this study.

disturbance) throughout the day until nightfall and made seven tows each of 30-min duration. The second wayline (double disturbance) was trawled over a period of 23 h. The distance of each haul was *c.* 2700 m, and fishing speed was *c.* 3 kn.

Four epifaunal samples were collected using a modified commercial dredge (rastell) from both the fished and adjacent unfished areas prior to and at time intervals after creating the trawling disturbance. The abundance of epifauna was ascertained by standardising the catch data to numbers per 1000 m². The rastell-dredge gear was towed for *c.* 15 min on each occasion to provide an adequate sample of the epifauna. After each catch was brought on deck, specimens were sorted into species and weighed with an *a posteriori* revision for weight loss in the laboratory. The mouth of the rastell-dredge consists of a rectangular metal frame 2 m × 0.4 m in dimension, to which a net with a 40-mm stretched mesh codend is fitted. The rastell-dredge is fitted with 16 tickler chains and behaves in an analogous way to a small survey beam trawl (Ramsay *et al.*, 1996).

The responses of demersal fish and benthic fauna to the effects of trawling were examined by comparing the biomass of the taxa caught in the first haul using the otter trawl with the mean of the first three hauls and last three hauls for each wayline. It is unlikely that the rastell-dredge would sample large fish species effectively, hence the catches from the otter trawl are most representative for these organisms. Differences in the total numbers of individuals and the abundance of representative species between fished and unfished areas, together with time elapsed after fishing, were determined using pair-wise *t*-tests.

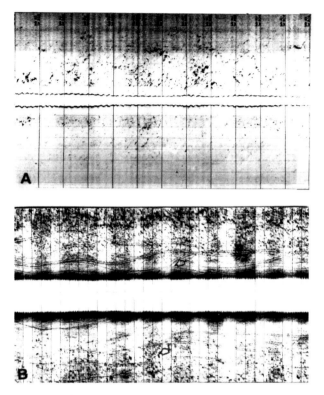

Fig. 8.2   Side-sonar sonographs of the waylines 1 and 2 prior to (A) and after (B) trawling.

Results

Trawl catches

A total of 89 species were caught during the seven consecutive hauls made on the single disturbance wayline (wayline 1). These comprised: 39 fish; 12 crustaceans; 19 molluscs; 10 echinoderms; 9 other invertebrate species; plant material. Similarly, a total of 84 species were caught during the 14 consecutive hauls made on the double-disturbance wayline (wayline 2). These comprised: 35 fish; 9 crustaceans; 17 molluscs; 10 echinoderms; 13 other invertebrate species; plant material. Of the 39 fish species caught on wayline 1, 36% of these decreased in abundance from the first set of three to the final three consecutive hauls. However, only *Cepola rubescens* ($P = 0.04$) and *Spicara* sp. ($P = 0.02$) decreased significantly in their abundance. On wayline 2, 35% of the species decreased after disturbance, but only the fish *Trigla lucerna* decreased significantly ($P = 0.003$). The crustacean *Squilla mantis* is the only species that increased significantly ($P = 0.01$).

On wayline 1 (Fig. 8.3), the crustaceans *Goneplax rhomboides*, *Medorippe lanata* and *Squilla mantis*, and the fishes *Gobius niger*, *Lesueurigobius friesii*, *L. suerii*,

Arnoglossus laterna and *Cepola rubescens* increased considerably in consecutive hauls. In contrast, the gurnard *Trigla lucerna* and sole *Solea vulgaris* decreased after creating the fishing disturbance. Similarly, on wayline 2 (Fig. 8.3), the following species showed a strong tendency to aggregate with increasing trawl disturbance: *Arnoglossus laterna*, *Cepola rubescens*, *Gobius niger*, *Squilla mantis* and *Liocarcinus*

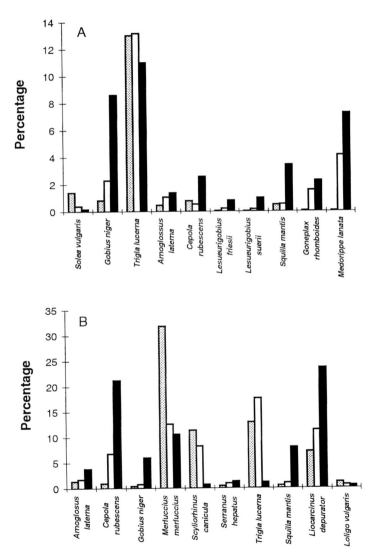

Fig. 8.3 Changes in the biomass of selected species at different stages during the commercial trawling process for (A) wayline 1 and (B) wayline 2. The shaded bars represent the biomass of taxa from the first haul, the open bars represent the mean of the first three hauls and the closed bars represent the mean biomass of the last three hauls made with the commercial otter trawl.

depurator. Both the fishes dogfish *Scyliorhinus canicula* and hake *Merlucius merlucius* decreased with increasing numbers of hauls along the wayline, despite the fact that both are predatory species. A large proportion of these two species were probably removed from the fished area because their large body size means they are retained more effectively by the meshes of the net. These results concur with those of previous studies and indicate that the observed short-term changes in the biomass of taxa within the trawled area are either attributed to the depletion of organisms or due to the aggregation of scavenging species.

Dredge samples

Changes in the abundance of mobile epifauna were detected on both waylines 1 and 2. Those species presented in Fig. 8.4 changed significantly in abundance with time. On wayline 1, the mean abundance of all the invertebrates and the fish *Lesueurigobius suerii* increased significantly immediately after fishing (t_0) relative to adjacent unfished areas, except for the deposit feeding gastropod *Aporrhais pespelicani*. Four days later (t_{102}), the abundance of these species was similar to that which occurred originally in the fished and unfished areas. On wayline 2, where the disturbance was double that of wayline 1, there was a short-term decrease in the abundance of crustaceans and fishes immediately after fishing was completed (t_0). Twenty-four hours later (t_{24}), an increase in the abundance of these taxa was detected. This was particular clear for the crab *Medorippe lanata*. The burrowing shrimp *Alpheus glaber* showed a similar response, although this was more delayed (t_{72}). In the case of the fishes, values after 24 h increased, but never exceeded those of the unfished areas.

Discussion

In common with other similar studies, scavenging species were attracted to the areas disturbed by the otter trawl. These scavenging species are presumably attracted by the odour of animals damaged or displaced by the trawl. These responses are relatively short-lived and last no more than several days. Not all species sampled by the commercial trawl demonstrated an aggregative response to trawl disturbance. For example, the important predator *Merluccius merluccius* decreased in abundance with increasing numbers of hauls. This species may behave exclusively as a predator, be unable to respond to the disturbance, or be removed by the commercial net so effectively that removal rate exceeded the immigration rate onto the areas of disturbed seabed. Similar behaviour was shown by the fishes *Scyliorhinus canicula*, *Solea vulgaris* and *Trigla lucerna*, which decreased in abundance within the disturbed areas.

Experimental trawling reduced the abundance of many epifaunal species. Thereafter, we observed over several days aggregation of epifaunal species that

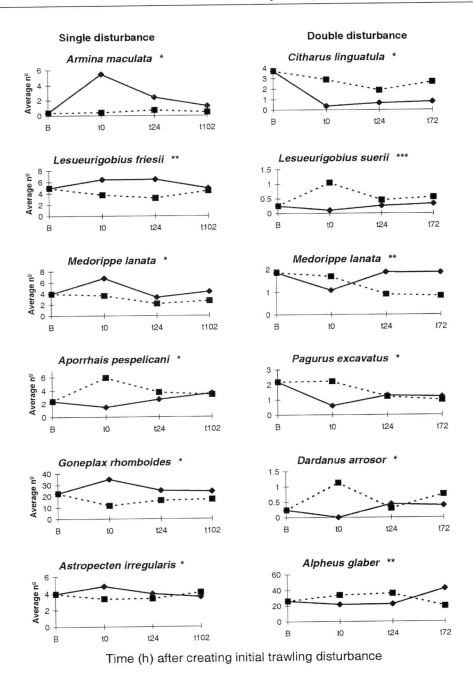

Time (h) after creating initial trawling disturbance

Fig. 8.4 Changes in the abundance of taxa on waylines 1 and 2 ascertained from samples collected with the rastell-dredge at time intervals before (**B**) and after (t0 = immediately after, t24 = 24 h after, t72 = 72 h after, t102 = 102 h after disturbance) creating the trawling disturbance. Significant differences occurred with both time and between fished (♦) and adjacent unfished (■) areas (*$P < 0.05$; **$P < 0.01$; ***$P < 0.001$).

was a typical response of mobile predators or scavengers. In addition, several hours after the initial disturbance was created, we observed black-marks over the trawled areas on the side-scan sonograph record. We interpreted these marks as fish marks that presumably related to fish attracted into the area of trawl disturbance (Kaiser & Spencer, 1994; Sánchez *et al.*, 1998).

The short-term responses of scavenging and some predatory species to trawl disturbance are now well established, although they may vary between different habitats (Kaiser & Spencer, 1994; Ramsay *et al.*, 1996a, b, 1997; Lindeboom & de Groot, 1998). Dietary analyses demonstrate that these scavengers are able to capitalise on animals killed or dug up by the passage of the trawl. However, the energetic benefits derived from this food source are short-lived and last for less than 3 days in all studies undertaken to date (Ramsay *et al.*, 1996a, b, 1997; Lindeboom & de Groot, 1998). Further research in the Mediterranean should address the possibility that energy subsidies created by fishing disturbance might cause shifts in benthic community structure such that they become dominated by populations of scavengers. These changes may be particularly apparent in populations of mobile epibenthic scavenging species, which are probably most food-limited in muddy seabed environments.

Acknowledgements

This research forms part of the contract 'Impact of bottom trawling on sediment and benthic communities in the NW Mediterranean' (Study No. 95/52, European Commission DG-XIV). The authors thank all the participants in the cruise 'Impact '97' for their help, as well as the skipper and crew of the FV *Monfi*. We also thank Sra F. Calderon for her help in the laboratory.

References

Carbonell, A., De Ranieri, S. & Martín, P. (1997) *Discards of the western Mediterranean trawl fleets. Final report to the European Commission*, MED 94/027, 142 pp.

Currie, D.R. & Parry, G.D. (1996) Effects of scallop dredging on a soft sediment community: a large-scale experimental study. *Marine Ecology Progress Series*, **134**, 131–50.

Dayton, P.K. & Hessler, R.R. (1972) Role of biological disturbance in maintaining diversity in the deep sea. *Deep-Sea Research*, **19**, 199–208.

Kaiser, M.J. & Ramsay, K. (1997) Opportunistic feeding by dabs within areas of trawl disturbance: possible implications for increased survival. *Marine Ecology Progress Series*, **152**, 307–10.

Kaiser, M.J. & Spencer, B.E. (1994) Fish scavenging behaviour in recently trawled areas *Marine Ecology Progress Series*, **112**, 41–9.

Kaiser M.J. & Spencer, B.E. (1996a) The effects of beam-trawl disturbance on infaunal communities in different habitats. *Journal of Animal Ecology*, **65**, 348–58.

Kaiser, M.J. & Spencer, B.E. (1996b) Behavioural responses of scavengers to beam trawl disturbance. In: *Aquatic Predators and their Prey* (eds S.P.R. Greenstreet & M.L. Tasker), pp. 117–23. Blackwell Science, Oxford.

Lindeboom H.J. & de Groot, S.J. (eds) (1998) *The effects of different types of fisheries on the North Sea and Irish Sea benthic ecosystems. NIOZ Report* 1998-1/*RIVO-DLO Report* C003/98. Netherlands Institute for Sea Research, Den Burg, Texel, The Netherlands.

Ramsay, K., Kaiser, M.J. & Hughes, R.N. (1996) Changes in hermit crab feeding patterns in response to trawling disturbance. *Marine Ecology Progress Series*, **144**, 63–72.

Ramsay, K., Kaiser, M.J. & Hughes, R.N. (1997) A field study of intraspecific competition for food in hermit crabs (*Pagurus bernhardus*). *Estuarine, Coastal and Shelf Science*, **44**, 213–20.

Ramsay, K., Kaiser, M.J. & Hughes, R.N. (1998) The response of benthic scavengers to fishing disturbance in different habitats. *Journal of Experimental Marine Biology and Ecology*, **224**, 73–89

Sánchez, P., Demestre, M., Palanques, A., Mas, J. & Kaiser, M.J. (1998) *Impact of bottom trawling on the sediments and benthic communities in the NW Mediterranean. Final Report to the European Commission*, DG XIV Study 95/52, 132 pp.

Tuck, I.D., Hall, S.J., Robertson, M.J., Armstrong, E. & Basford, D.J. (1998) Effects of physical trawling disturbance in a previously unfished sheltered Scottish sea loch. *Marine Ecology Progress Series*, **162**, 227–42.

Chapter 9
Food subsidies generated by the beam-trawl fishery in the southern North Sea

M. FONDS and S. GROENEWOLD

Netherlands Institute for Sea Research, NIOZ, Postbox 59, 1790 AB Den Burg, Texel, The Netherlands

Summary

1. The intensive beam-trawl fishery for sole and plaice in the southern North Sea produces large amounts of discard materials and much larger amounts of damaged fauna on the seabed. This material is rapidly consumed by opportunistic scavenging species, such as birds, crabs, starfish and fish. Damaged and exposed benthos is mainly consumed by fish, while discarded fish are mainly consumed by invertebrate scavengers. Trawling results in an increased rate of recycling of macro-benthic fauna and fish through the food web.

2. The balance between food generated by beam trawling and the potential food consumed by local populations of benthic carnivores and demersal fish was estimated for four different areas in the southern North Sea. On average, beam trawling an area once in the summer may generate $c.$ 127 g afdw (ash-free dry weight) 100 m^{-2}. This can be compared with a potential daily food consumption by benthic carnivores of $c.$ 13.2 g afdw 100 m^{-2}, 10.8 g by benthic invertebrates and 2.4 g by demersal fish. In winter, food production by beam trawling and potential daily food consumption by benthic carnivores is estimated to be lower: $c.$ 87 g generated compared with $c.$ 3.5 g consumption.

3. On average, beam trawling may generate $c.$ 180 g afdw 100 m^{-2}year^{-1} damaged benthos and approximately 15–38 g afdw 100 m^{-2}year^{-1} of discard fish, compared with a potential annual food demand of $c.$ 2450 g afdw 100 m^{-2}year^{-1} for benthic invertebrate carnivores and 550 g afdw 100 m^{-2}year^{-1} for demersal fish.

4. The annual amount of food supplied by beam trawling is approximately 7% of the maximum annual food demand of all common benthic predators considered together, which may help to maintain these populations but is insufficient to support further population growth.

5. While beam trawling undoubtedly increases food subsidies in the marine environment, it also removes large predators from the ecosystem. This may have led to higher growth rates of some fish and caused increases in the populations of small fish species such as dragonets, solenettes, scaldfish, lesser weever and gobies.

Keywords: demersal fishery, beam trawl, benthos, discards, scavengers.

Introduction

The beam-trawl fishery for sole (*Solea solea*) has increased considerably in the period 1960–1990 (Polet & Blom, 1998). Today, more than 440 vessels operate in the southern North Sea. Larger vessels (> 230 ships > 800 hp), each fishing with two 12-m-wide beam trawls at a speed of approximately 6 kn, are restricted to offshore areas outside the 12-mile zone. Since 1995, these vessels have also been excluded from a much wider protected area ('Plaice Box') off the Wadden Sea islands along the Dutch, German and Danish coast. Within these coastal areas, only smaller ships (300 hp) are allowed to trawl with two 4-m-wide beam trawls, which are usually towed at a speed of approximately 4 kn. About 450 shrimp beam trawlers also operate in the same area

The sole is a rather elongated and flexible flatfish, hence sole nets have a relatively small mesh size of 8 cm (stretched) in the codend to enable fishermen to catch sufficient animals to make a living (Beek *et al.*, 1983). This leads to a considerable amount of unwanted by-catch, mostly undersized plaice and dab, which are discarded during sorting of the catch (Beek *et al.*, 1990; Fonds, 1994a). Beam trawls used for sole fishing are equipped with at least 10 tickler chains in front of the groundrope, which may penetrate the sea bed to an average depth of *c.* 2–6 cm. This causes considerable direct mortality of benthic invertebrates within the trawl track (Bergman & Santbrink, this volume, Chapter 4) and may also lead to increased numbers of invertebrates in the discards, particularly starfish (*Asterias rubens* and *Astropecten irregularis*) and sea potatoes (*Echinocardium cordatum*).

In the southern North Sea, a variety of investigations have been carried out to estimate the general catch composition of beam trawls, the amount of discards produced, and the survival rate of different species in the discards. In addition, we also studied the direct mortality of benthic fauna due to beam trawling, the identity of carnivorous species feeding on discards and damaged or exposed benthos, and the importance of food produced by beam trawling for populations of these species (de Groot & Lindeboom, 1994; Lindeboom & de Groot, 1998). This chapter concentrates on the results of the last two topic areas.

Dead discarded material and damaged or exposed bottom fauna comprise a significant amount of organic matter. Sea birds following trawlers may consume around 10% of these discards, in particular the roundfish, smaller flatfish and offal (Camphuysen *et al.*, 1995). However, the remaining discards sink to the seabed where they become available for bottom-feeding scavengers and predators. This study investigated the identity of benthic animals that use this food source, and whether it is of any ecological importance for the food web in the benthic ecosystem.

Methods

Study locations

In the period between March 1992 and September 1995, 12 surveys were carried out with the research vessels *Tridens* and *Isis* (Directorate of Fisheries, Ministry of Agriculture, Nature Management and Fisheries, The Hague) at 14 locations in the southern part of the North Sea, from Helgoland and Weisse Bank to the Oystergrounds and south to the Dutch west coast (Fig. 9.1).

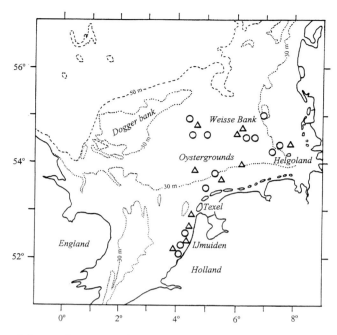

Fig. 9.1.   Map of the southern part of the North Sea showing the locations where repeated sampling of a trawled transect was carried out (○) and where baited traps were exposed on the sea floor (△).

Catch composition and catch efficiency

The catch composition was analysed for 126 'commercial' hauls of at least 1 h duration, using either 4-m or 12-m beam trawls (Craeymeersch *et al.*, 1998). Experimental trawling was carried out on several precisely defined strips (Fig. 9.1). In each of these areas, bottom fauna were sampled to measure their initial densities using a range of gear that included a Van Veen grab (sample size 0.18 m²), a benthos dredge Triple-D (sample size 25 m²) (Bergman & van Santbrink, 1994) and a fine-meshed 3-m beam trawl (sample size 1000–2000 m²). This enabled the estimation of the catch efficiency of commercial trawls and direct mortality of benthos in the trawl tracks (Bergman & van Santbrink, this volume, Chapter 4).

Identification of scavenging species

Repeated sampling on a trawled area

At each site (Fig. 9.1), a narrow strip of *c.* 2000 × 60 m was completely trawled (three times) with 12-m beam trawls. After trawling, the immigration of different species was followed over 1–3 days, by sampling regularly at 12- or 24-h intervals with a fine-meshed (1 cm stretched) 3-m beam trawl fitted with six tickler chains and meter wheels that measured precisely the distance covered during each haul. Changes in density of different species on the trawled strip were compared with changes on a nearby untrawled reference area. Species that showed a rapid increase, compared with initial or reference (untrawled) densities, were considered as potential scavengers. Their stomach contents were analysed to verify whether they were feeding on damaged and exposed fauna (Groenewold & Fonds, 1999).

Baited traps

In order to determine the feeding preferences of scavengers for certain discards, three types of traps, baited with different kinds of discard materials, were exposed for 2 days on the seabed at different localities (Fig. 9.1). Trap designs are described in Fonds *et al.* (1998). The traps were attached in series, with distances of about 5 m between consecutive traps, to 50-m-long steel chains that were stretched on the bottom between two heavy anchor weights. Each series always contained unbaited traps as a control. Trap catches were compared with densities of the same species in the local vicinity, which were estimated with the fine-meshed 3-m beam trawl. Trap catches in relation to different kinds of discard materials used as bait were analysed according to the methodology outlined by Krebs (1989). The selectivity of different kinds of bait in the traps for different species of scavengers is indicated by Manly's alpha index of preference. Values above the critical threshold indicate a significant positive selection of the bait. Food niche width of different species in the traps was estimated by Levin's index, 100 indicating not specialised and 0 most specialised (Krebs, 1989).

Food generated by beam trawling

Estimation of biomass of damaged benthic fauna in trawl tracks

At several stations in the southern North Sea direct mortality of invertebrates was estimated from the decrease in density after experimental trawling. Densities of benthic fauna were precisely estimated before and after trawling, by Van Veen grab, benthos dredge and fine-meshed 3-m beam trawl. In the estimates of total mortality of benthos, mortality of invertebrates in the discard are included (Bergman & van Santbrink, this volume, Chapter 4). In order to allow for calculations of biomass, samples of all common species were measured and weighed on board, while ash-free dry weight was later estimated in the laboratory. The total decrease in biomass (ash-free dry weight, afdw) following complete trawling of an area is considered to be equivalent to the amount of damaged and exposed benthos that is consumed by benthic carnivores.

Production of dead discards

The survival of different species in the discards of commercial hauls was estimated on board ship. Mortality on the sorting belt that carried the catch from the net hopper below deck was estimated. Living animals were stored into tanks with running sea water at ambient temperature to estimate additional mortality over a further 2 days (Fonds, 1994b). Total mortality is the sum of direct mortality on the sorting belt and the mortality observed in the survival tanks. Some species that showed very slow reactions to damage, such as whelks (*Buccinum undatum*) and quahogs (*Arctica islandica*), were taken to the laboratory to monitor their survival over a period of about 1 month (Witbaard & Klein, 1994; Mensink *et al*., in press).

Food consumption by benthic carnivores

Estimation of daily food consumption of abundant benthic carnivores

Measurements of metabolism, food consumption and growth of selected common species of benthic carnivores were carried out in the laboratory at different constant temperatures, in order to estimate their potential daily food consumption in the sea in summer and winter. The animals were kept singly or in groups, in small (10–100 l.) or large (700 l.) tanks, with running sea water at constant temperatures of 5, 10, 15 and 20°C. In separate experiments, oxygen consumption, food consumption and growth were measured over periods of about 1 month. The relationships of these parameters with the size of the animals and water temperature were estimated. Maximum daily food consumption rate was measured by feeding weighed excess rations and reweighing food left over the next day, together with a control to estimate the weight loss of food in sea water (cf. Fonds *et al*., 1992). In general, mussel meat (*Mytilus edulis*) was used as a standard food, although in some cases fish flesh (fillet of whiting) or shrimp (*Crangon crangon*) was used.

In situ *clearance rate of discard fish*

Dead discard fish were attached to 7-m long lines, weighted with lead and exposed on the sea bed at the same locations as the traps (Fig. 9.1). The fish were retrieved after 2 days and weight loss was measured. For comparison and as a control, bacterial decomposition of dead discard fish was measured on board over the same period in tanks with running sea water at ambient temperature. Weight loss of fish, exposed on the seabed, was corrected for the weight loss measured on board in controls. The rate of decay of discard fish (dab and whiting) in sea water was also measured in the laboratory in tanks with running sea water at constant temperatures of 5, 10 and 15°C.

Results

Catch composition and catch efficiency of beam trawls

The average catch composition of 12-m and 4-m beam trawls is presented in Table 9.1, together with data on average densities of all benthic animals (>1 cm in size) sampled in the same areas (Bergman *et al.*, 1998a, b). The commercial fishing gears caught mainly larger animals that were >4–5 cm diameter for invertebrates and >10–15 cm total length for fish (Fonds, 1994a). The majority of benthic animals in the southern North Sea are smaller, and the general catch efficiency for benthos appeared to be rather low: <1–3% (Craeymeersch *et al.*, 1998). Beam trawls catch mainly flatfish, while roundfish comprise <10% of the marketable fish. Discarded fish were represented mainly by undersized dab and plaice (>90%), and the total numbers discarded were about seven to 10 times the numbers of marketable fish, which is the equivalent of about twice as much weight as marketable fish.

Identification of scavenging benthic species

Repeated sampling on a trawled line

Repeated sampling with a fine-meshed beam trawl in areas disturbed with commercial trawl gear, showed that some species rapidly increased in numbers within 2 days, and some attained densities three to four times higher than the density measured prior to trawling or in an untrawled reference area (Table 9.2). This was particularly the case for dab (*Limanda limanda*), whiting (*Merlangius merlangus*), dragonets (*Callionymus lyra*) and swimming crabs (*Liocarcinus* spp.).

At four locations, from the western Dutch coast to Helgoland, fish were sampled for stomach content analysis before and after trawling a particular transect (Groenewold & Fonds, 1999). More than 1000 fish were examined, belonging to nine species: 139 dabs, 144 plaice (*Pleuronectes platessa*), 95 soles, 139 dragonets (*Callionymus lyra*), 93 tub gurnards (*Trigla lucerna*), 37 grey gurnards (*Eutrigla gurnardus*), 88 whiting, 20 lesser weevers (*Echiichthys vipera*) and 38 bull rout (*Myoxocephalus scorpius*). In general, fish from recently trawled areas had two to four times the amount of stomach contents of fish sampled from untrawled reference areas, and also ate a slightly wider spectrum of prey species: from three to eight prey species before trawling to four to 12 species after trawling. Soles and some predatory species (lesser weever and bull rout) showed no change in prey selection (mainly shrimps and gobies) nor in stomach fullness. However, the stomach contents of the other species sampled in recently trawled areas consisted mainly of benthic infauna damaged or exposed by trawling: particularly shellfish and the intestines of the sea potato (*Echinocardium cordatum*). Comparison of stomach contents of fish, sampled before and after trawling an area, indicated that the stomach fullness index (weight of stomach contents as % of fish weight) increased within 12 h after trawling, but rapidly decreased again to pre-trawling values 24–48 h after trawling.

Table 9.1 Catch composition and catch efficiency of 4-m and 12-m beam trawls in the southern North Sea. Comparison with densities of small fish species and benthic megafauna (>1 cm) estimated with a fine-meshed beam trawl and benthos dredge (Triple-D) in 1996 and 1997. The ratio of discarded materials to fish of marketable size is also shown

Area:	Shallow coastal area		Deeper offshore area		Coastal	Offshore	Coastal	Offshore
Trawl type:	4-m beam trawl		12-m beam trawl		Fine-meshed net		4-m beam trawl Catch efficiency	12-m beam trawl Catch efficiency
Number of hauls:	57		31		31	30	31	31
Catch per hectare (0.01 km²):	No.	(kg)	No.	(kg)	No.	No.	No. (%)	No. (%)
Sole (*S. solea*)	9	1.1	7	1.6	110	(<1)	8	–
Other marketable flatfish	11	3.1	11	3.0				
Marketable roundfish	2	0.4	1	0.4				
Discard flatfish	146	10.3	170	8.6	860	320	17	53
Discard roundfish	10	0.4	16	0.6	17	34	59	47
Small fish species					1540	540	<0.1	<0.2
Polychaetes (*Aphrodite*)	2	<0.1	37	0.9	100	3000	2	1
Crustaceans	182	2.5	116	2.0	14 000	14 000	1	1
Echinoderms	1042	35.2	1309	16.8	33 000	40 000	3	3
Molluscs	511	2.7	19	1.8	60 000	43 000	1	<1
All marketable fish	22	4.6	19	5.0				
All discard fish	156	10.7	186	9.2	2300	700	7	27
All invertebrates	1737	40.5	1383	19.5	107 100	100 000	<2	<2
Dead invertebrates	430	6.0	200	4.0				
Discard fish/market sole	17	9.7	26	5.8				
Discard fish/all market fish	7	2.3	10	1.8				
Dead inv./market sole	48	5.5	29	2.3				
Dead inv./all market fish	20	1.3	10	0.7				

Table 9.2 Evidence of immigration of demersal fish species and invertebrates into a newly trawled area. Numbers of stations where the species were observed, the number of stations where an increase in density was observed and the range of proportional increase in density observed at those stations

Species	No. of stations observed	Stations showing immigration	Range of proportional increase
Fish			
Sole	8	4	1–2
Plaice	9	7	1–2
Dab	9	8	1–4
Gadoids	9	3	3–4
Gurnards	7	2	1–4
Scaldfish	6	3	1–2
Dragonet	6	5	1–4
Hooknose	5	3	1
Weever	1	1	2
Gobies	6	3	1
Juvenile mullet	2	1	2
Invertebrates			
Starfish	8	2	1
Sandstar	5	1	1
Brittlestars	7	1	1
Swimming crab	9	6	1–3
Hermit crab	9	3	1
Shrimp	4	2	1–2

Baited traps

During five surveys, 370 traps were deployed at 14 locations (Fig. 9.1). The amphipod traps caught 48 000 amphipods, while the other traps caught more than 5500 larger animals belonging to 42 different species (>4000 crustaceans (12 species), *c.* 1200 echinoderms (six species), 140 fish (17 species), 100 molluscs (two species) and seven worms (two species)). Two species of small amphipods belonging to the family Lyssianidae (*Orchomene nana* (size 5–7 mm) and *Scopelocheirus hopei* (size 7–10 mm)) were caught in special amphipod traps baited with crushed crustaceans, at densities of hundreds to thousands of amphipods per trap. A large species of isopod (*Natatolana (Cirolana) borealis*) was caught in small numbers offshore in traps baited with fish.

For all trap catches taken together, crustaceans (*Liocarcinus* spp., *Pagurus bernhardus*) were dominant both in number (73%) and in weight (70%), followed by starfish (mainly *Asterias rubens*: numbers = 22%, weight = 11%) and fish (mainly gadoid fish: numbers = <3%, weight = 16%). Whelks (*Buccinum undatum*) at the offshore stations and the large edible crab (*Cancer pagurus*) occurred in smaller proportions in the trap catches.

For nine of the most abundant species, catches in traps were analysed in relation to different kinds of discard materials used as bait (Table 9.3). For each species, the area of attraction was estimated from the average number of a species caught in traps divided by the density of that species in the area. The traps were exposed with their openings into the main current direction, and scavengers were probably mainly attracted by the smell of the bait (cf. Sainte-Marie & Hargrave, 1987). The distance of attraction varied from 3 to 5 m for shrimps and brittle stars to more than 10 m for gadoid fish, hermit crabs, swimming crabs and the isopod *N. borealis*.

Most species caught in traps were not very selective in their choice of bait (Table 9.3). Swimming crabs showed some preference for fish bait, while whelks preferred molluscan flesh. The Lyssianid amphipods (*Orchomene nana*) showed a strong preference for crustacean bait, while the isopod *Natatolana borealis* was caught only in traps baited with fish. Crushed echinoderms or polychaetes (seamouse *Aphrodite aculeata*) appeared to be less attractive as bait, only gadoid fish showed a clear preference for seamice (Table 9.3).

Food generated by beam trawling

Biomass of damaged and exposed fauna within trawl tracks

Direct mortality of benthos in a trawled strip was estimated at several stations by sampling the benthos before and after trawling (Bergman & Santbrink, this volume, Chapter 4). In general, the direct mortality of benthos due to trawling was comparable for both 4-m and 12-m beam trawls. Species that live at a greater depth in the sediment (*Ensis* spp., *Lutraria lutraria*, *Callianasa subterranea*) and species that are protected by hard shells (*Donax vittatus*, *Chamelaea gallina*, *Turritella communis*, *Arctica islandica*) suffered lower mortality as compared other more vulnerable species (*Spisula* spp., *Angulus* spp., *Phaxas pellucidus*, *Dosinia lupinus*, *Gari fervensis*, *Echinocardium cordatum* and *Corystes cassivelaunus*). The seamouse (*Aphrodite aculeata*), in spite of its soft body, appeared to be fairly robust (8–16% mortality).

Ash-free dry weight, as a percentage of wet weight, was *c.* 10% for starfish and shellfish (with shells), *c.* 20% for crabs, *c.* 3–5% for sea urchins (*Echinocardium cordatum* and *Psammechinus miliaris*, respectively) and for worms *c.* 4% (*Lagis koreni* with tube) or 15% (*Nephthys* spp.). These values have been used to estimate the ash-free dry weight of food generated by beam trawling.

Production of dead discards

To estimate the total production of dead organic material as discards, the mortality of discarded animals must be calculated. Table 9.4 presents estimates for the most abundant species found in discards. The mortality of molluscs and crustaceans was

Table 9.3 Analysis of catches in traps baited with different kinds of discard materials

Species	Total No. in traps	Area of attraction (m²)	Critical value alpha	Manly's alpha values (×100) for different kinds of bait						Levin's index ×100 food niche
				No bait	Fish	Moll.	Crust.	Echin.	Polych.	
Shrimps (*Crangon crangon*)	95	9	25	11	22	31	36	–	–	83
Hermit crabs (*Pagurus bernhardus*)	291	135	16	2	25	18	19	23	13	79
Brittlestars (*Ophiura ophiura*)	86	8	16	13	35	25	7	20	0	61
Starfish (*Asterias rubens*)	920	27	16	1	27	30	27	9	6	61
Gadoid fish (juv. cod, whiting, bib)	28	281	16	0	22	22	25	0	31	58
Swimming crabs (*Liocarcinus* spp.)	555	118	16	2	47	23	7	9	12	46
Whelks (*Buccinum undatum*)	102	50	20	0	2	70	15	13	–	22
Amphipods (*Orchomene nana*)	42 000	–	16	0	0	0	85	13	2	7
Isopod (*Natatolana borealis*)	25	159	20	0	100	0	0	–	0	0

Manly's alpha values (×100) above the critical value indicate a positive selection of the bait. Levin's index (× 100) indicates food niche width : 0 = maximum specialisation, 100 = not specialised. The area of attraction is estimated as the average catch in traps compared with the average density in the same area.

Table 9.4   Mortality of some common species in the discards of commercial hauls with 4-m and 12-m beam trawls

Species	Coastal areas		Offshore areas	
	4-m beam trawl		12-m beam trawl	
	Mortality (%)	No. in trials	Mortality (%)	No. in trials
Fish				
Dab	98	138	> 99	87
Plaice	93	158	95	85
Flounder	74	68		
Sole	78	126	95	57
Turbot and brill	89	93		
Gurnards	> 96	123	100	20
Crustaceans				
Swimming crab	56	117	60	533
Masked crab	57	10	67	746
Edible crab	34	41	65	55
Hermit crab	10	27	25	224
Molluscs				
Prickly cockle			50	499
Quahog			88	99
Trough shells	33	27	26	134
Queen scallop			28	52
Whelk			55	
Echinoderms				
Starfish	4	200	6	765
Sandstar	7	88	11	610
Brittle star	9	153	21	1096
Polychaetes				
Sea mouse			6	388

in the order of 35–65%, and the mortality of starfish about 10%. Hermit crabs are protected by their shell and hence showed lower mortality (10–25%). The catch efficiency for quahogs was relatively low, and their total mortality in the trawl tracks was therefore assumed to be comparable with other mollusc species. The survival of whelks (*Buccinum undatum*) caught with a 12-m beam trawl, was estimated in the laboratory over 4–6 weeks and compared with the survival of whelks caught in baited traps (Mensink *et al.*, 1999). Whelks from traps showed little mortality, whereas the percentage mortality of whelks from the trawl was up to 55%. Some species in the discards showed less damage and had relatively high survival rates (starfish, sea mouse).

In general, the catch efficiency of beam trawls is relatively low for benthic invertebrates (Table 9.1). To estimate the total amount of food generated by beam trawling (Table 9.5), the amount of benthos killed in the discards was added to the amount of damaged benthos killed in the trawl tracks. Most roundfish in the discards, such as whiting, bib (*Trisopterus luscus*) or cod (*Gadus morhua*), were dead before they could be placed in survival tanks. In the rare case that live gurnards were found on the sorting belt, none subsequently survived. Survival experiments were also carried out with small fish species that pass through the trawl meshes. They were collected from very short hauls (1–2 min) made with a fine-meshed covering net. Their survival was at least 80%, which indicates that most of the small fish that escape through the meshes are not seriously damaged by the tickler chains. A high survival was also observed for sole and plaice collected from the same short hauls. When the high survival (*S*%) of these species in very short hauls is compared with survival estimates in hauls with commercial beam trawls for longer periods, as presented by van Beek *et al.* (1990), the percentage survival appears to decrease exponentially with increasing haul duration *t* (min) (Fonds, 1994b):

$$\text{Plaice}: \ S\% = 94e^{(-0.15t)}$$
$$\text{Sole}: \ S\% = 91e^{(-0.18t)}.$$

The chance of survival is still high (94% and 91% at $t = 0$) when the fish enter the net, which again indicates that they are relatively undamaged by the tickler chains (6–9% mortality). However, in commercial hauls of 1–2-h duration, the percentage survival of discarded flatfish is very low (van Beek *et al.*, 1990). Survival also depends on the catch composition in the codend, as large amounts of echinoderms can cause greater physical damage to fishes.

Food consumption by benthic carnivores

Laboratory measurements

Measurements of metabolism, maximum daily food consumption and/or maximum growth rate at different constant temperatures (5, 10, 15, 20°C) were carried out in the laboratory with starfish (*Asterias rubens*, *Astropecten irregularis*), brittle stars (*Ophiura ophiura* and *O. albida*), sea urchins (*Psammechinus miliaris*), hermit crabs (*Pagurus bernhardus*), swimming crabs (*Liocarcinus holsatus*), shore crabs (*Carcinus maenas*), shrimps (*Crangon crangon*), dab (*Limanda limanda*), juvenile whiting (*Merlangius merlangus*), cod (*Gadus morhua*) and bib (*Trisopterus luscus*). The average daily food consumption, with unlimited food rations, was correlated with size (weight) of the animals and with water temperature (Fonds *et al.*, 1998). For most species, oxygen consumption, food consumption and growth showed an exponential relationship with body weight, with a 'metabolic' weight exponent that varied between 0.65 and 0.85 for different species or at different temperatures (Fonds *et al.*, 1998; cf. Fonds *et al.*, 1992). In order to get a size-independent parameter for

Table 9.5 Estimates of daily food demand of common benthic carnivores, compared with the amount of food generated by a single trawling disturbance with 12-m beam trawls in four different areas. All values in g afdw 100 m^{-2}

Location (Fig. 9.1)	NW of Helgoland	Oyster-grounds	Weisze Bank	West of Holland	Average values
Water depth (m)	25	42	31	19	
Water temperature (°C)	17	15	12	19	16
% Silt in sediment	6	9	5–10	1	
Dates:	31-8-94	5-9-95	2-5-95	11-9-95	

Maximum daily food consumption at 15°C, in g afdw 100 m^{-2} estimated for the most abundant scavenging species in the area.

Liocarcinus	3.52	0.21	0.16	5.48	2.34
Pagurus	0.71	0.32	1.71	8.46	2.80
Crangon	0.22	0.02	0.53	0.93	0.43
Asterias	1.92	0.74	0.02	1.03	0.93
Astropecten	0.01	6.31	1.32	0.00	1.91
Ophiura	8.15	0.02	0.02	1.21	2.35
Buccinum	0.00	0.00	0.31	0.00	0.08
Dab	2.00	0.77	0.42	0.12	0.83
Plaice	3.84	0.05	0.01	0.10	1.00
Dragonet	1.04	<0.1	0.19	0.30	0.51
Whiting	0.10	0.08	0.25	0.10	0.13
Gurnards	0.17	0.02	0.05	0.00	0.06
Total consumption by invertebrates	14.5	7.6	4.1	17.1	10.8
Total consumption by fish	7.2	0.9	0.9	0.5	2.4
Consumption of all scavengers	21.7	8.5	5.0	17.6	13.2

Food generated by single passage of a beam trawl in g afdw 100 m^{-2}

Dead discard fish	40.7	5.8	21.6	8.1	19.1
Damaged benthos in trawl track	204.1	58.9	84.9	84.7	108.1
% Echinocardium	18	24	67	63	43
% Bivalves	69	26	15	25	34
% other taxa	13	50[a]	18	12	23
Total amount food generated:	244.8	64.7	106.5	92.8	127.2

Food supply for *n* days

All food for all scavengers	11.3	7.6	21.3	5.3	11.4
Invertebrates eating discard fish	2.8	4.4	7.8	0.5	3.9

[a]35% *Corystes*.

feeding and growth in terms of weight, all values were expressed as a proportion of the 'metabolic weight', with a mean weight exponent of 0.8 (weight exponent of 0.7 used for gadoid fish and shrimp). Temperature greatly affected metabolism, feeding and growth, and for most species increased exponentially from 5 to 15°C according to $e^{(bT)}$, with a value of the exponent of temperature b that varied between 0.1 and 0.2 for different species. Some species showed a similar increase, but at a higher temperature range from 10 to 20°C (e.g. swimming crab, shore crab and sole).

The conversion of food into body tissue was measured for some species fed at different ration levels. Food conversion efficiency appeared to be high for starfish (65–75%), lower for flatfish (45–55%) and lowest for crustaceans (30–40%). Laboratory measurements of maximum daily food consumption were then used to estimate the potential food consumption of the most abundant benthic carnivores and demersal fish in the sea (Table 9.5, Fonds *et al.*, 1998).

In situ *clearance rate of discard fish*

The weight loss of dead fish during decay, estimated in the laboratory with dab and whiting of different sizes at different temperatures, appeared to be strongly correlated with temperature and less dependent on size of the fish:

$$D = 23e^{(-0.076T)} \qquad (r^2 = 0.91, \ n = 110)$$

where D is the number of days from death to complete decay and T is temperature in °C (Groenewold & Fonds, in press). Complete decay, due to bacterial activity, took 7–8 days at 15°C and 16–18 days at 5°C, which sets a time limit for the availability of dead discard fish to benthic scavengers in summer and winter. Discard fish that were exposed on the seabed and retrieved after 2 days were always partly consumed by scavengers. Daily weight loss, corrected for the rate of decay, varied from about 0.9 g afdw in winter to 2.4–4.5 g afdw day^{-1} in spring and summer. The rate of clearance of discard fish, observed at the different stations, was usually higher for larger fish and positively correlated with the densities of crabs and starfish found in the local area.

Food generated by beam trawling and food consumed

Food produced by trawling compared with potential daily food consumption

The amount of food generated by trawling was estimated for four different stations where densities of benthos before and after the complete trawling of a narrow strip had been ascertained using the methods described earlier. Table 9.5 presents estimates of the production of dead discard materials and damaged fauna in the trawl tracks, after trawling an area once with 12-m beam trawls. Based on laboratory measurements of maximum daily food consumption of selected common benthic predators, the potential daily food consumption for populations of these benthic carnivores in the four selected areas was estimated in g afdw 100 m^{-2}. The total daily food consumption varied from 5 to 22 g afdw 100 m^{-2} and depended on abundance and species composition in each area. The average amount of food generated in the form of damaged benthos was about five to 10 times as much as the amount of food in the form of discarded fish. The total amount of food generated varied between 95 and 245 g afdw 100 m^{-2}, sufficient for about 5–21 days at the maximum feeding rate of the local populations of scavengers in summer. Analysis of stomach contents of fish, feeding in a recently trawled area, indicated that increases in fish stomach fullness lasted not more than 1–2 days, while discard fish exposed on the bottom in summer were almost completely consumed in about 2–3 days.

Annual amount of food generated by trawling

In summer, the average value of the potential daily food consumption by all benthic carnivores in the four areas studied, at 16°C, is c. 13 g afdw 100 m^{-2} day^{-1} (11 g by benthic carnivores and 2 g by demersal fish). The average amount of food generated by trawling is estimated at c. 108 g afdw 100 m^{-2} damaged benthos + 19 g afdw 100 m^{-2} discarded fish, sufficient for about 10 days at maximum feeding rate for the local populations of scavengers.

In winter, the rate of food consumption of the different species will be at least three to 10 times lower when considering the effect of temperature on feeding and growth. Assuming that the composition of the benthic fauna does not change very much, an average daily consumption of c. 3.5 g afdw 100 m^{-2} day^{-1} is estimated from the data in Table 9.5. The amount of food generated by trawling will also decrease because some benthic infaunal species (for example, sea potatoes *Echinocardium cordatum*) will become less active, burying deeper into the sediment beyond the reach of beam trawls. When sea potatoes are excluded from calculations of the production of damaged benthos by beam trawling in the winter period, the total amount is c. 68 g afdw 100 m^{-2}. If the amount of discarded fish is assumed to be similar as in summer (19 g afdw), the total amount becomes 87 g afdw 100 m^{-2}, sufficient for about 25 days' feeding by the local scavengers in winter at 5°C.

Food generated compared with food demand

The potential annual food demand can be estimated as the average consumption in summer and winter, multiplied by 365. This gives values, in g afdw m^{-2}, of c. 24.5 g for all invertebrates + 5.5 g for the demersal fish, i.e. 30 g afdw m^{-2} year^{-1}.

The average annual trawling intensity in the southern bight of the North Sea, south of 55° N where intensive beam trawling for sole takes place (Rijnsdorp et al., 1998), is estimated to vary between 50% and 250% of the surface area of different ICES rectangles. The average trawling intensity for the Dutch sector is estimated at 140% of the surface of the seabed (Bergman et al., 1998a, b). If it is assumed that the most intensively trawled areas are trawled completely twice each year, once in the summer and once in the winter, then the total amount of food generated by trawling in g afdw m^{-2}, will be 1.76 g of damaged benthos + 0.38 g of discard fish, i.e. 2.14 g m^{-2}. Hence, beam trawling may generate about 7% of the annual food demand of the entire scavenger population in such areas.

Discussion

After more than 30 years of intensive beam trawling in the southern North Sea, many large benthic invertebrates (*Alcyonium digitatum, Buccinum undatum, Tealia felina, Ostrea edulis*) have become scarce, and large slowly reproducing rays, skates and small shark species have become rare (Walker & Heessen, 1996; Philippart, 1998; Vooys & van der Meer, 1998). The present-day macrobenthic infauna consists predominantly of small species (brittlestars, small bivalves) or species that live deeper within the seabed (sea potatoes, razor shells).

From the relationship between the amount of discard fish and the catch of marketable sole (Table 9.1) the total annual amount of discards can be roughly estimated from the annual landings of sole. Based on annual landings of about 30 000 t (up to 1995), the annual production of discard fish by the sole fishery is estimated at *c*. 190 000 t and the amount of dead invertebrates at *c*. 85 000 t. The International Council for Exploration of the Sea (ICES) working group on the ecosystem effects of fishing activity (Anon., 1998) estimated that the annual amount of discards produced by beam trawlers over the period 1978–1990 was 153 600 t, compared with an average catch of 22 600 t of sole. This equates to 6.8 kg of discarded fish per kg of landed sole, comparable with estimates in Table 9.1. An annual production of discarded fish of about 150 000–190 000 t by the sole fishery in the southern North Sea (236 000 km^2), assuming 18% ash-free dry weight for flatfish, gives a value of approximately 0.11–0.15 g afdw m^{-2}, which is comparable with the estimate of 0.19 g m^{-2} in Table 9.5. However, this estimate is much lower than the annual value of 0.38 g afdw m^{-2} estimated for the more intensely trawled areas.

Identification of scavenging species

The combined information obtained by repeated trawling and from baited traps gives an indication of the benthic carnivores and demersal fish that show opportunistic scavenging behaviour in newly trawled areas. The trap catches may have been rather selective and affected by aggressive interactions between scavengers. In general, most scavenging species were not very selective in their choice of food. However, it is evident that scavenging amphipods show a distinct preference for damaged or discarded dead crustaceans. This has also been observed in the Clyde Sea (Nickell & Moore, 1992).

Food generated compared with food demand

The amount of damaged and exposed benthos generated in the trawl tracks, including the mortality of benthos in the discards, was at least four to 10 times greater than the amount of discarded fish that were produced by trawling. In summer most of this material is consumed within 2 days by fast-moving fish (dab, dragonet, gadoids), swimming crabs and shrimp, that rapidly immigrate into newly trawled areas. *In situ* underwater observations in the Irish Sea have shown that, while some invertebrates (starfish, hermit crabs) take a share of the damaged benthos (Evans *et al.*, 1996; Kaiser & Spencer, 1996; Ramsay *et al.*, 1996, 1997), fish appear to be major consumers of damaged benthos within newly trawled areas of the seabed (Kaiser & Spencer, 1994; Kaiser & Ramsay, 1997).

Dead discarded fish that sink to the bottom are consumed mainly by invertebrate scavengers such as swimming crabs, hermit crabs, starfish, sea urchins and whelks (Evans *et al.*, 1996; Fonds *et al.*, 1998; Ramsay *et al.*, 1998). Some larger roundfish (gurnards and possibly larger gadoids) may take some of the smaller discarded fish species (Fonds *et al.*, 1998). Consumption of larger dead fish may takes some time, and the scavengers will always have to compete with bacterial decay and consumption. In summer, the decay of dead discard fish will take 1 week; in winter, this may take much longer (2–3 weeks) owing to the lower sea-water temperatures. Dead fish that are frequently found in more intensely trawled areas in the southern North Sea are probably those discards that have not yet been consumed (personal observations).

Demersal fish immigrate rapidly into recently trawled areas to feed on the damaged and exposed fauna (Kaiser & Spencer, 1994; Kaiser & Ramsay, 1997; Fonds *et al.*, 1998; Groenewold & Fonds, in press). Some fast-moving invertebrates (crabs) may also utilise some of this food. However, if fish are the main consumers, the annual amount of damaged and exposed benthos generated by intensive beam trawling in the southern North Sea, would be approximately 30% of the annual food demand of demersal fish. If the dead discarded fish are consumed mainly by invertebrate scavengers, the total annual production by beam trawlers (15–38 g afdw 100 m^{-2}) is less than 2% of their annual food demand. Hence, demersal fish may benefit more than invertebrate carnivores from the large amounts of damaged benthos in trawled tracks. An increase in growth rate of several demersal fish species after the increase in beam trawl intensity in the southern North Sea (Veen, 1971, 1976; Beek, 1988; Millner & Whiting, 1996; Rijnsdorp & van Leeuwen, 1996) may have been partly related to an increase in availability of benthic infauna due to beam trawling. However, other factors such as the elimination of competitors and predators may well be equally, if not more, important than changes in food supply. For example, a long-term trend of increasing population density has been observed for several small benthic fish species in the North Sea (van Leeuwen *et al.*, 1994).

Greenstreet *et al.* (1997) estimated the balance between food production and food consumption by fish in the North Sea over the four quarters of the year for different feeding guilds (planktivores, benthivores, piscivores). Their estimates can be transformed into ash-free dry weight (13.6% of wet weight) and summed for the whole year. For the demersal fish that feed on benthos, this leads to the following estimates in gram ash-free dry weight per square metre per year: benthos production *c.* 18 g, annual consumption of benthos by demersal fish 5 g, 1 g consumption by demersal piscivores. For the area of the southern North Sea, where the beam-trawl fishery for sole is concentrated (Welleman, 1989, Rijnsdorp *et al.*, 1997), the total biomass and annual production of the macrobenthos have been estimated by Wilde *et al.* (1984) and Duineveld *et al.* (1990, 1991). For a biomass of about 8–15 g afdw m^{-2}, annual production is estimated at *c.* 20 g afdw m^{-2}, similar to the estimates of Greenstreet *et al.* (1997).

Our estimate of annual food demand of demersal scavenging fish of *c.* 5.5 g m^{-2} agrees fairly well with the estimates by Greenstreet *et al.* (1997). Of the 18–20 g of benthos production, 5 g is consumed by fish, which leaves only 13–15 g for other (invertebrate) carnivores. Compared with our estimate of an annual (maximum) food demand of invertebrate scavengers of *c.* 25 g, this suggests that they will get only half that amount. Hence, food for invertebrate carnivores in the benthic system is limited, and any food generated by trawling disturbance will be rapidly consumed.

In general, maintenance food requirements of fish and invertebrates are approximately 20–30% of their maximum rations (Fonds *et al.*, 1992, 1998). The addition of 7% or 10% of the annual food demand of all benthic scavengers together may assist the maintenance of their populations. It is unlikely that it will support an increase in scavenger population density.

It is possible that, in some areas, starfish and small fish species such as dragonets have increased together with the increasing trawling intensity (Philippart *et al.*, 1998; Ramsay *et al.*, this volume, Chapter 10). There is also evidence that the structure of assemblages of demersal invertebrates and fish in the southern North Sea have changed (Rogers *et al.*, 1998; Rumohr *et al.*, 1998). However, there is no clear evidence of a general increase in population density of scavenging species over the whole area of the southern North Sea (Buijs *et al.*, 1994).

References

Anon. (1998) Report of the working group on ecosystem effects of fishing activity. *International Council for the Exploration of the Sea*, C.M. 1998/ACFM/ACME:1, pp. 175–7.

van Beek, F.A. (1988) On the growth of sole in the North Sea. *International Council for the Exploration of the Sea*, C.M. 1988/G:24, 24 pp.

van Beek, F.A., Rijnsdorp, A.D. & van Leeuwen, P.I. (1983) Results of the mesh selection experiments on sole and plaice with commercial beamtrawl vessels in the North Sea in 1981. *International Council of the Exploration of the Sea*, C.M. 1983/B:16, 25 pp.

van Beek, F.A., van Leeuwen, P.I. & Rijnsdorp, A.D. (1990) On the survival of plaice and sole discards in the otter trawl and beam trawl fisheries in the North Sea. *Netherlands Journal of Sea Research*, **26**, 151–60.

Bergman, M.J.N. & van Santbrink, J.W. (1994) A new benthos dredge (Triple-D) for quantitative sampling of infauna species of low abundance. *Netherlands Journal of Sea Research*, **33**, 129–33.

Bergman, M.J.N., Ball, B., Bijleveld, C., Craaymeersch, J.A., Munday, B.W., Rumohr, H. & van Santbrink, J.W. (1998a) Direct mortality due to trawling. In: *The effects of different types of fisheries on the North Sea and Irish Sea benthic ecosystems* (eds H.J. Lindeboom & S.J. de Groot), pp. 167–84. *NIOZ Report* 1998-1/*RIVO-DLO Report* C003/98. Netherlands Institute for Sea Research, Den Burg, Texel, The Netherlands.

Bergman, M.J.N., Craaymeersch, J.A., Polet, H. & van Santbrink, J.W. (1998b) Fishing mortality in invertebrate populations due to different types of trawl fisheries in the Dutch sector of the North Sea. In: *The effects of different types of fisheries on the North Sea and Irish Sea benthic ecosystems* (eds. H.J. Lindeboom & S.J. de Groot), pp. 167–84. *NIOZ Report* 1998-1/*RIVO-DLO Report* C003/98. Netherlands Institute for Sea Research, Den Burg, Texel, The Netherlands.

Beukema, J.J. (1992) Expected changes in the Wadden Sea benthos in a warmer world: lessons from periods with mild Winters. *Netherlands Journal of Sea Research*, **30**, 73–9.

Beukema, J.J., Essink, K. & Michaelis, H. (1996) The geographic scale of synchronised fluctuation patterns in zoobenthos populations as a key to underlying factors: climate or man-induced. *ICES Journal of Marine Science*, **53**, 964–71.

Buijs, J., Craaymeersch, J.A., van Leeuwen, P.I. & Rijnsdorp, A.D. (1994) *De epi- en endofauna van de Nederlandse, Duitse en Deense kustzone, een analyse van 20 jaar bijvangst gegevens. BEON Report* 94-11, 63 pp. Netherlands Institute for Sea Research, Den Burg, Texel, The Netherlands.

Camphuysen, C.J., Calvo, B., Durinck, J., Ensor, K., Follestad, A., Furness, R.W., Garthe, S., Leaper, G., Skov, H., Tasker, M.L. & Winter, C.J.N. (1995) *Consumption of discards in the North Sea. NIOZ Report 1995-1*, 202 pp. Netherlands Institute for Sea Research, Den Burg, Texel, The Netherlands.

Craaymeersch, J.A., Ball, B., Bergman, M.J.N., Damm, U., Fonds, M., Munday, B.W. & van Santbrink, J.W. (1998) Catch efficiency of commercial trawls. In: *The effects of different types of fisheries on the North Sea and Irish Sea benthic ecosystems* (eds H.J. Lindeboom & S.J. de Groot), pp. 167–84. *NIOZ Report* 1998-1/*RIVO-DLO Report* C003/98. Netherlands Institute for Sea Research, Den Burg, Texel, The Netherlands.

Duineveld, G.C.A, de Wilde, P.A.W.J. & Kok, A. (1990) A synopsis of the macrobenthic assemblages and benthic ETS activity in the Dutch sector of the North Sea. *Netherlands Journal of Sea Research*, **26**, 125–36.

Duineveld, G.C.A., Kunitzer, A., Niermann, U., de Wilde, P.A.J. & Gray, J.S. (1991) The macrobenthos of the North Sea. *Netherlands Journal of Sea Research*, **28**, 53–65.

Evans, P.L., Kaiser, M.J. & Hughes, R.N. (1996) Behaviour and energetics of whelks, *Buccinum undatum* (L.), feeding on animals killed by beam trawling. *Journal of Experimental Marine Biology and Ecology*, **197**, 51–62.

Fonds, M. (1994a) Catch composition of 12-m beamtrawl and 4-m beamtrawl for sole fishing. In: *Environmental impact of bottom gears on benthic fauna in relation to natural resources management and protection of the North Sea* (eds S.J. de Groot & H.J. Lindeboom), pp. 95–130. *NIOZ Report* 1994-11/*RIVO-DLO Report* C026/94. Netherlands Institute for Sea Research, Den Burg, Texel, The Netherlands.

Fonds, M. (1994b) Mortality of fish and invertebrates in beam trawl catches and the survival chances of discards. In: *Environmental impact of bottom gears on benthic fauna in relation to natural resources management and protection of the North Sea* (eds S.J. de Groot & H.J. Lindeboom), pp. 131–46. *NIOZ Report* 1994-11/*RIVO-DLO Report* C026/94. Netherlands Institute for Sea Research, Den Burg, Texel, The Netherlands

Fonds, M., Cronie, R., Vethaak, A.D. & van der Puyl, P. (1992) Metabolism, food consumption and growth of plaice (*Pleuronectes platessa*) and flounder (*Platichthys flesus*) in relation to fish size and temperature. *Netherlands Journal of Sea Research*, **29**, 127–43.

Fonds, M., Groenewold, S., Hoppe, I., Kaiser, M.J., Munday, B.W. & Ramsay, K. (1998) Scavenger responses to trawling. In: *The effects of different types of fisheries on the North Sea and Irish Sea benthic ecosystems* (eds H.J. Lindeboom & S.J. de Groot), pp. 185–244. *NIOZ Report* 1998-1/*RIVO-DLO Report* C003/98. Netherlands Institute for Sea Research, Den Burg, Texel, The Netherlands.

Greenstreet, P.R., Bryant, A.D., Broekhuizen, N., Hall, S.J. & Heath, M.R. (1997) Seasonal variation in the consumption of food by fish in the North Sea and implications for food web dynamics. *ICES Journal of Marine Science*, **54**, 243–66.

Groenewold, S. & Fonds, M. (in press) The effects of discards and damaged benthos, produced by beam trawl fishery, for benthic scavengers in the southern North Sea. *ICES Journal of Marine Science*.

De Groot, S.J. & Lindeboom, H.J. (1994) *Environmental impact of bottom gears on benthic fauna in relation to natural resources management and protection of the North Sea. NIOZ Report* 1994-11/*RIVO-DLO Report* C026/94, 257 pp. Netherlands Institute for Sea Research, Den Burg, Texel, The Netherlands.

Heessen, H.J.L. (1996) Time-series data for a selection of forty fish species caught during the International Bottom Trawl Survey. *ICES Journal of Marine Science*, **53**, 1079–84.

Heessen, H.J.L. & Daan, N. (1996) Long-term trends in ten non-target North Sea fish species. *ICES Journal of Marine Science*, **53**, 1063–78.

Kaiser, M.J. & Ramsay, K. (1997) Opportunistic feeding by dabs within areas of trawl disturbance : possible implications for increased survival. *Marine Ecology Progress Series*, **152**, 307–10.

Kaiser, M.J. & Spencer, B.E. (1994) Fish scavenging behaviour in recently trawled areas. *Marine Ecology Progress Series*, **112**, 41–9.

Kaiser, M.J. & B.E.Spencer (1996) The effects of beam-trawl disturbance on infaunal communities in different habitats. *Journal of Animal Ecology*, **65**, 348–58.

Krebs, C.J. (1989) *Ecological Methodology*, 654 pp. HarperCollins, NewYork.

Van Leeuwen, P.I., Rijnsdorp, A.D. & Vingerhoed, B. (1994) Variations in abundance and distribution of demersal fish species in the coastal zone of the south-eastern North Sea between 1980 and 1993. *International Council for the Exploration of the Sea*, C.M. 1994/G:10.

Lindeboom, H.J. & de Groot, S.J. (eds) (1998) *The effects of different types of fisheries on the North Sea and Irish Sea benthic ecosystems. NIOZ Report* 1998-1/*RIVO-DLO Report* C003/98, 404 pp. Netherlands Institute for Sea Research, Den Burg, Texel, The Netherlands.

Mensink, B., Fischer, C.V., Cadée, G.C., Fonds, M., ten Hallers-Tjabbes, C.C. & Boon, J.P. (in press) Shell damage and mortality in the common whelk, *Buccinum undatum*, caused by beam trawl fishery. *Journal of Sea Research*.

Millner, R.S. & Whiting, C.L. (1996) Long-term changes in growth and population abundance of sole in the North Sea from 1940 to the present. *ICES Journal of Marine Science*, **53**, 1185–95.

Nickell, T.D. & Moore, P.G. (1992) The behavioural ecology of epibenthic scavenging invertebrates in the Clyde area: field sampling using baited traps. *Cahiers de biologie marine*, **32**, 353–70.

Philippart, C.J.M. (1998) Long-term impact of bottom fisheries on several by-catch species of demersal fish and benthic invertebrates in the southeastern North Sea. *ICES Journal of Marine Science*, **55**, 342–52.

Polet, H. & Blom, W. (1998) Size of bottom trawling fleets. In: *The effects of different types of fisheries on the North Sea and Irish Sea benthic ecosystems* (eds H.J. Lindeboom & S.J. de Groot), pp. 71–82. *NIOZ Report* 1998-1/*RIVO-DLO Report* C003/98. Netherlands Institute for Sea Research, Den Burg, Texel, The Netherlands.

Ramsay, K., Kaiser, M.J. & Hughes, R.N. (1996) Changes in hermit crab feeding patterns in response to trawling disturbance. *Marine Ecology Progress Series*, **144**, 63–72.

Ramsay, K., Kaiser, M.J., Moore, P.G. & Hughes, R.N. (1997) Consumption of fisheries discards by benthic scavengers: utilisation of energy subsidies in different marine habitats. *Journal of Animal Ecology*, **66**, 884–96.

Ramsay, K., Kaiser, M.J. & Hughes, R.N. (1998) The responses of benthic scavengers to fishing disturbance by towed gears in different habitats. *Journal of Experimental Marine Biology and Ecology*, **224**, 73–89.

Rijnsdorp, A.D. & van Leeuwen, P.I. (1996) Changes in growth of North Sea plaice since 1950, in relation to density, eutrophication, beam-trawl effort and temperature. *ICES Journal of Marine Science*, **53**, 1199–213.

Rijnsdorp, A.D., van Leeuwen, P.I., Daan, N. & Heessen, H.J.L. (1996) Changes in abundance of demersal fish species in the North Sea between 1906–1909 and 1990–1995. *ICES Journal of Marine Science*, **53**, 1054–62.

Rijnsdorp, A.D., Buijs, A.M., Storbeck, F. & Visser, E.G. (1998) Micro-scale distribution of beam trawl effort in the southern North Sea between 1993 and 1996, in relation to the trawling frequency of the seabed and the impact on benthic organisms. *ICES Journal of Marine Science*, **55**, 403–19.

Rogers, S.I., Rijnsdorp, A.D., Damm, U. & Vanhee, W. (1998) Demersal fish populations in the coastal waters of the UK and continental NW Europe from beam trawl survey data collected from 1990 to 1995. *Journal of Sea Research*, **39**, 79–102.

Rumohr, H., Ehrich, S., Knust, R., Kujawski, T., Philippart, C.J.M. & Schroeder, A. (1998) Long term trends in demersal fish and benthic invertebrates. In: *The effects of different types of fisheries on the North Sea and Irish Sea benthic ecosystems* (eds H.J.Lindeboom & S.J. de Groot), pp. 280–352. *NIOZ Report* 1998-1/*RIVO-DLO Report* C003/98. Netherlands Institute for Sea Research, Den Burg, Texel, The Netherlands.

Sainte-Marie, B. & Hargrave, B.T. (1987) Estimation of scavenger abundance and distance of attraction to bait. *Marine Biology*, **94**, 431–43.

de Veen, J.F. (1971) Veranderingen in de visstand in de Noordzee. *Wadden Bulletin*, **6**, 2–7.

de Veen, J.F. (1976) On changes in some biological parameters in North Sea sole (*Solea solea* L.). *Journal du Conseil International pour l'Exploration de la Mer*, **37**, 60–90.

de Vooys, C.G.N. & van der Meer, J. (1998) Changes between 1931 and 1990 in b-catches of 27 animal species from the southern North Sea. *Netherlands Journal of Sea Research*, **39**, 291–8.

Walker, P.A. & Heessen, H.J.L. (1996) Long-term changes in ray populations in the North Sea. *ICES Journal of Marine Science*, **53**, 1085–93.

Welleman, H. (1989) *Litteratuur studie naar de effecten van de bodem visserij op de bodem en het bodem leven. Report Netherlands Institute for Fisheries Research (RIVO-DLO)*, MO 89-201, 58 pp. Netherlands Institute for Fisheries Research, IJmuiden, The Netherlands.

Wilde, P.A.W.J., Berghuis, E.M. & Kok, A. (1984) Structure and energy demand of the benthic community of the Oysterground, central North Sea. *Netherlands Journal of Sea Research*, **18**, 143–59.

Witbaard, R. & Klein, R. (1994) Long-term trends in the effects of the southern North Sea beamtrawl fishery on the bivalve mollusc *Arctica islandica* L. *ICES Journal of Marine Science*, **51**, 99–105.

Chapter 10
Impact of trawling on populations of the invertebrate scavenger *Asterias rubens*

K. RAMSAY[1], M.J. KAISER[1], A.D. RIJNSDORP[2], J.A. CRAEYMEERSCH[3] and J. ELLIS[4]

[1]*School of Ocean Sciences, University of Wales – Bangor, Menai Bridge, Gwynedd, LL59 5EY, UK*
[2]*Netherlands Institute for Fisheries Research (RIVO-DLO), Haringkade 1, PO Box 68, NL-1970 AB IJmuiden, The Netherlands*
[3]*Netherlands Institute for Fisheries Research (RIVO-DLO), Postbus 77, 640 AB Yerseke, The Netherlands*
[4]*The Centre for Environment, Fisheries and Aquaculture Science, Lowestoft Laboratory, Pakefield Road, Lowestoft, Suffolk, NR33 0HT, UK*

Summary

1. The relationship between starfish numbers and fishing effort is a quadratic, meaning that, as fishing effort increases, starfish numbers also increase until they reach a turning point, after which starfish numbers decline as fishing effort further increases.
2. This relationship, although significant, is fairly weak, suggesting that other factors must strongly influence starfish numbers.
3. Until we know more about the ecology and population dynamics of starfish populations, it will be difficult to determine the exact extent of the impact of beam trawling on starfish populations.

Keywords: starfish populations, fishing effort, NAO.

Introduction

In almost every fishery world-wide, man is increasing the input of carrion to benthic communities through fisheries activities. Beam-trawl fisheries generate carrion in two ways: firstly, by the practice of discarding unwanted material from the catch, and secondly, by killing benthic fauna on the seabed, crushing some animals with the heavy parts of the trawl whilst others are damaged as they pass through the meshes of the net (Lindeboom & de Groot, 1998). The survival of discarded animals varies between different species; some fish such as dragonets *Callionymus lyra* are extremely vulnerable and nearly 100% die after capture, whilst others such as the starfish *Asterias rubens* are less susceptible to physical damage and less than 6% die (Kaiser & Spencer, 1995; Lindeboom & de Groot, 1998). A proportion of discards are consumed by seabirds (Furness *et al.*, 1988; Hudson & Furness, 1988); the number taken varying according to the species discarded, as most seabirds eat roundfish in preference to flatfish or invertebrates (Hudson & Furness, 1988). Those discards not eaten by seabirds will sink through the water column, where some may be eaten by cetaceans and fishes (Hill & Wassenburg, 1990); the remainder sink to the seabed where any moribund animals are available as food for benthic scavengers. Beam

151

trawls are specifically designed to penetrate the surface layers of the sediment in order to increase the catch efficiency for sole and other flatfish, and this disturbance results in damage to non-target benthic animals such as the burrowing heart urchin *Echinocardium cordatum* (Bergman & Hup, 1992; Ramsay *et al.*, 1998) and the bivalve molluscs *Ensis* sp. (Eleftheriou & Robertson, 1992; Gaspar *et al.*, 1994) and *Acanthocardia echinata* (Ramsay *et al.*, 1998). Those species with brittle skeletal structures, that live in tubes or that have poor regenerative abilities are most vulnerable to this sort of damage (MacDonald *et al.*, 1996). In addition to the animals killed by the passage of the trawl many more will be damaged or disturbed, resulting in increased vulnerability to predators (Kaiser & Spencer, 1994; Ramsay *et al.*, 1996). Thus, beam trawls, as well as other fishing gears, can produce an additional food resource for scavengers.

Recent studies have shown that benthic scavengers feed on both fisheries discards and on animals damaged *in situ* by fishing gears (Wassenburg & Hill, 1987, 1990; Berghahn, 1990; Hill & Wassenburg, 1990; Kaiser & Spencer, 1994, 1996; Ramsay *et al.*, 1996, 1997, 1998). Some of these studies demonstrated that scavengers can gain more food by foraging in areas disturbed by fishing activity than in nearby unfished areas. Wassenburg & Hill (1987) found that sand crabs *Portunus pelagicus* sampled from an area whilst fishing was in progress had more food in their stomachs than crabs sampled from the same area some days later, when fishing was prohibited. They attributed this dietary increase to the higher availability of carrion discarded by fishing boats. Similarly, Ramsay *et al.* (1996) demonstrated that hermit crabs *Pagurus bernhardus* sampled from an experimentally trawled area had more food in their stomachs than those sampled from an adjacent unfished (control) area. Diver observations of fished and adjacent unfished areas have shown that the proportion of starfish *Asterias rubens* feeding was higher in the fished area for up to 44 h after fishing had taken place (Ramsay *et al.*, 1998). The starfish were observed to feed on damaged and displaced bivalves, gastropods, crustaceans and echinoderms.

It has been suggested that, by increasing the amount of carrion available to scavengers, fishing may lead to an expansion in the size of some scavenger populations (Furness, 1984; Furness *et al.*, 1988; Berghahn, 1990; Wassenburg & Hill, 1990; Kaiser & Spencer, 1994, 1996; Ramsay *et al.*, 1996, 1998). To date, the strongest evidence for such population changes comes from a study of seabird populations in the North Sea, which demonstrated that some populations of scavenging seabirds expanded concurrently with an increase in fishing effort (Furness, 1984; Furness *et al.*, 1988). However, one cannot assume that increasing the food supply available to a population will automatically lead to an increase in its size. Firstly, the animals in question must have the ability to find and utilise this food source. Secondly, life history will have an important role to play, especially when species have two very separate stages in their life cycle (e.g. most benthic species which have a planktonic larval stage and a benthic adult stage), in which case there is no guarantee that subsidising the food supply to the adult stage will have any effect on the population density. Lastly, many benthic scavengers are themselves damaged by fishing activities, leading to fishing mortality, which may counteract any benefits gained from extra food availability.

There are two ways of tackling the question of whether increased amounts of fisheries-generated carrion can lead to an expansion of scavenger populations. The first is to estimate the total amount of extra energy produced by fishing in the form of carrion and to relate this to the energy requirements of existing scavenger populations (Fonds & Groenewold, this volume, Chapter 9). If the extra energy produced by fishing is not a significant proportion of the latter, it would seem unlikely that fishing could lead to an increase in scavenger populations other than on a localised scale. We have adopted an alternative approach, which is to look for evidence that fishing has altered the size of scavenger populations by comparing the density of a selected scavenger species with fishing effort.

The life-history strategy, susceptibility to damage by fishing activities and feeding behaviour of scavengers will affect the probability that an enhanced carrion supply will lead to a population expansion. Species that are likely to benefit most are those that do not suffer high fishing mortality and also gain a true energetic benefit from feeding on fisheries-derived carrion or other animals damaged by fishing. In addition to this, an increase in population size would only be expected to occur when a population is food-limited. A species that is heavily dependent on carrion as its food source would also seem more likely to expand its population in response to fishing disturbances. However, with the possible exception of one amphipod species *Orchomene nanus* (Moore & Wong, 1995), benthic scavengers appear to be highly opportunistic feeders, switching readily from predation to scavenging according to the availability of different food types (Britton & Morton, 1994).

The starfish *Asterias rubens* is an opportunistic scavenger that has been shown to gain extra food by foraging in fished areas (Ramsay *et al.*, 1998) and also feeds on fisheries discards (Ramsay *et al.*, 1997; Lindeboom & de Groot, 1998). Fishing mortality is very low (less than 6%) (Fonds, 1994; Kaiser & Spencer, 1995; Lindeboom & de Groot, 1998). Populations of *A. rubens* that appear to be food limited have been described by Vevers (1949) and Nichols & Barker (1984). In addition, *A. rubens* is ubiquitous in European waters and is commonly found in large numbers, making it more useful for a study of a widespread phenomenon such as fishing activity. There is also some evidence to suggest that numbers of *A. rubens* have increased in the southern North Sea over the past 80 years, which may be linked to the effects of beam trawling (Lindeboom & de Groot, 1998).

Methods

The Netherlands Institute for Fisheries Research (RIVO-DLO) has carried out benthic surveys in the south-eastern North Sea (off the coast of The Netherlands) since 1973. These surveys collected information regarding the abundance of various benthic species, including starfish *Asterias rubens*. Benthic species are collected using a 6-m beam trawl with four tickler chains and a 40-mm stretched codend mesh towed for 15 min at a speed of 3.5 kn. Data are also available concerning the number of hours fished per year by beam trawlers (the fishing gear that causes most seabed disturbance in this area of the North Sea; Anon., 1995) in standardised blocks known as ICES rectangles. These effort data are derived from those reported as a statutory requirement to the Ministry of Agriculture, Nature Management and Fisheries, Netherlands. These values have been used to estimate the area of seabed that is swept by beam trawls per year (see Kaiser *et al.* (1996) for a description of the methodology used for these calculations). The relationship between fishing effort and starfish numbers was examined for all rectangles for the years between 1973 and 1996 (Fig. 10.1).

Another dataset was obtained from the Centre for Environment, Fisheries and Aquaculture Science (CEFAS) and consisted of numbers of starfish caught at 88 sites in the English Channel and Celtic Sea. The sampling gear was a commercial-type 4-m beam trawl, towed for 30 min. The data used were for the year of 1998 only and thus differ from the Netherlands data in that there can be no time-series effects.

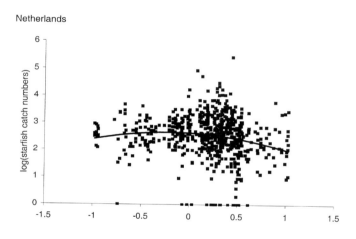

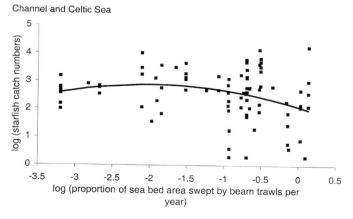

Fig. 10.1 The relationship between fishing effort and starfish numbers.

Effort data were also available for these sites and were converted to area swept by beam trawls per year as described above. These effort data were extracted for UK beam trawlers and foreign beam trawlers landing their catch in UK ports. Foreign beam trawlers landing their catch in their home country were not included in the dataset, and this will result in an underestimation of fishing effort, as Belgian and Dutch beam trawlers commonly work in the Channel and Celtic Sea. Scallop dredging is also a significant form of disturbance in parts of this area. It seems reasonable, however, to assume that the underestimation is consistent across the stations.

The relationship between fishing effort and starfish numbers was explored at both sites using generalised linear models. A time lag of 2 years was applied to the data, e.g. starfish numbers for 1996 were compared with effort data for 1994. Vevers (1949) suggests that, in favourable conditions, *A. rubens* may grow to an arm radius size of 8–9 cm within 1 year; a size class that could easily be caught in the beam trawls used in this study. In the areas studied, *A. rubens* generally spawn in April or May (Nichols & Barker, 1984) and gonad development begins the preceding November. Nutrient reserves in the pyloric caeca are an important source of energy for the process of gametogenesis and therefore food supply in the summer preceding spawning (when nutrients are deposited in the pyloric caeca) will be an important factor determining fecundity (Jangoux & van Impe, 1977; Oudejans *et al.*, 1979; Barker & Nichols, 1983). This implies that a lag of 2 years could be expected between an enhanced food supply and increased starfish catches. A preliminary analysis carried out on the Netherlands data showed that a 2-year time lag resulted in a higher level of significance ($P < 0.0001$) than a 1-year lag ($P = 0.0006$).

Temperature data were added to the Netherlands model, using the North Atlantic Oscillation (NAO) data (Dickson & Brander, 1993; Tunberg & Nelson, 1998). These index values are calculated as the difference between the normalised mean winter atmospheric pressure anomalies at Ponta Delgadas (Azores) and Akureyri (Iceland) and have been shown to be a reasonable predictor of subsurface temperatures in the North Sea (Tunberg & Nelson, 1998).

Results

At both sites, the relationship that gave the best fit was a quadratic (Fig. 10.1, Table 10.1), given by the following equations:

$$\text{Netherlands: log(starfish numbers)} = -0.4\log(\text{area swept})^2$$
$$-0.1\log(\text{area swept}) + 2.6$$

$$\text{Channel and Celtic Sea: log(starfish numbers)} = -0.2\log(\text{area swept})^2$$
$$-0.8\log(\text{area swept}) + 2.1.$$

The turning point of each equation (i.e. the point at which the starfish numbers stop increasing with increasing fishing effort and start decreasing) was calculated by

Table 10.1   Analysis of variance (ANOVA) for the relationship between fishing effort and starfish numbers using the following model: $\log_{10}$(starfish numbers) $= a(\log_{10}$(proportion of the seabed area swept per year))$^2 + b \log_{10}$(proportion of the seabed area swept per year) $+ c$

	df	Sum of squares	Mean square	F value	P > F
Netherlands 1973–1996					
Model	2	11.8	5.88	9.73	0.0001
Error	674	407.5	0.60		
Total	676	419.2			
Channel and Celtic Sea 1998					
Model	2	5.6	2.8	3.6	0.0318
Error	85	66.8	0.8		
Total	87	72.5			

solving the differential of the equation. The turning points were 82% and 1% of the seabed area swept per year for the Netherlands and the Channel and Celtic Sea data, respectively.

Starfish mortality due to fishing was also calculated for fishing effort values of 82% swept (the turning point for the Netherlands data) and 1000% (close to the maximum for the Netherlands data). Starfish caught by commercial beam trawls have a mortality rate of between 0% and 6% (Fonds, 1994; Kaiser & Spencer, 1995; Lindeboom & de Groot, 1998) and a beam trawl catches between 12% and 39% of starfish that occur in a fished area (Bergman & Santbrink, 1994; Santbrink & Bergman, 1994). This suggests that when 82% of the seabed is swept per year 10–32% of starfish in the rectangle are caught in that year (assuming an even distribution of starfish and fishing effort throughout the rectangle). Of these starfish that are caught, at least 94% will survive (Kaiser & Spencer, 1995; Lindeboom & de Groot, 1998), and therefore the percentage of starfish killed (%caught × mortality rate of 6%) will be between 0.6% and 1.9%. When fishing effort increases to a level where the whole seabed is swept on average 10 times per year, the estimated mortality of starfish due to fishing is between 7% and 23%.

Adding temperature data (NAO index) to the Netherlands model resulted in a significant improvement in the model (Table 10.2, $F_{2,672} = 4.52$, $P = 0.011$). However, although the model is significantly improved, the addition of temperature data explains only a further 1.3% of the variance.

Discussion

For both geographical locations, the relationship between fishing effort and starfish density is significant but has a high error sum of squares, which suggests that factors other than fishing activity may play an important part in determining the size of starfish populations. These factors may be a mixture of anthropogenic and natural

Table 10.2 Analysis of variance (ANOVA) for the relationship between fishing effort, temperature and starfish numbers using the following model: $\log_{10}$(starfish numbers) = a ($\log_{10}$(proportion of the seabed area swept per year))^2 + b $\log_{10}$(proportion of the seabed area swept per year) + c + d(NAO index) + e(NAO index × $\log_{10}$(proportion of the seabed area swept per year))

	df	Sum of squares	Mean square	*F* value	*P* > *F*
Netherlands					
Model	4	17.2	4.3	7.17	0.0001
Error	672	402.1	0.6		
Total	676	419.2			

effects, such as eutrophication and climate. The addition of temperature data to the Netherlands model showed that temperature does influence starfish populations, although a large percentage of the variance remained unexplained, suggesting that other factors are also important. It is possible that the NAO data are not the best data to use, as temperature at specific times of the year may be important, e.g. temperature during the time of metamorphosis. In order to complete a robust analysis on these data, temporal trends in starfish numbers should be examined in relation to a variety of environmental factors known to influence starfish populations. However, it is difficult to allow for temporal trends in the present model since we have only limited knowledge of which environmental factors are likely to influence starfish populations. Nevertheless, the data suggest that trawling activity does have some influence on starfish populations. According to the models, at low levels of fishing effort, starfish numbers increase as effort increases. However, after a threshold level, starfish numbers decrease with increasing fishing effort. It is surprising that the threshold (or turning point) of each equation is so different in the two different areas. This may be due to inaccuracies in recording or calculating fishing effort (as discussed in the Methods section) and, whilst it seems safe to assume that the effort data are fairly consistent within areas, it may be inadvisable to make comparisons between areas. The effort data for the southern North Sea are believed to be reasonably accurate and have to some extent been verified by comparisons with overflight datasets (Jennings *et al.*, 1999).

The models suggest that at low levels of fishing intensity, the extra food made available by fishing could result in an expansion of starfish populations. However, as fishing effort intensifies further, other deleterious impacts of fishing, such as depletion of natural prey items and starfish mortality due to fishing, may cause a reduction in the size of starfish populations.

In order to examine the feasibility of this hypothesis, one firstly needs to examine the effect of food availability on starfish populations. Starfish from prey-rich areas have larger ovaries and, presumably, higher fecundity (Vevers, 1949; Nichols & Barker, 1984). In areas where food is limited, subsidies from fishing activities may increase the fecundity of the starfish population. If starfish populations are never food-limited, the addition of extra food through fishing activities would have no effect. However, starfish, in common with most other scavengers, are extremely plastic in their food requirements, and can survive at least 17 months with virtually no food (Vevers, 1949), but can also consume up to 25% of their own body weight in 24 h (author's unpublished data). Under conditions of excess food, starfish can grow at a rate of more than 1 cm per month (expressing size in terms of radius length, from the mouth to the end of the longest arm) (Vevers, 1949). Starfish have been recorded with arm lengths in excess of 30 cm (Vevers, 1949), although a size of around 10 cm is more common (Hayward & Ryland, 1990). It is possible that the submaximal size of starfish in many areas may be an indication of food limitation.

If a starfish population was food-limited, the addition of extra food through fishing activities would be expected to result in an increase in body size and fecundity. The magnitude of the response would depend on the amount of food made available by fishing. Fonds (in Lindeboom & de Groot, 1998) calculated that, throughout the southern North Sea, the carrion produced from fishing activities would support only 9% of the scavenger populations present. However, it may be that a spatial scale encompassing the whole of the southern North Sea is misleading and that, in certain localised areas, the food input from fishing activities could be sufficient to increase the fecundity of a starfish population. Nevertheless, an increase in the fecundity of a starfish population will not necessarily lead to an increase in population size. At present, little is known about the pre-adult stages of the starfish life cycle and the factors regulating population size. If survival through the larval stage and metamorphosis is strongly density dependent, or is controlled by some environmental factor, increased egg production is unlikely to be related to adult population size. Juvenile mortality is known to be high in many benthic invertebrates, often exceeding 90% (Gosselin & Qian, 1997) and this could also play an important part in the regulation of starfish populations. The mechanism of starfish larval dispersal is also unknown, and it is possible that increased larval production by one population may influence settlement some considerable distance away (Morgan, 1995), while not affecting the original population.

According to the model, at higher intensities of fishing, other factors that reduce starfish numbers apparently become more important. At the turning point of the Netherlands model, the percentage of starfish killed was calculated to be between 0.6% and 1.9%. It seems unlikely that this level of mortality due to fishing will have a large impact on the population. Even when fishing effort increases to a level where the whole seabed is swept on average 10 times per year (the maximum for these data), the estimated mortality of starfish due to fishing was between 7% and 23%, which still seems low for a highly fecund species. However, the effort data used in the model are at a scale of ICES rectangles, and within each such rectangle the distribution of fishing effort will be extremely patchy (Rijnsdorp *et al.*, 1998). If starfish aggregate in the most heavily fished areas, the fishing mortality could be far higher than that calculated above.

The fishing mortality of non-target species that are the prey of starfish may be another factor leading to the reduction of starfish numbers. Intensive fishing activity tends to lead to a decrease in numbers of long-lived species, e.g. the bivalve *Arctica islandica* L. (Bergman & Hup, 1992), and fishing also damages other bivalve species, e.g. *Ensis* spp. (Eleftheriou & Robertson, 1992; Gaspar *et al.*, 1994). Therefore, it is possible that by depleting populations of bivalves, fishing is reducing prey availability for starfish. If the supply of food to the adult stage of the starfish life cycle is an important factor controlling these populations, a reduction in prey availability could be one cause of a decrease in starfish numbers at high fishing intensities. However, although trawling removes the larger bivalve species, it may also result in an increase in the abundance of smaller, short-lived benthic species, such as cirratulid polychaetes (Tuck *et al.*, 1998). The diet of *A. rubens* has not been well studied, mainly owing to the difficulty of identifying the stomach contents. However, Anger *et al.* (1977) showed that *A. rubens* do indeed prey upon smaller bivalves and polychaetes, as well as the larger bivalves often thought of as the staple diet of starfish. If starfish are readily able to feed on the small short-lived species that increase in abundance in response to fishing disturbance, the depletion of larger bivalves due to intensive fishing effort would be unlikely to influence starfish numbers.

As well as removing prey species, fishing can also remove predators. The effect of this on populations of *Asterias rubens* is likely to be small, however, as few animals pre-date on this species, with the exception of large asteroideans such as *Crossaster papposus* (which themselves are unlikely to suffer high fishing mortality), cod *Gadus morhua* (which probably take relatively small numbers of small individuals) and spider crabs *Hyas araneus* (which can remove limbs). Fishing can also alter the habitat of an area (Brown, 1989) rendering it unsuitable for recolonisation by many of the original species, although this has only been documented in a few fairly unusual communities, such as horse mussel *Modiolus modiolus* beds. As starfish can be found in almost any habitat, it seems likely that they could recolonise most areas even after they have been altered.

We conclude that the relationship between fishing effort and starfish numbers may be partly attributable to the opposing effects of increased availability of carrion at low fishing intensities and decreased natural prey availability and/or increased fishing mortality at higher fishing intensities. The cause of the relationship could be inferred with more certainty if more information were available on the link between starfish nutritive requirements and population dynamics. It also seems clear that other factors must influence starfish numbers and these may prove to be more important than the supply of extra food derived from fishing activities. It is possible that temporal trends in starfish numbers may confound the analysis; for example, echinoderm numbers in the North Sea have been increasing over the last 40 years and the reasons for this increase are not known (Lindley *et al.*, 1995). A greater knowledge of the ecology of starfish populations would help to resolve these questions and, in particular, more information is needed concerning mortality during the egg, larval and juvenile stages of the starfish life cycle.

References

Anger, K., Rogal, U., Schriever, G. & Valentin, C. (1977) In-situ investigations on the echinoderm *Asterias rubens* as a predator of soft-bottom communities in the western Baltic Sea. *Helgoländer wiss Meeresunters*, **29**, 439–59.

Anon. (1995) *Report of the study group on ecosystem effects of fishing activities. International Council for the Exploration of the Sea cooperative research report* No. 200. ICES, Copenhagen, Denmark.

Barker, M.F. & Nichols, D. (1983) Reproduction, recruitment and juvenile ecology of the starfish *Asterias rubens* and *Marthasterias glacialis. Journal of the Marine Biological Association of the United Kingdom*, **63**, 745–65.

Berghahn, R. (1990) On the potential impact of shrimping on trophic relationships in the Wadden Sea. In: *Trophic Relationships in the Marine Environment, Proceedings of the 24th European Marine Biology Symposium* (eds M. Barnes & R.N. Gibson), pp. 130–40. Aberdeen University Press, Aberdeen.

Bergman, M.J.N. & Hup, M. (1992) Direct effects of beam trawling on macrofauna in a sandy bottom sediment in the southern North Sea. *ICES Journal of Marine Science*, **49**, 5–11.

Bergman, M.J.N. & van Santbrink, J.W. (1994) Direct effects of beam trawling on macrofauna in sandy areas off the Dutch coast. In: *Environmental impact of bottom gears on benthic fauna in relation to natural resources management and protection of the North Sea* (eds S.J. de Groot & H.J. Lindeboom), pp. 179–208. *NIOZ Report* 1994-11/*RIVO-DLO Report* C026/94. Netherlands Institute for Fisheries Research, Den Burg, Texel, The Netherlands.

Britton, J.C. & Morton, B. (1994) Marine carrion and scavengers. *Oceanography and Marine Biology Annual Review*, **32**, 369–434.

Brown, R.A. (1989) Bottom trawling in Strangford Lough: problems and policies. In: *Distress Signals, Signals from the Environment in Policy and Decision Making, 3rd North Sea Seminar* 1989, pp. 117–27

Dickson, R.R. & Brander, K.M. (1993) Effects of a changing windfield on cod stocks of the North Atlantic. *Fisheries Oceanography*, **2**, 124–53.

Eleftheriou, A. & Robertson, M.R. (1992) The effects of experimental scallop dredging on the fauna and physical environment of a shallow sandy community. *Netherlands Journal of Sea Research*, **30**, 289–99.

Fonds, M. (1994) Mortality of fish and invertebrates in beam trawl catches and the survival chances of discards. In: *Environmental impact of bottom gears on benthic fauna in relation to natural resources management and protection of the North Sea* (eds S.J. de Groot & H.J. Lindeboom), pp. 131–46. *NIOZ Report* 1994-11/*RIVO-DLO Report* C026/94. Netherlands Institute for Fisheries Research, Den Burg, Texel, The Netherlands.

Furness, R.W. (1984) Seabird–fisheries relationships in the north-east Atlantic and North Sea. In: *Marine Birds: their Feeding Ecology and Commercial Fisheries Relationships* (eds D.N. Nettleship, G.A. Sanger & P.F. Springer), pp. 162–9. Canadian Wildlife Service Special Publication Dartmouth, Nova Scotia.

Furness, R.W., Hudson, A.V. & Ensor, K. (1988) Interactions between scavenging seabirds and commercial fisheries around the British Isles. In: *Seabirds and Other Marine Vertebrates: Competition, Predation and other Interactions* (ed. J. Burger), pp. 240–68. Columbia University Press, Columbia.

Gaspar, M.B., Richardson, C.A. & Monteiro, C.C. (1994) The effects of dredging on shell formation in the razor clam *Ensis siliqua* from Barrinha, southern Portugal. *Journal of the Marine Biological Association of the United Kingdom*, **74**, 927–38.

Gosselin, L.A. & Qian, P.-Y. (1997) Juvenile mortality in benthic marine invertebrates. *Marine Ecology Progress Series*, **146**, 265–82.

Hayward, P.J. & Ryland, J.S. (1990) *The Marine Fauna of the British Isles and North-west Europe.* Oxford University Press, New York.

Hill, B.J. & Wassenberg, T.J. (1990) Fate of discards from prawn trawlers in Torres Strait. *Australian Journal of Marine and Freshwater Research*, **41**, 53–64.

Hudson, A.V. & Furness, R.W. (1988) Utilization of discarded fish by scavenging seabirds behind whitefish trawlers in Shetland. *Journal of Zoology*, **215**, 151–66.

Jangoux, M. & van Impe, E. (1977) The annual pyloric cycle of *Asterias rubens* L. (Echinodermata: Asteroidea). *Journal of Experimental Marine Biology and Ecology*, **30**, 165–84.

Jennings, S., Alvsvåg, J., Cotter, A.J., Ehrich, S., Greenstreet, S.P.R., Jarre-Teichmann, A., Mergardt, N., Rijnsdorp, A.D. & Smedstad, O. (1999) Fishing effects in northeast Atlantic shelf seas: patterns in fishing effort, diversity and community structure. III International fishing effort in the North Sea: an analysis of temporal and spatial trends. *Fisheries Research.*

Kaiser, M.J. & Spencer, B.E. (1994) Fish scavenging behaviour in recently trawled areas. *Marine Ecology Progress Series*, **112**, 41–9.

Kaiser, M.J. & Spencer, B.E. (1995) Survival of by-catch from a beam trawl. *Marine Ecology Progress Series*, **126**, 31–8.

Kaiser, M.J. & Spencer, B.E. (1996) Behavioural responses of scavengers to beam trawl disturbance. In: *Aquatic Predators and their Prey* (eds S.P.R. Greenstreet & M.L. Tasker), pp. 116–23. Blackwell Science, Oxford.

Kaiser, M.J., Hill, A.S., Ramsay, K., Spencer, B.E., Brand, A.R., Veale, L.O., Prudden, K., Rees, E.I.S., Munday, B., Ball, B. & Hawkins, S.J. (1996) Benthic disturbance by fishing gear in the Irish Sea: a comparison of beam trawling and scallop dredging. *Aquatic Conservation of Marine and Freshwater Ecosystems*, **6**, 269–85.

Lindeboom, H.J. & de Groot, S.J. (1998) *The effects of different types of fisheries on the North Sea and Irish Sea benthic ecosystems. NIOZ Report* 1998-1/*RIVO-DLO Report* C003/98, 404 pp. Netherlands Institute for Sea Research, Den Burg, Texel, The Netherlands.

Lindley, J.A., Gamble, J.C. & Hunt, H.G. (1995) A change in the zooplankton of the central North Sea (55° to 58°N): a possible consequence of changes in the benthos. *Marine Ecology Progress Series*, **119**, 299–303.

MacDonald, D.S., Little, M., Eno, N.C. & Hiscock, K. (1996) Towards assessing the sensitivity of benthic species and biotopes in relation to fishing activities. *Aquatic Conservation of Marine and Freshwater Ecosystems*, **6**, 257–68.

Moore, P.G. & Wong, Y.M. (1995) *Orchomene nanus* (Krøyer) (Amphipoda: Lysianassoidea), a selective scavenger of dead crabs: feeding preferences in the field. *Journal of Experimental Marine Biology and Ecology*, **192**, 35–45.

Morgan, S.G. (1995) Life and death in the plankton: larval mortality and adaptation. In: *Ecology of Marine Invertebrate Larvae* (ed. L. McEdward), pp. 279–321. CRC Press, Boca Raton, FL.

Nichols, D. & Barker, M.F. (1984) A comparative study of reproductive and nutritional periodicities in two populations of *Asterias rubens* (Echinodermata: Asteroidea) from the English Channel. *Journal of the Marine Biological Association of the United Kingdom*, **64**, 471–84.

Oudejans, R.C.H.M., van der Sluis, I. & van der Plas, A.J. (1979) Changes in the biochemical composition of the pyloric caeca of female seastars, *Asterias rubens*, during their annual reproductive cycle. *Marine Biology*, **53**, 231–8.

Ramsay, K., Kaiser, M.J. & Hughes, R.N. (1996) Changes in hermit crab feeding patterns in response to trawling disturbance. *Marine Ecology Progress Series*, **144**, 63–72.

Ramsay, K., Kaiser, M.J., Moore, P.G. & Hughes, R.N. (1997) Consumption of fisheries discards by benthic scavengers: utilisation of energy subsidies in different marine habitats. *Journal of Animal Ecology*, **66**, 884–96.

Ramsay, K., Kaiser, M.J. & Hughes, R.N. (1998) The responses of benthic scavengers to fishing disturbance by towed gears in different habitats. *Journal of Experimental Marine Biology and Ecology*, **224**, 73–89.

Rijnsdorp, A.D., Buys, A.M., Storbeck, F. & Visser, E.G. (1998) Micro-scale distribution of beam trawl effort in the southern North Sea between 1993 and 1996 in relation to the trawling frequency of the sea bed and the impact on benthic organisms. *ICES Journal of Marine Science*, **55**, 403–19.

Van Santbrink, J.W. & Bergman, M.J.N. (1994) Direct effects of beam trawling on macrofauna in a soft bottom area in the southern North Sea. In: *Environmental impact of bottom gears on benthic fauna in relation to natural resources management and protection of the North Sea* (eds S.J. de Groot & H.J. Lindeboom), pp. 147–78. *NIOZ Report* 1994-11/*RIVO-DLO Report* C026/94. Netherlands Institute for Fisheries Research Den Burg, Texel, The Netherlands.

Tuck, I.D., Hall, S.J., Robertson, M.R., Armstrong, E. & Basford, D.J. (1998) Effects of physical trawling disturbance in a previously unfished sheltered Scottish sea loch. *Marine Ecology Progress Series* **162**, 227–42.

Tunberg, B.G. & Nelson, W.G. (1998) Do climatic oscillations influence cyclical patterns of soft bottom macrobenthic communities on the Swedish west coast. *Marine Ecology Progress Series*, **170**, 85–94.

Vevers, H.G. (1949) The biology of *Asterias rubens* L.: growth and reproduction. *Journal of the Marine Biological Association of the United Kingdom*, **28**, 165–87.

Wassenberg, T.J. & Hill, B.J. (1987) Feeding by the sand crab *Portunus pelagicus* on material discarded from prawn trawlers in Moreton Bay, Australia. *Marine Biology*, **95**, 387–93.

Wassenberg, T.J. & Hill, B.J. (1990) Partitioning of material discarded from prawn trawlers in Moreton Bay. *Australian Journal of Marine and Freshwater Research*, **41**, 27–36.

Chapter 11

Seabirds and commercial fisheries: population trends of piscivorous seabirds explained?

C.J. CAMPHUYSEN[1] and S. GARTHE[2]

[1]Netherlands Institute for Sea Research, NIOZ, PO Box 59, 1790 AB Den Burg, Texel, The Netherlands
[2]Institut für Meereskunde, Abt. Meereszoologie, Düsternbrooker Weg 20, 24105 Kiel, Germany

Summary

1. Most North Sea seabirds have expanded their breeding range and have increased in number over the last century. While the relaxation of persecution, egg collecting and exploitation of seabirds are probably the most important factors, it has often been suggested that commercial fisheries have influenced these trends. This paper examines three important effects of fisheries on North Sea seabirds: (1) the provision of offal and discards as an artificial and additional source of food for seabirds; (2) overfishing of large predatory fish in commercial fisheries; and (3) overfishing of small fish in industrial fisheries.

2. The amount of discards and offal currently discharged in commercial fisheries could potentially support over 6 million North Sea seabirds. However, the species with the most dramatic range expansion and increase in breeding numbers, the northern fulmar, was shown to rely on discards to a substantially lower extent than anticipated on the basis of earlier studies. Only in summer could discards and offal explain an estimated 50% of the overall food consumption by these seabirds. Some of the large *Larus* gulls used discards to a considerable extent, particularly in winter. In all cases, it appeared that non-breeders were most frequently using discards and offal, while nesting birds made a greater effort to feed on more natural resources. It has been shown that several species experienced poor breeding success when proportionally large amounts of discards occurred in chick diets. black-legged kittiwakes largely ignored discards as a food supply over most of the breeding season, while preferring to capture sandeels and other small fish.

3. It has been suggested that the gross overfishing of large predatory fish over the last century has led to increases in the survival and stocks of young fish. While there is circumstantial evidence, there are few factual data to indicate that seabirds have profited from this newly established and abundant food resource. Several examples, however, have shown that overfishing of certain stocks leads to declines in the reproductive output of seabirds, suggesting that their food supply is not necessarily super-abundant.

4. Catches by industrial fisheries of sand eel, sprat, herring and Norway pout have increased dramatically over the last 40 years. A major crash of sand-eel stocks around Shetland in the late 1980s could not be attributed to industrial fisheries with any certainty, and remains an area of controversy. If industrial fisheries in the North Sea continue to increase their catches, stock collapses may be expected in analogy with similar fisheries elsewhere in the world. The west and north-western parts of the North Sea are most heavily prosecuted by industrial fisheries. These are the regions where most seabirds breed and where sandeels form their staple food in summer.

5. While fish stocks, seabird numbers, the energetic requirements of seabirds and the distribution of birds at sea are now generally well known, the foraging strategies of seabirds have so far received insufficient attention. As a result, fluctuations in the availability of prey are largely unknown and any shift in fishing effort or changes in the fishery policy could have unexpected side effects on seabirds.

Keywords: *Fulmarus glacialis*, *Morus bassanus*, *Catharacta skua*, *Larus fuscus*, *Larus argentatus*, *Rissa tridactyla*, *Uria aalge*, industrial fisheries, commercial fisheries, discards, offal, consumption, population trends, seabird community.

Introduction

In the 20th century, most species of seabirds in the North Sea have greatly increased in numbers, established new colonies, and/or expanded their breeding range (Cramp *et al.*, 1974; Becker & Erdelen, 1987; Lloyd *et al.*, 1991; Arts, 1993; Hüppop, 1996). Over the last 15 years, the growth in some of these populations has ceased (Thompson *et al.*, 1996, 1997, 1998; van Dijk *et al.*, 1997a, b, c, 1998), but seabird numbers in the North Sea are generally at a historically high level (Hunt & Furness, 1996). These changes have led to speculation as to what might have caused these tendencies. Until the beginning of the 20th century, seabirds were heavily exploited in most of their range (Croxall *et al.*, 1984; Evans & Nettleship, 1985; Mearns & Mearns, 1998) and, in fact, some populations are still harvested (Nørrevang, 1986; Beatty, 1992). Coulson (1963) argued that the increase of black-legged kittiwakes *Rissa tridactyla* in Britain was due to the relaxation of the exploitation. On the other hand, it has been argued, particularly with reference to the spread of the northern fulmar *Fulmarus glacialis* in north-western Europe, that seabirds benefited from the rich, new food supply made available, first by offal from whaling and later from commercial (whitefish) trawlers (Fisher, 1952). Later authors suggested that more complex trophic interactions might have occurred. The cropping of large piscivorous predators and cannibalistic species (e.g. cod *Gadus morhua*) by large-scale commercial fisheries, benefit seabirds by increasing the abundance of small fish (Sherman *et al.*, 1981; Dunnet *et al.*, 1990; Hunt & Furness, 1996). As such, even commercial 'overfishing' might have been beneficial for seabirds.

Seabirds interact with commercial fisheries in a number of ways (Furness, 1982; Camphuysen, 1990). Seabirds and fisheries may compete for fish, and seabirds might reduce commercial catches by eating large quantities of fish that would otherwise be available to man. Conversely, commercial fisheries might overexploit fish stocks such that the food resources of seabirds become depleted. Fisheries probably always have greater effects on seabirds rather than vice versa (Hunt & Furness, 1996). There is little doubt that the production of discards (unwanted by-catch of small fish, unmarketable species of fish and benthic invertebrates) and offal (discarded waste of gutted marketable fish) in commercial fisheries is of great significance for some species of seabirds. Discards and offal benefit a group of 'scavenging' seabirds, by 'offering' prey that would otherwise be unavailable and out of reach for these birds (Camphuysen *et al.*, 1995; Garthe *et al.*, 1996).

Over the last 30 years, considerable data has been collected with regard to breeding numbers, interannual variations in breeding success, prey selection, seasonal shifts in diets and foraging behaviour (Furness & Hunt, 1996; Anon., 1998). The exploitation of discards and offal, and the spatial and temporal patterns in the distribution at sea of North Sea seabirds in relation to the distribution of fishing activities, have been studied in much detail during two consecutive European Commission funded projects (Camphuysen *et al.*, 1993, 1995; Garthe *et al.*, 1996). The objective of this chapter is to evaluate the main interactions that occur between seabirds and fisheries in the North Sea, particularly in the light of recent population trends and local declines in reproductive success. We evaluate the consumption of discards and offal by seabirds throughout the year and throughout the North Sea. Further, we examine trends in populations and breeding success in non-scavenging seabirds, and examine the patterns in the context of food supply and interactions with commercial and/or industrial fisheries.

Methods

Studies of scavenging seabirds were conducted on-board fisheries research vessels and commercial fishing vessels, mainly between 1988 and 1995 (Furness *et al.*, 1988, 1992; Hudson & Furness, 1988, 1989; Hudson, 1989; Garthe, 1992, 1993; Camphuysen *et al.*, 1993, 1995; Garthe & Hüppop, 1993, 1994, 1996; Garthe *et al.*, 1996; Camphuysen & Garthe, 1997). In these projects, a combination of experimental discarding, counts of scavenging seabirds at the ship's stern, counts of seabirds at sea and counts of fishing vessels at sea were used to quantify the consumption of discards by seabirds. In addition, the dominance hierarchies among seabirds at the stern of fishing vessels were studied to establish prey choice and consumption rates, and to relate overall seabird numbers at sea to the numbers of birds attracted and sustained by (commercial) fisheries. The methodology used in these studies is explained in considerable detail in Camphuysen *et al.* (1995). Subregions used for the data analyses were derived from the standard International Council for the Exploration of the Sea (ICES) subdivisions of the North Sea (Fig. 11.1).

The distribution and abundance of seabirds at sea was established during ship-based surveys between 1985 and 1998 in the North Sea (Stone *et al.*, 1995). In these routine surveys, birds are recorded as either swimming or flying in 10-min periods, following methods first formulated by the Canadian Wildlife Service and published by Brown *et al.* (1975), described in detail by Tasker *et al.* (1984).

The energetic requirements of seabirds were calculated following Bryant & Furness (1995), where the basal metabolic rate (BMR, W) of seabirds is estimated on the basis of the mass of birds: BMR $= 2.30M^{0.774}$ kJ day^{-1} (where M equals the bird's body weight in kg). Following Anon. (1994) we have conservatively assumed a total metabolic rate of 3.9 BMR during the breeding season and 2.5 BMR outside the breeding season.

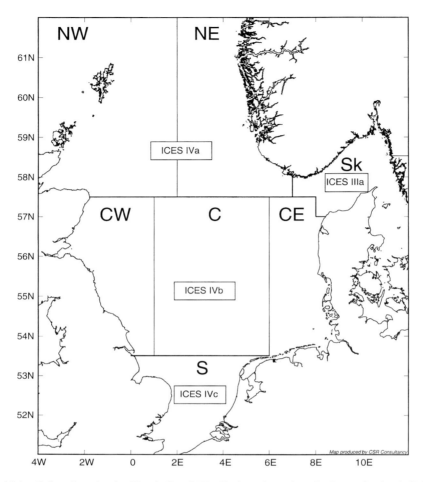

Fig. 11.1   Subregions in the North Sea (NW–S), based on the ICES standard subdivisions.

Results

Recent population trends of seabirds in the North Sea

Populations of most species of seabirds in the western and north-western North Sea have increased dramatically, particularly over the last 100 years (Fig. 11.2). The northern fulmar (which did not breed on the British Isles apart from an ancient colony in the Outer Hebrides on St Kilda) colonised Britain around the turn of the century, and has established new colonies throughout the area over the last 100 years (Fisher, 1952; Cramp *et al.*, 1974; Lloyd *et al.*, 1991). Today, the northern fulmar is one of Britain's most numerous seabirds. Northern gannets *Morus bassanus* (Nelson, 1978), great skuas *Catharacta skua* (Furness, 1987), most *Larus* gulls (Harris, 1970) and until quite recently black-legged kittiwakes (Coulson, 1983) have increased in

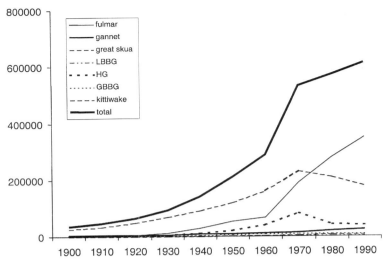

Fig. 11.2 Numbers of pairs of scavenging seabirds breeding on North Sea coasts (NE Britain). Numbers are interpolated from census data to give figures for every tenth year, from 1900 to 1990 (after Furness, 1992). LBBG, lesser black-backed gull; HG, herring gull; GBBG, greater black-backed gull.

numbers in their North Sea colonies. Similar trends were reported for non-scavenging seabirds such as common guillemots *Uria aalge*, razorbills *Alca torda*, and puffins *Fratercula arctica* (Stowe & Harris, 1984; Harris, 1984; Lloyd *et al.*, 1991). Most populations of seabirds are currently at their highest historical levels (NW and CW in Table 11.1).

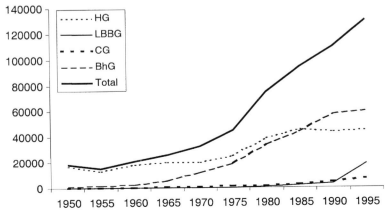

Fig. 11.3 Numbers of breeding *Larus* gulls in the German Bight, 1950–1995. HG, herring gull; LBBG, lesser black-backed gull; CG, common gull; BhG, black-headed gull.

Table 11.1 Numbers of seabirds nesting in five subregions around the North Sea (summary of late 1980s, early 1990s estimates (after Hunt & Furness, 1996) slightly modified with more recent data). Units are occupied sites (OS), breeding pairs or breeding individuals (indiv.)

Species	NW IVa west	NE IVa east	CW IVb west	CE IVb east	S IVc	Total	Census units
Northern Fulmar	294 150	0	12 600	35	700	307 485	OS
Great Cormorant	1485	18	703	0	265	2471	pairs
European Shag	13 485	1755	4565	0	0	19 805	pairs
Northern Gannet	21 650	0	22 150	5	0	43 805	pairs
Great skua	7300	5	0	0	0	7305	pairs
Lesser black-backed gull	2600	25 500	2200	15 800	3250	49 350	pairs
Herring gull	41 850	34 050	40 450	96 300	24 500	237 150	pairs
Great black-backed gull	9900	14 500	30	0	0	24 430	pairs
Black-legged Kittiwake	206 600	3000	200 000	3300	2550	415 450	pairs
Arctic tern	55 950	8630	5350	4710	85	74 725	pairs
Common tern	1160	39 800	1730	14 400	4380	61 470	pairs
Sandwich tern	1120	1500	5590	14 690	7645	30 545	pairs
Guillemot	507 485	440	167 610	4900	0	680 435	indiv.
Razorbill	54 540	300	18 260	16	0	73 116	indiv.
Black guillemot	20 845	2890	3	0	0	23 738	indiv.
Atlantic Puffin	124 290	21 695	79 975	0	0	225 960	indiv.
Total	1 364 410	154 083	561 216	154 156	43 375	2 277 240	4 554 480

In the eastern and southern half of the North Sea, several species of *Larus* gulls have established populations that subsequently showed exponential growth. *Larus* gulls in the German Bight increased from 18 000 pairs in 1950 to 130 000 pairs in 1995, a nearly 10-fold increase in breeding numbers (Fig. 11.3). Herring and lesser black-backed gulls (*Larus argentatus* and *L. fuscus*) nesting in The Netherlands increased from less than 2500 pairs in 1900 to nearly 120 000 pairs in 1996, with a particularly sharp increase since the late 1960s (Fig. 11.4; Spaans, 1998a, b). From the island of Helgoland in the German Bight, formerly an important breeding station for seabirds, northern fulmar and northern gannet have established new colonies this century, while the local populations of black-legged kittiwakes and common guillemots have increased considerably since the late 1960s (Fig. 11.5; Hüppop, 1996). Also, we may safely conclude that the numbers of seabirds nesting in this part of the North Sea have not been as high at any time in our human (written) history (CE and S in Table 11.1).

In the north-eastern part of the North Sea, at least in recent years, similar patterns of increase and colonisation along the Norwegian west coast have been reported in black-legged kittiwakes, northern fulmars and northern gannets (Brun, 1971, 1972; Toft, 1983; Barrett & Vader, 1984). The total numbers of seabirds at sea in the North Sea are considerably greater than these breeding numbers indicate, but total numbers are subject to species-specific seasonal fluctuations. From ship-based

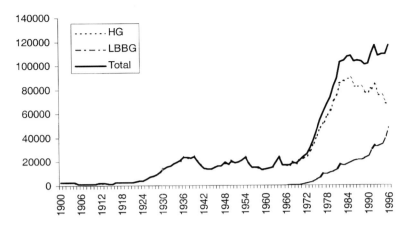

Fig. 11.4 Numbers of herring gulls and lesser black-backed gulls nesting in The Netherlands since 1900.

surveys in the North Sea, it has been estimated that the total number of seabirds in the North Sea range from *c.* 4 million (June) to over 8 million (February and late summer to early autumn; Fig. 11.6). In summer, well over 50% of these birds are found in the northern and north-western North Sea (subregions NW and CW).

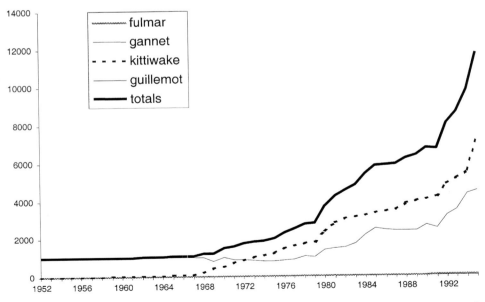

Fig. 11.5 Numbers of seabirds breeding on Helgoland since the late 1960s. The lines for fulmars and guillemots run so close to the *x* axis that it is difficult to see any variation in these species.

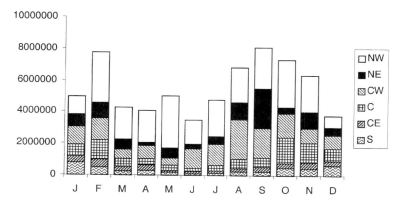

Fig. 11.6 Numbers of seabirds at sea in the North Sea (12 commonest species), based on monthly mean densities in each subregion (Fig. 11.1) (after Anon., 1994).

Energetic requirements of seabirds in the North Sea

The estimated annual energy requirements of the more common species of seabirds in the North Sea are estimated to be just over 3900×10^9 kJ (Anon., 1994; Table 11.2). Populations of the northern fulmar, northern gannet, herring gull, great black-backed gull *Larus marinus*, kittiwake and guillemot together require 89% of all the energetic requirements. The first five of these species obtain a substantial amount of their food from the discards of fishing vessels. In contrast, the energetic requirements of the three commoner terns (totalling over 300 000 breeding individuals in the North Sea) and of cormorant *Phalacrocorax carbo* and shag *P. aristotelis* are almost negligible.

Discards and offal as an artificial source of food

Camphuysen *et al.* (1995) identified the following seabirds as common scavengers at fishing vessels in the North Sea: northern fulmar, northern gannet, great skua, herring gull, lesser black-backed gull, great black-backed gull and black-legged kittiwake. When not feeding at fishing vessels, the fulmar is a surface feeder, specialising in catching zooplankton, squid and small fish. All other seabirds are mainly piscivorous when at sea, although some of the gulls have terrestrial feeding modes. Gannets are deep-plunge diving seabirds, the great skua is a surface feeder, while all gulls are shallow-plunge diving species or surface feeders.

Consumption rates of scavenging seabirds were measured aboard fisheries research vessels during sessions of experimental discarding and on-board commercial fishing vessels. Overall consumption rates for offal (liver and guts of processed fish), discarded roundfish (usually roundfish below the minimum landing size and non-target species), flatfish and benthic invertebrates varied per season and in different

Table 11.2 Mass (g), basal (BMR; kJ day^{-1} and W) and field metabolic rate (FMR; W) of some common seabirds in the North Sea. FMR (1) = 3.9 BMR (breeding season), FMR (2) = 2.5 BMR (non-breeding season; according to assumptions in Anon. (1994) and estimated annual energy requirements (kJ 10^9 and %)

Species	Mass (g)	BMR (W)	FMR (W) (1)	FMR (W) (2)	Annual requirement (kJ × 10^9 %)	
Fulmarus glacialis	820	5	19	12	1095	28.1
Sula bassana	3000	13	51	33	273	7.0
Phalacrocorax carbo	2750	12	48	31	6	0.2
Phalacrocorax aristotelis	2000	10	37	24	47	1.2
Catharacta skua	1430	7	29	18	21	0.5
Larus fuscus	860	5	19	12	70	1.8
Larus argentatus	1150	6	24	16	451	11.6
Larus marinus	1830	9	35	22	301	7.7
Rissa tridactyla	405	3	11	7	307	7.9
Sterna sandvicensis	240	2	7	5	2	0.1
Sterna hirundo	125	1	4	3	3	0.1
Sterna paradisaea	110	1	4	3	3	0.1
Uria aalge	860	5	19	12	1024	26.3
Alca torda	710	4	17	11	100	2.6
Cepphus grylle	375	3	10	7	8	0.2
Fratercula arctica	380	3	10	7	108	2.8

parts of the North Sea (Table 11.3). Consumption rates were particularly high in winter.

Consumption rates by seabirds of different prey items in the discards offered ranged from 6% for benthic invertebrates to nearly 90% for offal (Table 11.4). Camphuysen (1993) indicated that both the consumption of flatfish and benthic invertebrates, less preferred food for the scavengers at beam trawlers in the southern North Sea, increased when competition was high, but was negligible when the number of scavengers was low in proportion to the discard supply. The relatively high consumption rates for these less-preferred discard types in an area where the potential demand (numbers of seabirds at sea) exceeds the resource (amounts discarded km^{-2}) suggests that competition for these resources may be greatest in the northern and western North Sea (Table 11.5).

The northern fulmar is one of the more common scavengers behind fishing vessels in the North Sea, and this species has shown the most spectacular increase in numbers and breeding range in the North Sea of any seabird (Fisher, 1952; Lloyd *et al.*, 1991). The expansion of the fulmar has frequently been attributed mainly to its success as an offal scavenger, first associated with whalers in the arctic and subarctic,

Table 11.3   Consumption rates (%) of discards and offal taken by scavenging seabirds in six subregions (all seasons combined) and in each of the four seasons (all subregions combined) in the North Sea (modified from Camphuysen *et al.* (1995) and Garthe *et al.* (1996))

	Roundfish	Flatfish	Offal	Benthic invertebrates
Season				
Winter	92	35	100	17
Spring	76	22	94	8
Summer	70	10	94	3
Autumn	82	20	97	3
Subregion				
NW	90	28	99	9
NE	89	41	98	3
CW	84	32	92	1
C	75	14	90	1
CE	63	10	54	3
S	71	8	100	4

Table 11.4   Total amounts of discards and offal in the North Sea (estimates for 1990), and the overall consumption of discards and offal by seabirds (after Garthe *et al.*, 1996)

	Amount discarded	Consumed	% Consumed
Roundfish	262 000 t	206 000 t	79
Flatfish	300 000 t	38 000 t	13
Offal	62 800 t	55 000 t	88
Benthic invertebrates	150 000 t	9 000 t	6

Table 11.5   The energetic requirements of seabirds at sea and the energetic content of available discards (after Garthe *et al.*, 1996)

Subregion	Energetic requirements of scavengers ($\times 10^6$ kJ)	Energetic equivalents of discards and offal ($\times 10^6$ kJ)	Index of required/available discards	Potential shortage or surplus if discards were only prey
NW	899 000	463 000	1.9	shortage
NE	312 000	456 000	0.7	surplus
CW	364 000	269 000	1.4	shortage
C	322 000	659 000	0.5	surplus
CE	120 000	698 000	0.2	surplus
S	175 000	871 000	0.2	surplus
Total	2 192 000	3 216 000	0.6	surplus

and later in association with commercial fishing fleets. Several studies have demonstrated a strong affinity between fulmars and fishing vessels (Rees, 1963; Wahl & Heinemann, 1979; Tasker *et al.*, 1987). From studies of scavenging seabirds at whitefish trawlers around Shetland, fulmars were ranked at the apex of the interspecific dominance hierarchy, and were able to obtain the best quality food (i.e. fish livers or the entire offal; Furness *et al.*, 1988; Hudson & Furness, 1988, 1989). Tasker *et al.* (1987) concluded that fishing activities were an important determinant in fulmar distribution at sea. More recent studies, often on different scales, were less conclusive and suggested that, although fulmars were evidently attracted by fishing vessels, the spatial distribution of the main fisheries and fulmars apparently did not match very well (Camphuysen *et al.*, 1995; Stone *et al.*, 1995). Also, recent investigations have shown that the position in the dominance hierarchy at fishing vessels was not so high in other parts of the North Sea, as compared with studies around Shetland (Camphuysen, 1993, 1994; Garthe, 1993; Garthe & Hüppop, 1993, 1994).

In light of these findings, Camphuysen & Garthe (1997) re-examined the distribution and scavenging habits of northern fulmars in the North Sea on the basis of a large set of recently gathered data. These data included all available discards experiments ($n = 841$), the European Seabirds at Sea database version 2.0 (1979–1993), systematic snapshot counts of fishing vessels in the North Sea (see Camphuysen *et al.*, 1995, for details), and a literature study of fulmar diets. On the basis of fleet composition and landings statistics (ICES working group reports), the availability of discards and offal per km^2 was calculated for each subregion in the North Sea (Fig. 11.1). Spearman rank correlation coefficients showed that negative correlations occurred for northern fulmar densities at sea versus the availability of offal (t km^{-2}) (R_S –0.770, $P < 0.05$), roundfish discards (R_S –0.428, n.s.), flatfish discards (R_S –0.942, $P < 0.001$), discarded benthic invertebrates (R_S –0.942, $P < 0.001$) and all fishery waste combined (R_S –0.828, $P < 0.05$). There was no correlation between numbers of fishing vessels at sea and densities of northern fulmars at this scale, but the numbers of fulmars observed at the stern were highly correlated with the overall densities of northern fulmars at sea (R_S –0.893, $P < 0.001$). These results suggested strongly that commercial fisheries were not the prime determinant of fulmar distribution at sea. Further results indicated that fulmars had a rather northerly, offshore distribution, mainly over waters with a rather strong Atlantic influence, suggesting that their natural (zooplanktonic) prey was a much more important parameter influencing their distribution at sea.

As fulmars are at the apex of the dominance hierarchy when in competition with other seabirds, we were able to calculate the probability (p) of a fulmar successfully obtaining a discarded fish or offal in the competition with other scavengers (Furness *et al.*, 1988; Hudson & Furness, 1988, 1989; incidentally, all studies were conducted around Shetland where fulmars are super-abundant):

$$p = \frac{y}{y + ax^b}$$

where y is the number of northern fulmars and x is the number of other scavengers near the fishing vessel. The parameter a indicates the 'relative competitive strength' (RCS) of another scavenger, and b indicates whether the RCS is related to the absolute number of other scavengers. Reformulating the model in terms of a logit model (generalised linear models; Nelder & Wedderburn, 1972; McCullagh & Nelder, 1989) results in:

$$\log \frac{p}{(1-p)} = \log\left(\frac{1}{a}\right) + \log\left(\frac{f}{1-f}\right) + (1-b)\log(x)$$

where f is the fraction of fulmars present ($y/(y+x)$). The results of 245 experiments in which offal was discarded showed that fulmars took food in line with the predictions made on the basis of their numbers in relation to other scavengers. This indicated that the probability that a fulmar consumed offal was positively related to the relative abundance of fulmars at the trawl. With roundfish discards (723 experiments), however, fulmars consumed considerably less than expected on the basis of their numbers at the trawl. The probability that fulmars obtained roundfish in competition with other species was particularly low in autumn and winter and also in the subregions NW and CW. Fulmars were less likely to obtain offal in competitive interactions in autumn and in CW, whereas fulmars consumed considerably more of the discarded offal in summer than their relative abundance at trawls would have predicted.

The distinct decrease in overall consumption rates of fulmars in autumn and winter coincided with a sudden and dramatic increase in herring gulls in the North Sea (including the offshore waters in the north). Herring gulls became very abundant scavengers at fishing vessels all over this area. Fulmars appeared to be rather ill adapted as scavengers, particularly around trawlers that move while fish are sorted and discarded. Fulmars are stiff-winged and not very manoeuvrable in flight, which is a serious disadvantage in competitive situations with more agile gulls. This study concluded that less than 50% of the energetic requirements of fulmars may be satisfied by offal and roundfish discards, and that in autumn and winter only one-tenth to one-fifth of the northern fulmars in the North Sea could be sustained by fishery waste (Camphuysen & Garthe, 1997).

A detailed analysis of the scavenging habits of some of the larger, commoner *Larus* gulls (great black-backed gull, lesser black-backed gull and herring gull) was more complex. Although all three species breed in the North Sea, the great black-backed gull occurs year-round, the lesser black-backed gull is a summer visitor and the herring gull is very scarce at sea in (late) summer and invades this area in winter with approximately a million birds that breed in Scandinavia. The great black-backed gull is an effective scavenger, while mainly acting as a kleptoparasite of smaller species (only equalled by the northern gannet and the great skua). The lesser black-backed gull and the herring gull are the dominant species at the south-eastern North Sea beam trawl fisheries. For the *Larus* group of gulls, the general distribution patterns at sea were roughly opposite to that of the northern fulmar. The *Larus* gulls had a predominantly southerly, nearshore distribution, mainly over mixed waters with a small Atlantic influence, suggesting that the proximity of land for roosting sites was an important parameter influencing their distribution at sea. In the south-eastern North Sea, the distribution of fishing vessels was an important underlying parameter explaining *Larus* gull distribution.

We undertook a similar probability analysis (as above), for herring gulls and great black-backed gulls, which showed that, in winter, these birds obtain considerably more discards than expected from their relative numbers at the stern of trawlers (C.J. Camphuysen & J. van der Meer, unpublished data). These results indicate that northern fulmars are outcompeted in the presence of large numbers of other scavengers. Again, within the confines of the North Sea study area, the correlations between the numbers of gulls at sea and the presence of fishing vessels were weak or absent. There is very little doubt, however, that most gulls exploit fishery waste to a considerable extent, particularly in the south-eastern North Sea. Preliminary calculations of the numbers of gulls at sea that are potentially sustained by fishery waste alone suggest that at least 25% (up to 80% at certain times of the year) of the diet of these birds consists of offal and roundfish discards (Garthe, 1992; Hüppop & Garthe, 1993; Garthe & Hüppop, 1994; Camphuysen *et al.*, 1995; Garthe *et al.*, 1996; Camphuysen & Garthe, unpublished data).

Although discards and offal may be an important additional source of food, the reproductive output of seabirds that consume fishery waste almost exclusively is not necessarily very high. Hamer *et al.* (1991) demonstrated that the chick growth index of great skuas on Foula (Shetland Islands) declined considerably when more than 50% of the prey delivered by the parents were comprised of discards and offal. On the other hand, relatively high growth rates were observed when discards and offal comprised a smaller proportion (20–30%) of chick diets. Similarly, Spaans *et al.* (1994) showed that the reproductive output of lesser black-backed gulls was high in years when herring dominated chick diets and the use of fishery waste was comparatively low. Black-legged kittiwakes nesting on the east coast of Britain largely ignored discards as a food supply over most of the breeding season, and preferred to consume sandeels *Ammodytes* spp. and other small fish (Camphuysen, 1995).

Studies of scavenging seabirds around fishing vessels have given detailed insights into the dominance hierarchies among seabirds, suggesting that measures to reduce the amount of discards would affect the lower ranked species first. For example, Furness (1992) suggested that an increase in trawl net mesh size would increase the mean length of discarded fish. This in turn would lower the handling efficiency of smaller scavenging seabirds that are less well adapted to consuming large fish. A total ban of discards and offal would obviously lead to major changes in seabird diets. A prompt introduction of such a ban would inevitably increase the predation risk for smaller birds from large seabirds such as great black-backed gulls and great skuas.

The effect of commercial fisheries on seabirds

Other common piscivorous seabirds breeding in the North Sea are cormorant, shag, Arctic tern *Sterna paradisaea*, common tern *S. hirundo*, Sandwich tern *S. sandvicensis*, common guillemot, razorbill, black guillemot, and puffin. Of these, the cormorant and shag are foot-propelled pursuit diving seabirds, the terns are shallow plunge divers, while the auks are all wing-propelled pursuit diving species.

Most piscivorous seabirds feed mainly on small shoaling fish, such as herring *Clupea harengus*, sprat *Sprattus sprattus* and sandeels. Sandeels are particularly important prey in the breeding season, whereas clupeoids and gadoids become more important during the winter period. The ICES working group on seabird ecology (Anon., 1994) estimated the annual consumption of 18 seabird species, based on seabird abundance estimates and energetic requirements (as described earlier) at 600 000 t per annum. Included in this estimate are nearly 200 000 t of sandeels, 30 000 t of sprat and young herring, 22 000 t of gadoid fish, 13 000 t of mackerel, 13 000 t of large herring, 71 000 t of offal, 109 000 t of discards and 146 000 t of 'other prey'. These estimates were based to a considerable extent on diet studies conducted during summer, and it is likely that the consumption of sandeels in winter (Oct–Mar, 40 000 t) was significantly overestimated. Estimates of offal and discards consumption were respectively slightly higher and considerably lower than estimates of the annual consumption of these by Garthe *et al.* (1996) based on a discards projects conducted in 1993 and 1994 (Camphuysen *et al.*, 1993, 1995). There is little doubt, however, that the total consumption of small fish by seabirds (sandeel, sprat, young herring, small gadoids) is in the range of 250 000 t per annum (about one-tenth of that consumed by large predatory fish; Anon., 1994).

It is thought that stocks of immature fish in the North Sea have increased, thanks to the overfishing of larger size classes of piscivorous fish (Andersen & Ursin, 1977; Daan, 1978, 1980; Furness, 1982; Jones, 1983; Furness & Monaghan, 1987; Daan *et al.*, 1990). Similarly, congruent shifts in sandeel abundance in the western and eastern North Atlantic ecosystems were also explained by predator removal by fisheries (Sherman *et al.*, 1981). While these shifts in age structure, species composition and length distribution of North Sea fish are largely beyond doubt, the conclusion that seabirds must have benefited is largely speculative and certainly untested. Although the increase in the amount of small fish intuitively must have resulted in a larger resource for piscivorous seabirds, the availability of prey may not have changed quite that dramatically. Although seabirds consume only a small proportion of the immature fish in the North Sea, shifts in the availability of small fish may negatively affect seabirds even when the fish stock itself is hardly affected (see below). A clear example of a negative effect of commercial fisheries off the Norwegian coast is the reduced reproductive output of puffins nesting on the Lofoten islands over a 20-year period following the collapse of local herring (Barth, 1978; Lid, 1981; Vader, 1988; Vader *et al.*, 1989; Anker-Nilssen, 1992).

The effects of industrial fisheries on seabirds

One result of the ecosystem changes caused by the overfishing of large fish has been the development of industrial fisheries (those targeting small fish, not meant for human consumption), particularly for sandeels and Norway pout *Trisopterus esmarkii* (Madsen, 1978; Anon., 1984; Gauld *et al.*, 1986). Industrial fisheries in the North Sea have increased substantially since the early 1960s. In particular, the sandeel fishery is currently the largest fishery in the North Sea. This fishery has caused concern among conservationists because many top predators (seals, cetaceans and seabirds) rely on sandeels as the main part of their diet (Furness, 1990). Seabirds may compensate for shortages of prey during the breeding season by working harder, by allocating more time to foraging, switching prey, abandoning nesting attempts, or ultimately may starve to death. As a result, the direct effects of fisheries impacts on prey availability can easily be overlooked. Most seabirds are K-strategists in terms of their life histories (large and long-lived species, living in predictable environments, with a slow rate of colonisation, a low reproductive rate, low density independent mortality, with rather stable populations; Campbell & Lack, 1985). Over their long life span, seabirds invest in few, well-provisioned young and conserve energy reserves for themselves. While seabird populations are now monitored all over Europe, several direct responses of seabirds to food shortages, such as the reduced feeding rates of chicks, longer feeding trips or shifts in the diet, are seldom studied on a regular basis.

During the 1980s, the breeding success of several seabirds at Shetland declined markedly, coincident with a marked decline in landings of sandeels from an industrial fishery that operated close to the Shetland coast (Heubeck, 1988, 1989; Goodlad, 1989; Monaghan *et al.*, 1992a, b). The seabird species affected most severely were surface feeders that fed predominantly on O-group sandeels (Arctic terns, black-legged kittiwakes and great skuas). The kleptoparasitic Arctic skuas *Stercorarius parasiticus* were also affected, as was the puffin, a pursuit diving species. The diet of northern gannets changed from predominantly sandeels to clupeoid and gadoid fish. After a few years of poor breeding success, significant declines occurred in breeding populations of Arctic terns, black-legged kittiwakes and common guillemots. The numbers of Arctic terns increased dramatically again in 1991, coincident with the appearance of a large sandeel year class.

These population shifts led to suggestions that the industrial fishery competed for the same resource as the seabirds and was responsible for the decline in their food supply. However, studies carried out by the Scottish Office indicated that the decline in sandeel abundance was the result of poor recruitment to the Shetland stock and, hence, that natural fluctuations in sandeel survivorship were the main cause of the decline in prey availability for seabirds breeding at Shetland (Anon., 1994). The suggestion that 'natural fluctuations' cause disaster in a stock that is heavily fished is not new (Berrill, 1997), and there is still considerable controversy over the impact of the Shetland fishery (Monaghan, 1992). World-wide, there are many examples of fisheries that were not closed at times when an assumed overabundant resource showed signs of depletion, and the sandeel fishery in the North Sea is not the best example of a well-managed industry. More recently, Danish industrial sand-eel catches off the Scottish east coast declined spectacularly, while concomitantly the seabird populations in the Firth of Forth show the first signs of food shortages (S. Greenstreet and S. Wanless, personal communication). The collapse of the Barents Sea capelin *Mallotus villosus* stock was thought to have occurred as a result of recruitment failure and cod predation. This also coincided with the establishment of an industrial fishery that was not regulated adequately as catches declined (Hamre, 1988; Tjelmeland, 1989; Vader *et al.*, 1989; Hopkins & Nilssen, 1991).

Discussion

It is clear that the overall numbers of seabirds have increased spectacularly over the last century. In some cases this began in either the 18th or 19th century, but few populations are studied sufficiently to reach such conclusions. Nature conservation is a 20th-century development. As a result, the mass slaughter of birds, even of species that were generally considered as pests, has come to a halt. Some authors largely explain the dramatic increase in numbers of seabirds simply by referring to relaxation of the immense scale of exploitation and persecution of seabirds prior to the 20th century (Coulson, 1963). While this might indeed have been an important factor, it should be realised that many terrestrial or freshwater species suffered equal or at least similar losses (Mearns & Mearns, 1998), while an increase and expansion of the breeding range, as witnessed in the slowly reproducing seabirds, is unrivalled. Moreover, the gannetry on Sula Sgeir in the Outer Hebrides, although still exploited in a sustainable manner, has increased at very much the same rate as all the other colonies in the British Isles (Nelson, 1978). Spaans (1998a, b) analysed the rise and fall of herring gull and lesser black-backed gull populations in The Netherlands and concluded that the relaxation of persecution must have been an important force driving the spectacular increase of these birds, although it could not explain all trends. When breeding colonies became protected, however, all populations increased. The increase stopped when culling was introduced to 'manage' gull numbers for the protection of other birds. Culling (in the late 1960s) was not exercised such that populations declined, but when full protection measures were taken in the 1970s, populations exploded to unprecedented levels and peaked in the 1980s (herring gull) or 1990s (lesser black-backed gull).

Many authors commenting on the increase in seabirds have suggested that man has supported these trends by providing extra sources of food (fishery waste, garbage). Gulls, fulmars, gannets and a variety of other seabirds cannot be blamed for exploiting the ample supply of fishery waste. These birds and their scavenging behaviour are highly visible to the layman, but scavenging should not be overestimated in the absence of less easily gathered data on natural feeding. There is mounting evidence that northern fulmars, for example, are highly specialised seabirds, exploiting a wide range of natural prey, most notably zooplankton. When feeding on such resources, these birds disperse over vast areas of ocean and become 'invisible'. The analysis outlined in this chapter clearly suggested that offal and discards, albeit an important prey for fulmars, were additional prey rather than staple foods, and perhaps of significance in particular for non-breeding individuals. The feeding ecology of none of the scavenging seabirds in the North Sea has been studied sufficiently to reach conclusions as to what proportion of these birds can be sustained by discards and offal. In addition, it is not clear what would happen if this additional source of food were reduced or reduced by future management measures.

The conclusion that seabirds, in general, have profited from overfishing is not yet supported by the available data. Prey availability for seabirds could also be reduced if large predatory fish decline in numbers and therefore fail to drive fish towards the surface during their feeding bouts. There is no linear correlation between fish stocks and the numbers or amounts of fish available to seabirds. Even when the suggestion that seabirds have profited from fisheries activities may hold some truth, simply repeating the statement without additional information is unfounded. Furness & Monaghan (1987) suggested a shift in the energy flow through parts of the North Sea ecosystem. From a situation in the late 19th century, where seabirds and particularly large predatory fish shared a common resource of sandeels, sprats, pout and herring (and ultimately zooplankton), modern intensified food fisheries had enlarged the resource for seabirds of sandeels, sprats and pout by removing predatory fish (and herring). In a further possible scenario for the 1990s, they proposed that the development of large-scale industrial fisheries would then remove a substantial part of the small fish biomass such that seabird populations would come under pressure through lack of food.

Industrial fisheries are currently of the greatest concern for seabird biologists in the North Sea. While catches continue to increase and the fishery is not tightly regulated in comparison with other fisheries, and while the effect of these fisheries on fish stocks are largely unknown, most species of seabirds and marine mammals are under threat of population decline in the North Sea. During the breeding season, sandeels are staple food for many seabirds and marine mammals and it seems only a matter of time before serious conflicts arise. We recommend more detailed studies of the effect of industrial fisheries on North Sea fish stocks and, in particular, of the effect of industrial fisheries in areas where large numbers of seabirds breed and rely on sandeels (e.g. Firth of Forth, Scottish east coast, Orkney and Shetland). The numbers, distribution and energetic requirements of seabirds are well known, but their foraging strategies, the aggregative and functional responses of feeding seabirds, and factors influencing prey availability are still largely unstudied. As a result, most statements on the effect of fisheries on seabirds are largely speculative. As a consequence, any change in fishery policy could have unexpected and perhaps undesirable side effects.

Acknowledgements

We would like to thank our colleagues for stimulating discussions and for the data collected during the EC-funded discards projects 1992–95: Frank Albers, Per Andell, Rob Barrett, Peter Becker, Belen Calvo, Gilles Chapdelaine, Kenny Ensor, Olaf Flore, Arne Follestad, Bob Furness, Simon Greenstreet, Stefan Groenewold, Jens Lund Hansen, George Hunt, Ommo Hüppop, Eduard Koopman, Genevieve Leaper, Mardik Leopold, Jaap van der Meer, Bill Montevecchi, Max Nitske, Henk Offringa, Claire Pollock, Jan Seys, Henrik Skov, Carolyn Stone, Andrew Stronach, Mark Tasker, Uwe Walter, Sarah Wanless, Andy Webb, Richard White, Chris Winter and Olaf Zeiske.

References

Andersen, K.P. & Ursin, E. (1977) A multispecies extension to the Beverton and Holt theory of fishing, with accounts of phosphorous circulation and primary production. *Meddelelser Danmarks Fisk og Havundersøkelser*, 7, 319–435.

Anker-Nilssen, T. (1992) Food supply as a determinant of reproduction and population development in Norwegian Puffins *Fratercula arctica*. Ph.D. thesis, University of Trondheim, 150 pp.

Anon. (1984) Report of the industrial fisheries working group. *International Council for the Exploration of Sea*, C.M. 1984/Assess 12, 148 pp.

Anon. (1994) Report of the study group on seabird/fish interactions. *International Council for the Exploration of Sea*, C.M. 1994/L:3, 119 pp.

Anon. (1998) Report of the Working Group on Seabird Ecology. Oceanography Committee. *International Council for the Exploration of Sea*, CM 1998/C:5, Ref. E, 37pp.

Arts, F. (1993) *Kustbroedvogeldatabase DGW/SOVON. Voorlopig SOVON rapport 93/08*. SOVON, Beek-Ubbergen.

Barrett, R.T. & Vader, W. (1984) The status and conservation of breeding seabirds in Norway. In: *Status and Conservation of the World's Seabirds* (eds J.P. Croxall, P.G.H. Evans & R.W.Schreiber), pp. 323–33. *Technical Publication* No. 2. ICBP, Cambridge.

Barth, E.K. (1978) Lundetragedien på Røst. *Fauna (Oslo)*, 31, 2734.

Beatty, J. (1992) *Sula – the Seabird-Hunters of Lewis*, 143 pp. Michael Joseph, London.

Becker, P.H. & Erdelen, M. (1987) Die Bestandsentwicklung von Brutvogeln der deutschen Nordseeküste seit 1950: Aspekte für den Artenschutz. *Berichte Deutschen Sektion Internationalen Rat für den Vogelschutz*, 26, 63–73.

Berrill, M. (1997) *The Plundered Seas – Can the World's Fish be Saved?* 208 pp. Sierra Club Books, San Francisco.

Brown, R.G.B., Nettleship, D.N., Germain, P., Tull, C.E. & Davis, T. (1975) *Atlas of Eastern Canadian Seabirds*, 220 pp. Canadian Wildlife Service, Bedford Institute for Oceanography, Dartmouth.

Brun, E. (1971) Populasjonsendringer hos noen sjøfuglarter i sør-norge. *Sterna*, 10, 35–56.

Brun, E. (1972) Establishment and population increase of the Gannet *Sula bassana* in Norway. *Ornis Scandinavica*, 3, 27–38.

Bryant, D.M. & Furness, R.W. (1995) Basal metabolic rates of North Atlantic seabirds. *Ibis*, 137, 219–26.

Campbell, B. & Lack, E. (1985) *A Dictionary of Birds*, 670 pp. T. & A.D. Poyser, London.

Camphuysen, C.J. (1990) *Fish stocks, fisheries and seabirds in the North Sea. Technisch Rapport Vogelbescherming* No. 5, 120 pp. Vogelbescherming Nederland, Zeist.

Camphuysen, C.J. (1993) Foerageermogelijkheden voor zeevogels in de boomkorvisserij: een verkennend onderzoek. *Sula*, **7**, 81–104.

Camphuysen, C.J. (1994) Flatfish selection by herring gulls *Larus argentatus* and lesser black-backed gulls *Larus fuscus* scavenging at commercial beamtrawlers in the southern North Sea. *Netherlands Journal of Sea Research*, **32**, 91–8.

Camphuysen, C.J. (1995) Kittiwakes *Rissa tridactyla* in the North Sea: pelagic ecology, fisheries relationships and feeding strategies. *Limosa*, **68**, 123.

Camphuysen, C.J. & Garthe, S. (1997) Distribution and scavenging habits of Northern Fulmars in the North Sea. *ICES Journal of Marine Science*, **54**, 654–83.

Camphuysen, C.J., Ensor, K., Furness, R.W., Garthe, S., Hüppop, O., Leaper, G., Offringa, H. & Tasker, M.L. (1993) *Seabirds feeding on discards in winter in the North Sea. Final report to the European Commission, study contr.* 92/3505, *NIOZ Report* 1993-8, 142 pp. Netherlands Institute for Sea Research, Den Burg, Texel, The Netherlands.

Camphuysen, C.J., Calvo, B., Durinck, J., Ensor, K., Follestad, A., Furness, R.W., Garthe, S., Leaper, G., Skov, H., Tasker, M.L. & Winter, C.J.N. (1995) *Consumption of discards by seabirds in the North Sea. Final report to the European Commission, study contract BIOECO/93/10, NIOZ Report* 1995-5, 202 pp. Netherlands Institute for Sea Research, Den Burg, Texel, The Netherlands.

Coulson, J.C. (1963) The status of the Kittiwake in the British Isles. *Bird Study*, **10**, 147–79.

Coulson, J.C. (1983) The changing status of the Kittiwake *Rissa tridactyla* in the British Isles, 1969–1979. *Bird Study*, **30**, 9–16.

Cramp, S., Bourne, W.R.P. & Saunders, D. (1974) *The Seabirds of Britain and Ireland*, 287 pp. Collins, London.

Croxall, J.P., Evans, P.G.H. & Schreiber, R.W. (1984) *Status and Conservation of the World's Seabirds. Technical Publication* No. 2, 778 pp. International Council for Bird Preservation, Cambridge.

Daan, N. (1978) Changes in cod stocks and cod fisheries in the North Sea. *Rapports et Procés-Verbaux des Réunions, Conseil International pour l'Exploration de la Mer*, **172**, 39–57.

Daan, N. (1980) A review of replacement of depleted stocks by other species and the mechanisms underlying such replacement. *Rapp. P.-v. Réunions Conseil Internationale pour l'Exploration de la Mer*, **177**, 405–21.

Daan, N., Bromley, P.J., Hislop, J.R.G. & Nielsen, N.A. (1990) Ecology of North Sea fish. *Netherlands Journal for Sea Research*, **26**, 343–86.

van Dijk, A.J., Hustings, F., Sierdsema, H. & Meijer, R. (1997a) *Kolonievogels en zeldzame broedvogels in Nederland in 1995. SOVON Monitoringrapport* 1997/06, 63 pp. SOVON, Beek-Ubbergen.

van Dijk, A.J., Hustings, F., Sierdsema, H. & Verstrael, T. (1997b) Kolonievogels en zeldzame broedvogels in Nederland in 1992–1993. *Limosa*, **70**, 11–26.

van Dijk, A.J., Hustings, F., Sierdsema, H. & Verstrael, T. (1997c) Kolonievogels en zeldzame broedvogels in Nederland in 1994. *Limosa*, **70**, 101–12.

van Dijk, A.J., Boele, A., Zoetebier, D. & Meijer, R. (1998) *Kolonievogels en zeldzame broedvogels in Nederland in 1996. SOVON Monitoringrapport* 1998/07, 70 pp. SOVON, Beek-Ubbergen.

Dunnet, G.M., Furness, R.W., Tasker, M.L. & Becker, P.H. (1990) Seabird ecology in the North Sea. *Netherlands Journal for Sea Research*, **26**, 387–425.

Evans, P.G.H. & Nettleship, D.N. (1985) Conservation of the Atlantic Alcidae. In: *The Atlantic Alcidae* (eds Nettleship D.N. & T.R. Birkhead), pp. 428–88. Academic Press, London/New York.

Fisher, J. (1952) *The Fulmar. Collins New Naturalist Series*, Facsimile 1984, 496 pp. Collins, London.

Furness, R.W. (1982) Competition between fisheries and seabird communities. *Advances in Marine Biology*, **20**, 225–307.

Furness, R.W. (1987) *The Skuas*, 363 pp. T. & A.D. Poyser, Calton.

Furness, R.W. (1990) A preliminary assessment of the quantities of Shetland sandeels taken by seabirds, seals, predatory fish and the industrial fishery in 1981–83. *Ibis*, **132**, 205–17.

Furness, R.W. (1992) *Implications of changes in net mesh size, fishing effort and minimum landing size regulations in the North Sea for seabird populations. Report to JNCC and Scottish Office*, 62 pp. Applied Ornithology Unit, Department of Zoology, University of Glasgow, Glasgow.

Furness, R.W. & Monaghan, P. (1987) *Seabird Ecology*. Blackie, Glasgow.

Furness, R.W., Hudson, A.V. & Ensor, K. (1988) Interactions between scavenging seabirds and commercial fisheries around the British Isles. In: *Seabirds & Other Marine Vertebrates: Competition, Predation and Other Interactions* (ed. J. Burger), pp. 240–68. Columbia University Press, New York.

Furness, R.W., Ensor, K. & Hudson, A.V. (1992) The use of fishery waste by gull populations around the British Isles. *Ardea*, **80**, 105–13.

Garthe, S. (1992) *Quantifizierung von Abfall und Beifang der Fischerei in der südöstlichen Nordsee und deren Nutzung durch Seevögel*, 111 pp. Diplomarbeit Mathematisch-Naturwissenschaftlichen Fakulteit, Institut für Meereskunde, Christian-Albrechts University, Kiel.

Garthe, S. (1993) Quantifizierung von Abfall und Beifang der Fischerei in der südöstlichen Nordsee und deren Nutzung durch Seevögel. *Hamburger avifaunistischen Beiträge*, **25**, 125–237.

Garthe, S. & Hüppop, O. (1993) Gulls and fulmars following ships and feeding on discards at night. *Ornis Svecica*, **3**, 159–61.

Garthe, S. & Hüppop, O. (1994) Distribution of ship-following seabirds and their utilization of discards in the North Sea in summer. *Marine Ecology Progress Series*, **106**, 1–9.

Garthe, S. & Hüppop, O. (1996) Nocturnal scavenging by gulls in the southern North Sea. *Colonial Waterbirds*, **19**, 232–41.

Garthe, S., Camphuysen, C.J. & Furness, R.W. (1996) Amounts of discards in commercial fisheries and their significance as food for seabirds in the North Sea. *Marine Ecology Progress Series*, **136**, 1–11.

Gauld, J.A., McKay, D.W. & Bailey, R.S. (1986) The current state of the Industrial Fisheries. *Fishing News*, 27 June 1987, 30–1.

Goodlad, J. (1989) Industrial fishing in Shetland waters. In: *Seabirds and Sandeels: Proceedings of a Seminar*, Lerwick, Shetland, 15–16 October 1988 (ed. M. Heubeck), pp. 50–9. Shetland Bird Club, Lerwick.

Hamer, K.C., Furness, R.W. & Caldow, R.G.W. (1991) The effects of changes in food availability on the breeding ecology of Great Skuas *Catharacta skua* in Shetland. *Journal of Zoology, London*, **223**, 175–88.

Hamre, J. (1988) Some aspects of the interrelation between the herring in the Norwegian Sea and the stocks of capelin and cod in the Barents Sea. *International Council for the Exploration of the Sea* C.M. 1988/H:42, Pelagic Fish Committee, Reference G.

Harris, M.P. (1970) Rates and causes of increase of some British gull populations. *Bird Study*, **17**, 325–35.

Harris, M.P. (1984) *The Puffin*, 224 pp. T. & A.D. Poyser, Calton.

Heubeck, M. (1988) Shetland's seabirds in dire straits. *British Trust for Ornithology News*, **158**, 1–2.

Heubeck, M. (ed.) (1989) *Seabirds and Sandeels: Proceedings of a Seminar*, Lerwick, Shetland, 15–16 October 1988, 81 pp. Shetland Bird Club, Lerwick.

Hopkins, C.C.E & Nilssen, E.M. (1991) The rise and fall of the Barents Sea capelin (*Mallotus villosus*): a multivariate scenario. *Polar Research*, **10**, 535–46.

Hudson, A.V. (1989) Interspecific and age-related differences in the handling time of discarded fish by scavenging seabirds. *Seabird*, **12**, 40–4.

Hudson, A.V. & Furness, R.W. (1988) Utilization of discarded fish by scavenging seabirds behind white fish trawlers in Shetland. *Journal of Zoology, London*, **215**, 151–66.

Hudson, A.V. & Furness, R.W. (1989) The behaviour of seabirds foraging at fishing boats around Shetland. *Ibis*, **131**, 225–37.

Hunt, G.L. & Furness, R.W. (eds) 1996. Seabird/fish interactions, with particular reference to seabirds in the North Sea. *International Council for the Exploration of the Sea Cooperative Research report* No. 216, 87 pp. International Council for the Exploration of the Sea, Copenhagen.

Hüppop, O. (1996) Die Brutbestände Helgoländer Seevögel von 1952 bis 1995. *Ornithologisch Jahresbericht Helgoland*, **6**, 72–5.

Hüppop, O. & Garthe, S. (1993) Seabirds and fisheries in the southeastern North Sea. *Sula*, **7**, 9–14.

Jones, R. (1983) The decline in herring and mackerel and the associated increase in other species in the North Sea. In: *Proceedings Expert Consultation to Examine Changes in Abundance and Species Composition of Neritic Fish Resources* (eds Sharp G.D. & Csirke J.), *Fisheries and Agriculture Organisation, Fisheries Reports*, **291**, 507–19.

Lid, G. (1981) Reproduction of the puffin on røst in the Lofoten Islands in 1964–1980. *Fauna Norvegica Serie C, Cinclus*, **4**, 30–9.

Lloyd, C., Tasker, M.L. & Partridge, K. (1991) *The Status of Seabirds in Britain and Ireland*, 355 pp. T. & A.D. Poyser, London.

Madsen, K.P. (1978) The industrial fisheries in the North Sea. *Rapports et Procés-Verbaux des Réunions, Conseil International pour l'Exploration de la Mer*, **172**, 27–30.

McCullagh, P. & Nelder, J.A. (1989) *Generalized Linear Models*, 2nd edn, 511 pp. Chapman and Hall, London.

Mearns, B. & Mearns, R. (1998) *The Bird Collectors*. Academic Press, San Diego.

Monaghan, P. (1992) Seabirds and sandeels: the conflict between exploitation and conservation in the northern North Sea. *Biodiversity and Conservation*, **1**, 98–111.

Monaghan, P., Uttley, J.D. & Burns, M.D. (1992a) Effect of changes in food availability on reproductive effort in Arctic terns *Sterna paradisaea*. *Ardea*, **80**, 71–81.

Monaghan, P., Uttley, J.D., Walton, P., Wanless, S., Hamer, K. & Burns, M.D. (1992b) The influence of changes in sand-eel availability on breeding seabirds. In: *Proceedings of the Seabird Group Conference 'European Seabirds'*, Glasgow 27–29 March 1992 (ed. M.L. Tasker), pp. 17–18. The Seabird Group, Sandy.

Nelder, J.A. & Wedderburn, R.W.M. (1972) Generalized linear models. *Journal of the Royal Statistical Society A*, **135**, 370–84.

Nelson, J.B. (1978) *The Gannet*, 336 pp. T. & A.D. Poyser, Berkhamsted.

Nörrevang, A. (1986) Traditions of sea bird fowling in the Faroes: an ecological basis for sustained fowling. *Ornis Scandinavica*, **17**, 275–81.

Rees, E.I.S. (1963) Marine birds in the Gulf of St. Lawrence and Strait of Belle Isle during November. *Canadian Field-Naturalist*, **77**, 98–107.

Sherman, K., Jones, C., Sullivan, L., Smith, W., Berrien, P. & Ejsymont, L. (1981) Congruent shifts in sand-eel abundance in western and eastern North Atlantic ecosystems. *Nature*, **291**, 487–9.

Spaans, A.L. (1998a) Breeding Lesser Black-backed Gulls *Larus graellsii* in The Netherlands during the 20th century. *Sula*, **12**, 175–84.

Spaans, A.L. (1998b) The herring gull *Larus argentatus* as a breeding bird in The Netherlands during the 20th century. *Sula*, **12**, 185–98.

Spaans, A.L., Bukacinska, M., Bukacinski, D. & van Swelm, N.D. (1994) *The relationship between food supply, reproductive parameters and population dynamics in Dutch Lesser Black-backed Gulls Larus fuscus: a pilot study. IBN Research Report* 94/9, 58 pp. Instituut voor Bos- en Natuuronderzoek, Arnhem.

Stone, C.J., Webb, A., Barton, C., Ratcliffe, N., Reed, T.C., Tasker, M.L., Camphuysen, C.J. & Pienkowski, M.W. (1995) *An Atlas of Seabird Distribution in North-west European Waters*, 326 pp. Joint Nature Conservation Committee, Peterborough, UK.

Stowe, T.J. & Harris, M.P. (1984) Status of guillemots and razorbills in Britain and Ireland. *Seabird*, **7**, 5–18.

Tasker, M.L., Jones, P.H., Dixon, T.J. & Blake, B.F. (1984) Counting seabirds at sea from ships: a review of methods employed and a suggestion for a standardized approach. *Auk*, **101**, 567–77.

Tasker, M.L., Webb, A., Hall, A.J., Pienkowski, M.W. & Langslow, D.R. (1987) *Seabirds in the North Sea*, 336 pp. Nature Conservation Council, Peterborough, UK.

Thompson, K., Brindley, E. & Heubeck, M. (1996) *Seabird Numbers and Breeding Success in Britain and Ireland, 1995. UK Nature Conservation* No. 20, 55 pp. Joint Nature Conservation Committee, Peterborough, UK.

Thompson, K., Brindley, E. & Heubeck, M. (1997) *Seabird Numbers and Breeding Success in Britain and Ireland, 1996. UK Nature Conservation* No. 21, 55 pp. Joint Nature Conservation Committee, Peterborough, UK.

Thompson, K., Brindley, E. & Heubeck, M. (1998) *Seabird Numbers and Breeding Success in Britain and Ireland, 1997. UK Nature Conservation* No. 22, 60 pp. Joint Nature Conservation Committee, Peterborough, UK.

Tjelmeland, S. (1989) Cod–capelin interactions in the Barents Sea: a new light on past management of Capelin. In: *Abstracts and Scientific Papers & Posters of the Symposium on Multispecies Models Relevant to Management of Living Resources*, The Hague, The Netherlands, 2–4 October 1989. *International Council for the Exploration of the Sea, MSM Symposium*, No. 20.

Toft, G.O. (1983) Changes in the breeding seabird population in Rogaland, S.W. Norway, during 1949–1979. *Fauna Norvegica Serie C. Cinclus*, **6**, 8–13.

Vader, W. (1988) A challenge for science. *Naturopa*, **60**, 25–6.

Vader, W., Anker-Nilssen, T., Bakken, V., Barrett, R. & Strann, K.B. (1989) Regional and temporal differences in breeding success and population development of fish-eating seabirds in Norway after collapses of herring and capelin stocks. *Transactions 19th IUGB Congress*, Trondheim 1989, pp. 143–50.

Wahl, T.R. & Heinemann, D. (1979) Seabirds and fishing vessels: co-occurrence and attractions. *Condor*, **81**, 390–6.

Part 4
Long-term changes associated with fishing

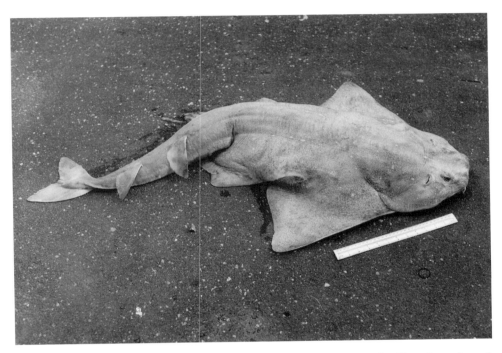

The angel shark *Squatina squatina* is a large (1.5–2.0 m long) bottom-dwelling elasmobranch which appears to have undergone a decline in abundance and is now only occasionally caught in British waters. This mature male is shown against a 30 cm ruler. (Reproduced with the permission of Stuart Rogers.)

Chapter 12

Distribution of macrofauna in relation to the micro-distribution of trawling effort

J.A. CRAEYMEERSCH[1], G.J. PIET[1], A.D. RIJNSDORP[1] and J. BUIJS[2]

[1]*Netherlands Institute for Fisheries Research (RIVO-DLO), PO Box 68, 1970 AB IJmuiden, The Netherlands*
[2]*Netherlands Institute of Ecology – Centre for Estuarine and Coastal Ecology, PO Box 140, 4400 AC Yerseke, The Netherlands*

Summary

1. Information on the micro-scale distribution of fishing activities on the Dutch Continental Shelf was derived from automated position registration systems. This enabled a better assessment of their impact on the benthic fauna.
2. A direct gradient analysis points to a globally significant difference in species composition between intensively fished and less heavily fished locations. It is, however, very likely that the major part of these differences is not related to differences in trawling effort but to differences in environmental factors.
3. Differences in fishing effort between areas best explained the differences that occurred in spionid worm densities.

Keywords: benthic community, fishing effort, spionid worms.

Introduction

Benthic invertebrates comprise a large proportion of the catch of mobile demersal fishing gears, such as beam trawls, that target flatfishes in the North Sea. The weight of by-catch of in- and epifaunal species is often several times the amount of marketable fish in the catch. The species composition of the by-catch depends largely on the faunal composition at the trawling site, but tends to be dominated by starfish, heart urchins and crabs (Lindeboom & de Groot, 1998). Of those species that are discarded into the sea, a fraction will not survive the capture and sorting process. This mortality is species dependent: <10% for starfish and brittlestars, 50–70% for most crustaceans and almost 90% for the bivalve *Arctica islandica*. However, the catch efficiency of commercial trawls for these species is low and therefore the overall mortality is very low when expressed as a percentage of the initial density of these animals. A larger fraction of the mortality occurs in the trawl path, because many animals that are not caught in the net are damaged or killed by the fishing gear as it passes over the seabed. Thus, the total direct mortality varies from 10% from 80%, and fragile or superficial living species experience the highest mortalities (Lindeboom & de Groot, 1998; Bergman & van Santbrink, this volume, Chapter 4).

The long-term impact of bottom fisheries on a particular species will depend on a combination of the direct mortality at each fishing event, the distribution of the fishing effort, the distribution of that species, and its life-history characteristics such as longevity and fecundity. Long-living species with a low fecundity will be affected more than short-living species with high fecundity. Benthic scavengers may benefit from the additional food supply from discards or animals damaged in the trawl path (Kaiser & Spencer, 1994; Lindeboom & de Groot, 1998; Collie *et al.*, 1997; Groenewold & Fonds, unpublished data; Fonds & Groenewold, this volume, Chapter 9; Ramsay *et al.*, this volume, Chapter 10). The longer-term effects of fisheries on the benthic communities may be evaluated from long-term trends in benthos or by-catch data (Schroeder & Knust, unpublished data; Philippart, 1998). They may also be inferred from comparisons between fished and unfished areas (Hall *et al.*, 1993; Lindeboom & de Groot, 1998; Tuck *et al.*, 1998).

Since 1993, the spatial distribution of the Dutch beam trawl fleet has been studied on a scale of 1×1-n mile squares (Rijnsdorp *et al.*, 1998; Piet *et al.*, unpublished data). Previously, information on fishing activities was limited to a scale of approximately 30×30 n miles (ICES rectangles). Rijnsdorp *et al.* (1998) showed that, in eight of the most heavily fished ICES rectangles of the North Sea, 47–71% of the surface area was trawled one to five times a year, 9–44% less than once a year, and 0–4% between 10% and 50% of the year. This detailed information on the distribution of fishing effort offers a novel opportunity to compare the benthic fauna of areas subjected to different levels of fishing disturbance. This chapter reports a first evaluation of differences in the macrobenthic infauna with respect to differences in fishing effort. First, we will focus on changes in species composition along a gradient in fishing effort. Secondly, we will focus on differences in the density of Spionidae, a polychaete family of which many species are known to be opportunists, often being the first species to colonise disturbed sediments (Gudmundsson, 1985).

Methods

Trawling effort

The micro-scale distribution of the beam trawling activities of a representative samples of the Dutch fleet has been studied since 1993 (Rijnsdorp *et al.*, 1998). In the period 1993–1995, 25 beam-trawl vessels (24 of them with an engine power > 300 hp or ≈221 kW) were equipped with an automated position registration system (APR), which records their position every 6 min at an accuracy of ±180 m. In 1996, another six vessels (five of them with an engine power <300 hp) were equipped with an APR. By calculating vessel speed, it is possible to estimate the fishing position of the sampled vessels, assuming a fishing speed of respectively 4.3 and 6 n miles h^{-1} for ships with an engine power of less and more than 300 hp, respectively. For the present study, the total number of APR data recorded in each 1×1-n mile rectangle during a 4-year period (1-4-1993 until 31-3-1997) were used.

Macrobenthos

During the last decade, the macrobenthic infauna on the Dutch Continental Shelf was surveyed in great detail. During the period 1985–1993, samples were taken from about 800 stations. All surveys were carried out in spring and used the same type of sampling equipment. In most surveys, a Reineck boxcorer was used and sometimes Van Veen grabs. The contents of the boxcorer or grab was washed over a sieve with round holes of 1 mm diameter. The work resulted in an atlas describing the occurrence and distribution of the most common species, together with relevant information on their ecology (Holtmann *et al.*, 1996a). The reader should refer to this atlas for more information on the different surveys and their methodology.

Statistical analyses

For the purpose of this study, sub-areas were selected for further analyses, based on the following conditions. First, the fishing effort within an area was not randomly distributed in space. Secondly, the faunal composition within an area had to be more or less similar. Based on the macrobenthos data mentioned above, the Dutch Continental Shelf can be divided into four sub-areas (Holtmann *et al.*, 1996b): the southern part of the Dogger Bank, the Oysterground, the southern offshore area and the coastal area. This spatial pattern has been consistent over time (Holtmann *et al.*, 1998). The coastal area, generally fished by smaller vessels (<300 hp), was excluded from this study because, prior to 1996, only a single small vessel was sampled and, secondly, different gear types might be used by smaller vessels during the year. On the relatively small Dogger Bank area, macrobenthic data were too sparse to allow further analysis. Finally, two sub-areas, one situated in the Oysterground and one in the offshore area, were chosen for detailed analysis. Figure 12.1 gives the location of the infauna sampling locations (79 samples in the offshore area, 129 samples in the Oysterground area) superimposed by the fishing effort data for those locations.

The relationship between the (log-transformed) total density of spionids and fishing effort was measured by a linear regression, and the significance tested. The few stations without spionids (two and 15 in the offshore and Oysterground area, respectively) were excluded from the analyses.

To determine whether there is a relationship between fishing effort distribution and infaunal community structure, a direct gradient analysis was performed. In a direct gradient analysis, the species composition is directly related to measured environmental variables: the first axis of the ordination is constructed in such a way as to optimise explicitly the fit to the supplied environmental data (ter Braak & Prentice, 1988). In a partial canonical ordination, the effect of one or more covariables can be factored out. The result is an ordination of the residual variation in the species data that remains after fitting the effects of the covariables. This is especially interesting in our study, as we are not interested in environmental variation but wanted to focus on the species' responses to fishing disturbance. Here,

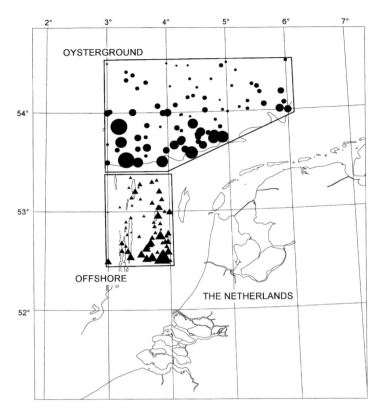

Fig. 12.1 Location of the sample locations (●, Oysterground area; ▲, offshore area) with superimposed symbols that are scaled to represent increasing fishing effort.

we removed possible effects of depth and sediment characteristics (silt content, median grain size). We used two ordination methods: partial canonical correspondence analysis (CCA) and partial redundancy analysis (RDA). RDA, the canonical form of principal components analysis (PCA), is based on a linear response model between species and environmental variables (e.g. as expected in short segments of ecological gradients); CCA is based on a unimodal response model. RDA and CCA allow measurement of the amount of variation in the species data that can be explained by the environmental variables. The significance of this relationship can be tested by a Monte Carlo permutation test. More information about these techniques can be found in Jongman *et al.* (1987), ter Braak (1988a), ter Braak & Prentice (1988), Borcard *et al.* (1992) and Palmer (1993). Macrobenthic abundance data were log-transformed prior to analysis. Species found at less than 10 stations were excluded from the analyses, resulting in 50 and 92 species in the offshore and Oysterground areas, respectively. All analyses were done with the CANOCO program of ter Braak (1988b), version 3.10.

Results

The (log-transformed) total density of spionids significantly increased with increasing fishing effort, both in the offshore area ($R^2 = 0.120$; $P < 0.001$) and in the Oysterground area ($R^2 = 0.034$; $P = 0.027$) (Fig. 12.2).

Figure 12.3 shows the ordination diagrams (site scores determined by weighted averaging) resulting from the partial RDA and partial CCA analyses. In these plots, stations with a similar species composition are closest to each other, stations that are dissimilar in species composition are furthest apart. Stations represented by larger circles experienced a higher level of fishing disturbance. In the offshore area, fishing effort accounted for 2.0% (RDA) and 2.2% (CCA) of the variance remaining after removing the effect of depth and sediment characteristics (silt content, median grain size). The Monte Carlo permutation test on the first axis was significant at probability levels of <0.01 and 0.01, respectively. In the Oysterground area, fishing effort accounted for 0.9% (RDA) and 1.5% (CCA) of the residual variance. For both RDA and CCA analyses, the Monte Carlo permutation test on the first axis was significant at a probability level of <0.01. The covariables explained 6–10% of the total variance in the species data.

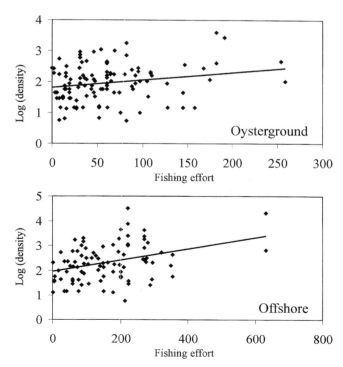

Fig. 12.2 Relationship between fishing effort and density (ind m^{-2}) of spionids.

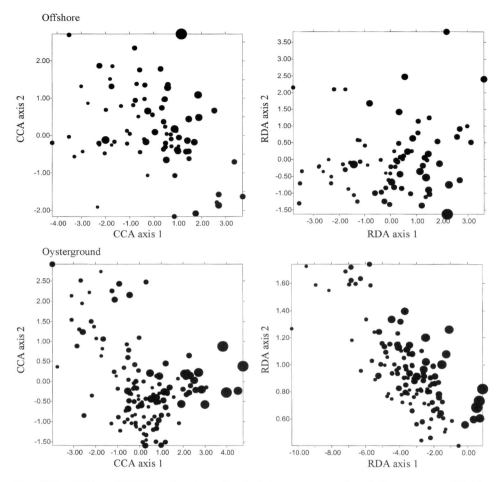

Fig. 12.3 CCA and RDA ordinations. Symbol size represents the relative amount of fishing effort.

Discussion

The assessment of the impact of trawling on the benthic fauna of the North Sea has previously been seriously hampered by the low resolution of information on the spatial distribution of fishing effort (ICES subrectangles). Estimated trawling frequencies (Welleman, 1989; Lindeboom, 1995) have been based on the assumption that fishing effort in a rectangle was homogeneously distributed. In reality, however, this is not the case, and fishing activities tend to be concentrated in small areas, as is shown by Rijnsdorp *et al.* (1998) for the Dutch fleet. Data for the benthic fauna of the Dutch Continental Shelf data are available on a fine spatial resolution (Holtmann *et al.*, 1996) and, thus, we had the opportunity to investigate whether the small-scale patterns in fishing effort are reflected in the benthic community.

The present study certainly has some limitations. First, we only considered infaunal species, while epifaunal taxa may be more vulnerable to fishing disturbance. Secondly, the most vulnerable species may have already decreased earlier this century. Finally, our analysis might have suffered from the fact that the benthos data and the effort data did not cover the same period. During the period 1993–1996, however, the micro-distribution of the sampled vessels showed a remarkable consistency (Rijnsdorp *et al.*, 1998). In a study of all data from the Dutch Continental Shelf in the period 1986–1997, Holtmann *et al.* (1998) found that the species compositions in the 1986 and 1990 observations were different from later years. However, most of the data used in this study were from 1991 to 1993. Thus, our implicit assumption that the trawling effort distribution within each area was similar in the period that macrobenthic sampling took place is most likely not violated.

The results of our study at first seem to point to a significant impact of bottom trawling. The total density of spionids increases with increasing fishing disturbance. In addition, there is a significant, although small, relationship between the species composition at each site and fishing effort that occurs at that site.

The direct gradient analyses point to an overall significant difference in species composition between intensively fished and less heavily fished locations. In the offshore area, stations with a high abundance of species such as *Pseudocuma longicornis* and *Urothoe brevicornis* are situated on the left side of the ordination diagram, and stations with a high abundance of the species *Sigalion mathildae* and *Atylus swammerdami* are situated on the right side of the diagram. In the Oysterground area, stations with a high abundance of species such as *Sigalion mathildae*, *Ampelisca brevicornis* and *Arctica islandica* are located on the left side, and stations with a high abundance of *Urothoe poseidonis* on the right side of the diagram. The fishing effort gradient tends to run from left to right, suggesting that population sizes of, for example, *A. islandica* are decreasing in response to increasing fishing pressure, while those of, for example, *A. swammerdami* are increasing. *Ampelisca brevicornis* is a tube-dwelling amphipod. The tubes may rise above the surrounding bottom surface (e.g. see Fig. 6B in Valente *et al.*, 1992). Females reproduce once and carry the eggs in their marsupium. It is clear that bottom trawling will destroy the tubes. Although it is possible that only a part of the population is affected, as *A. brevicornis* can enter the water column (Klein *et al.*, 1975), frequent trawling will certainly inhibit re-colonisation by this species.

The quahog *Arctica islandica* only occurs north of the 30-m depth contour. The direct effects of beam trawling on populations of *A. islandica* have been illustrated clearly by Witbaard & Klein (1995) and Witbaard (1997). In the Oysterground area, few specimens have undamaged shells, and their mortality rate seems to be higher than that in the northern North Sea or western Atlantic. Witbaard & Klein (1994) observed that an increase in the occurrence of scars in the shells of *A. islandica* was noted in the 1980s coincident with an increase in the number of larger fishing vessels in the Dutch sector of the North Sea. The quahog is a slow-growing species that lives for many years and has irregular recruitment (Rees & Dare, 1993), which makes the population very sensitive to disturbance.

Thus, for the two species mentioned above, the observed negative relationship with fishing effort is to be expected. For *A. islandica*, however, most of the specimens collected were very small (0-group year class). Thus, the differences in distribution might also be the result of differences in spatfall. For the other species mentioned above, the relationship with fishing effort is hard to explain. *Atylus swammerdami*, *Pseudocuma longicornis*, *Urothoe brevicornis* and *U. poseidonis* are highly mobile and free swimming (Watkin, 1939; Jones, 1976; Cattrijsse *et al.*, 1993; Holtmann *et al.*, 1996a) and hence are not vulnerable to sediment disturbance. Moreover, these species have a short life cycle and at least one of them, the cumacean *P. longicornis*, breeds twice a year. Thus, their life-cycle strategy ensures a long-term survival. Surprisingly, *A. swammerdami* and *U. poseidonis* are found in higher densities in the most intensively trawled areas, while *P. longicornis* and *U. brevicornis* have higher abundances in the less heavily trawled areas. The relationship between the occurrence of certain species and trawl disturbance is clearly more complex that would first appear.

It is notable that, in the offshore area, the densities of the polychaete *Sigalion mathildae* are higher at the most intensively trawled locations, while, in the Oysterground area, their abundances are lowest at these locations. As *S. mathildae* lives 15–20 cm beneath the sediment surface, we would not expect it to be sensitive to fishing disturbances.

For most benthic species, however, it is difficult to indicate whether they are likely to be vulnerable or not. Data on the life-history parameters are either lacking or are scattered through often grey literature. But the data on the species discussed above suggest that, in the areas studied, the relationship between fishing effort and community structure may be largely correlative and not causal. The patterns found are, therefore, likely to be due to one or more environmental variables not included in our study. Structural differences in the environment may affect the fishing strategies of fishers either directly or indirectly, i.e. some parts of the studied areas might be more attractive to fish than others. Thus, as noted by Hall *et al.* (1993), the fact that 'unfished' areas are usually unfished is precisely because they differ from real fishing grounds, further complicates the interpretation of such studies. When the same environmental factors determine both species composition and fishing effort distribution, it is probably impossible to discriminate between environmental impact and fishery impact. However, if the environmental actors are factored out in a partial canonical analysis, the effect of fishing effort will at least be underestimated. While this may have occurred in our study by partialling out the effects of depth and grain size, nevertheless, the first axis in the canonical analyses with fishing effort as the only environmental variable, but without covariables, did not explain a larger part of the variance compared to the partial analyses.

Opportunistic species, among them many members of the polychaete family Spionidae, are characterised by high growth rates, a short life span, a low reproductive age and a large reproductive output. These characteristics enable them to adapt rapidly to environmental perturbation and quickly re-colonise disturbed habitats (Grassle & Grassle, 1974; Gudmundsson, 1985). For several areas of the North Sea, an increase in the abundance of opportunistic species has been reported and has been explained as an effect of eutrophication, pollution and fisheries (Rachor, 1990; Kröncke, 1992; Gray *et al.*, 1996). However, benthic communities often have such high natural variability that it is often not possible to distinguish between the different influences acting on that community. This is certainly true for highly dynamic areas such as the offshore area, that are already characterised by relatively opportunistic species (Holtmann *et al.*, 1998). Nevertheless, a positive relationship between spionid density and fishing effort was also found here.

Another important factor causing a shift in the benthic community from low productivity, long-lived species to high productivity, short-lived species is eutrophication (Rachor, 1990). The importance of both eutrophication and beam trawling in the benthic community shift can be inferred from changes in growth of plaice and sole (Rijnsdorp & van Beek, 1991; Rijnsdorp & van Leeuwen, 1996). The overall changes in growth suggest increased food availability in the southern North Sea. They can be broken down into effects of eutrophication (decreasing with increasing distance from the shore) on juveniles and of beam trawling (of minor importance within the 12-mile zone) on older fishes in deeper water, whereas both factors may play a role during an intermediate life phase. Therefore, in the sub-areas studied, differences in the fishing effort are the most likely factor to explain differences in spionid densities.

Acknowledgements

This work was financially supported by BEON (Dutch co-operation on Policy-Linked Ecological Research on the North Sea and the Wadden Sea). The authors would like to thank Magda Bergman, Jan van Santbrink and Peter Herman for their co-operation throughout the study. An anonymous referee made helpful comments.

References

Borcard, D., Legendre, P. & Drapeau, P. (1992) Partialling out the spatial component of ecological variation. *Ecology*, **73**, 1045–55.

ter Braak, C.J.F. (1988a) CANOCO – an extension of DECORANA to analyse species-environment relationships. *Vegetatio*, **75**, 159–60.

ter Braak, C.J.F. (1988b) Partial canonical correspondence analysis. In: *Classification and Related Methods of Data Analysis* (ed. H.H. Bock), pp. 551–8. Elsevier Science, Amsterdam.

ter Braak, C.J.F. & Prentice, I.C. (1988) A theory of gradient analysis. *Advances in Ecological Research*, **18**, 272–318.

Cattrijsse, A., Mees, J. & Hamerlynck,O. (1993) The hyperbenthic Amphipoda and Isopoda of the Voordelta and the Westerschelde estuary. *Cahier de Biologie Marine*, **34**, 187–200.

Collie, J.S., Escanero, G.A. & Valentine, P.C. (1997) Effects of bottom fishing on the benthic megafauna of Georges Bank. *Marine Ecology Progress Series*, **155**, 159–72.

Gudmundsson, H. (1985) Life history patterns of polychaete species of the family Spionidae. *Journal of the Marine Biology Association of the United Kingdom*, **65**, 93–111.

Grassle, J.F. & Grassle, J.P. (1974) Opportunistic life histories and genetic systems in marine benthic polychaetes. *Journal of Marine Research*, **32**, 253–84.

Hall, S.J., Robertson, M.R., Basford, D.J. & Heaney,S.D. (1993) The possible effects of fishing disturbance in the northern North Sea: an analysis of spatial patterns in community structure around a wreck. *Netherlands Journal of Sea Research*, **31**, 201–8.

Holtmann, S.E., Groenewold, A., Schrader, K.H.M., Asjes, J., Craeymeersch, J.A., Duineveld, G.C.A., van Bostelen, A.J. & van der Meer, J. (1996a) *Atlas of the zoobenthos of the Dutch Continental Shelf.* Ministry of Transport, Public Works and Water Management, North Sea Directorate, Rijkswijk, The Netherlands.

Holtmann, S.E., Belgers, J.J.M., Kracht, B. & Daan, R. (1996b) *The macrobenthic fauna in the Dutch sector of the North Sea in 1995 and a comparison with previous data. NIOZ Report*, 1996-8. Netherlands Institute for Sea Research, Den Burg, Texel, The Netherlands.

Holtmann, S.E., Duineveld, G.C.A., Mulder, M. & de Wilde, P.A.W.J. (1998) *The macrobenthic fauna in the Dutch sector of the North Sea in 1997 and a comparison with previous data. NIOZ Report* 1998-5. Netherlands Institute for Sea Research, Den Burg, Texel, The Netherlands.

Jones, N.S. (1976) *British Cumaceans. Synopses of British Fauna (New Series)* No. 7. Academic Press, London.

Kaiser, M.J. & Spencer, B.E. (1994) Fish scavenging behaviour in recently trawled areas. *Marine Ecology Progress Series*, **112**, 41–9.

Klein, G., Rachor, E. & Gerlach, S.A. (1975) Dynamics and productivity of two populations of the benthic tube-dwelling amphipod *Ampelisca brevicornis* (Costa) in Helgoland Bight. *Ophelia*, **14**, 139–59.

Lindeboom, H.J. (1995) Protected areas in the North Sea: an absolute need for future marine research. *Helgoländer Meeresuntersuchungen*, **49**, 591–602.

Lindeboom, H.J. & de Groot, S.J. (eds) (1998) *The effects of different types of fisheries on the North Sea and Irish Sea benthic ecosystems. NIOZ Report* 1998-1/*RIVO-DLO Report* C003/98. Netherlands Institute for Sea Research, Den Burg, Texel, The Netherlands.

Olsgard, F. & Gray, J.S. (1995) A comprehensive analysis of the effects of offshore oil and gas exploration and production on the benthic communities of the Norwegian continental shelf. *Marine Ecology Progress Series*, **122**, 277–306.

Palmer, M.W. (1993) Putting things in even better order: the advantages of canonical correspondence analysis. *Ecology*, **74**, 2215–30.

Philippart, C.J.M. (1998) Long-term impact of bottom fisheries on several by-catch species of demersal fish and benthic invertebrates in the south-eastern North Sea. *ICES Journal of Marine Science*, **55**, 342–52.

Rachor, E. (1990) Changes in sublittoral zoobenthos in the German Bight with regard to eutrophication. *Netherlands Journal of Sea Research*, **25**, 209–14.

Rees, H.L. & Dare, P.J. (1993) *Sources of mortality and associated life-cycle traits of selected benthic species: a review. Fisheries Research Data Report* No. 33. Ministry of Agriculture, Fisheries and Food, Lowestoft, UK.

Rijnsdorp, A.D. & van Beek, F.A. (1991) Changes in growth of plaice *Pleuronectes platessa* L. and sole *Solea solea* (L.) in the North Sea. *Netherlands Journal of Sea Research*, **27**, 441–57.

Rijnsdorp, A.D. & van Leeuwen, P.I. (1996) Changes in the growth of North Sea plaice since 1950 in relation to density, eutrophication, beam-trawl effort, and temperature. *ICES Journal of Marine Science*, **53**, 1199–213.

Rijnsdorp, A.D., Buys, A.M., Storbeck, F. & Visser, E.G. (1998) Micro-scale distribution of beam trawl effort in the southern North Sea between 1993 and 1996 in relation to the trawling frequency of the sea bed and the impact on benthic organisms. *ICES Journal of Marine Science*, **55**, 403–19.

Tuck, I., Hall, S.J., Robertson, M.R., Armstrong, E & Basford, D.J. (1998) *Marine Ecology Progress Series*, **162**, 227–42.

Valente, R.M., Rhoads, D.C., Germano, J.D. & Cabelli, V.J. (1992) Mapping of benthic enrichment patterns in Narragansett Bay, Rhode Island. *Estuaries*, **15**, 1–17.

Watkin, E.E. (1939) The pelagic phase in the life history of the amphipod genus *Bathyporeia*. *Journal of the Marine Biology Association of the United Kingdom*, **23**, 467–81.

Welleman, H. (1989) *Literatuurstudie naar de effecten van de bodemvisserij op de bodem en het bodemleven. Report Netherlands Institute for Fisheries Research (RIVO-DLO), MO* 89-201, 58 pp. Netherlands Institute for Fisheries Research, IJmuiden, The Netherlands.

Witbaard, R. (1997) Tree of the sea. The use of the internal growth lines in the shell of *Arctica islandica* (Bivalvia, Mollusca) for the retrospective assessment of marine environmental change. Ph.D. dissertation, Groningen.

Witbaard, R. & Klein, R. (1994) Long-term trends on the effects of the southern North Sea beamtrawl fishery on the bivalve mollusc *Arctica islandica* (L.) (Mollusca, Bivalvia). *ICES Journal of Marine Science*, **51**, 99–105.

Chapter 13
Long-term changes in North Sea benthos: discerning the role of fisheries

C.L.J. FRID and R.A. CLARK

Dove Marine Laboratory, University of Newcastle, Cullercoats, North Shields, NE30 4PZ, UK

Summary

1. Fishing occurs at the scale of ocean basins and has been going on for millennia. The scale and intensity of fishing has expanded in the last 100 years with the mechanisation of the fleet and the development of better navigational and vessel technology.
2. Fishing activities interact with the benthos through direct mortality of benthos as by-catch and net damaged organisms and inputs of organic matter in the form of carcasses and offal, and indirectly through alterations in sediment characteristics, altered sediment–water column fluxes, and changes in predation rates through changed abundance and size structure of populations of predatory fish.
3. Separating the effects of fishing from other long-term sources of variation in benthic communities is difficult. However, application of a precautionary approach to ecosystem management would suggest that action needs to be taken when there is sufficient weight of evidence.
4. Current data suggest reduced abundances of long-lived bivalves and increased abundances of scavenging crustacea and sea stars in the German Bight, and altered benthic community composition on at least some fishing grounds. There are also likely to have been major changes in the predation pressure applied by fish to the benthos. This suggests that both direct and indirect effects are manifested in the most intensively fished areas of the North Sea.
5. Managers must recognise that a healthy ecosystem is a requirement and aim of existing international agreements and a prerequisite for healthy fish populations. To date, fisheries management has failed adequately to protect the target species, we should now seek methods that also provide protection to the wider ecosystem and its functions.

Keywords: benthos, long term, indirect effects, fishing, direct mortality, fish predation.

Introduction

Ecosystems are continually changing, and these changes are driven by a multiplicity of factors. Some of these concern the internal dynamics of multispecies systems (see De Angelis & Waterhouse, 1987), while others are the result of abiotic extrinsic factors such as climate. Interactions between a diversity of internal and external factors and the fact that the responses to most are non-linear results in the complex behaviour of biological systems over time (De Angelis & Waterhouse, 1987).

The multiplicity of factors that affect ecosystems makes it difficult to assess both the relative importance of any single factor (their influence may itself be variable with time), anthropogenic or otherwise to the dynamics of the system. The most straightforward approach is the demonstration of changes in biological systems that occur in phase with changes in presumed causative factors (Pearson & Rosenberg, 1986; Buchanan & Moore, 1986b; Aebischer *et al.*, 1990; Austen *et al.*, 1991; Taylor *et al.*, 1992; Frid *et al.*, 1996; Frid & Huliselan, 1996). The existence of such correlations are not, however, powerful tools for identifying causative factors (Underwood, 1990, 1992, 1996). For small-scale phenomena, experimental approaches in which a single factor is manipulated in a controlled manner provide the strongest inferences regarding causality. However, such experimental approaches are on the scale of a few square metres at most, whereas marine systems cover areas of hundreds of thousands of hectares. Thus, such experimental investigations are not practical at both the large temporal and spatial scales of many marine systems (Mann & Lazier, 1991). Fishing has occurred in the coastal seas of the world for thousands of years and, as such, the ecological impacts of fishing may be both wide ranging and long term.

Impacts of fishing

The effect of fisheries on target species is well established (Pope & Macer, 1996; Rijnsdorp & Millner, 1996). Recent studies (see Bergman & van Santbrink, this volume, Chapter 4) demonstrate that the direct mortality caused by beam trawling, estimated as the total mortality associated with one fishing event, was species dependent and varied from 10% to 40% in gastropods, starfish, crustaceans and annelid worms, from 10% to 50% for the sea urchin *Echinocardium cordatum* and the masked crab *Corystes cassavelaunus*, and from 30% to 80% for a number of bivalves. At the population level, the mortality imposed by the trawl fishery will depend on the level of direct mortality, the trawling frequency and the overlap in spatial distribution between the fishery and the benthic organisms.

All net fisheries result in catches composed of target and non-target species, and individuals of the target species that are unmarketable, owing to size or other considerations. This material is generally returned to the sea as 'discards'. In addition, the processing of the catch at sea generates additional offal in the form of heads, guts etc. Seabirds feed on discarded fish and offal world-wide (Hill & Wassenberg, 1990; Wassenberg & Hill, 1990). Studies in the North Sea show that birds primarily take offal and discarded round fish and smaller proportions of flatfish, cephalopods and benthic invertebrates when these are discarded (Camphuysen *et al.*, 1993, 1995; Camphuysen & Garthe, this volume, Chapter 11). Discards that are not consumed at the surface may be taken in the water column by pelagic fish and marine mammals (Hill & Wassenberg, 1990; Wassenberg & Hill, 1990) or by benthic scavengers (Kaiser & Spencer, 1994; Ramsay *et al.*, 1996, 1998).

Although indirect ecosystem effects of fishing are less obvious than the direct effects, they may be more important in structuring benthic communities (Kneib, 1991). The indirect effects of fishing include changes in nutrient cycling caused by physical disturbance of the sediment–water interface and the addition of labile organic matter (discards and offal) to the system, the continued transfer of fixed carbon from the marine environment to the terrestrial system (Camphuysen *et al.*, 1995) and the changes in the food chain arising from manipulation of the density and size structure of the target populations (ICES, 1998). The benthos play an important role in remineralisation and release of nutrients to the water column (Rowe *et al.*, 1975). The rate of this remineralisation is critically dependent on the oxidation state of the sediment (Prins & Smaal, 1990; Sørensen, 1978). Physical disruption by towed bottom gears redistributes sediments and temporally alters the redox state of the system.

Exploitation of fish stocks has altered the abundance of fish in the seas and, frequently, the size composition of the fish populations (Pope *et al.*, 1988; Pope & Macer, 1996). Marine communities frequently exhibit size-structured food webs and these changes are therefore likely to lead to changes in the quantities and types of prey consumed. This phenomenon is often referred to as 'trophic cascades' and is well established for limnetic systems (Carpenter, 1988). In the marine environment, the greater difficulty of carrying out the necessary experimental studies means that the existence of trophic cascades has not yet been established, although a number of properties of marine communities suggest that they may occur (ICES, 1998).

Some of the most dramatic examples of ecosystem changes arising from fisheries exploitation of the controlling predators are the changes in the abundance and predation of krill in the Antarctic following exploitation of the baleen whales (Dayton *et al.*, 1995). Other examples that demonstrate a cascade of effects through the ecosystem following heavy fishing mortality or destructive fishing practices include the changes induced in the Californian kelp forests following hunting of the sea otters (Simenstad *et al.*, 1978; Estes, 1996), the changes in the intertidal areas of Chile induced by the removal of predators by fishers (Moreno *et al.*, 1986) and the changes induced by the rapid development of the demersal fishery in the Gulf of Thailand (Pauly, 1988). Yet these studies provide only indirect evidence of cause and effect and rely on inferential arguments and deductive reasoning, as it is difficult to establish the presence of true control areas for comparative purposes (Dayton *et al.*, 1995).

Taking account of the patchy distribution of beam trawl effort (e.g. Rijnsdorp *et al.*, 1998), studies of the annual direct fishing mortality rates on benthic invertebrates in the southern North Sea were estimated at between 7% and 45% of the individuals (de Groot & Lindeboom, 1994). Compared with the estimated percentage of the benthic production that is consumed by fish predators (~45%), the estimated direct fishing mortality rates are the same or lower. The combination of direct fishing mortality rates and indirect changes in predation pressure support the hypothesis that intensive trawling may have caused shifts in benthic assemblages from large, slowly reproducing species to small species with a high reproductive rate, e.g. polychaete species. The proliferation of polychaetes in the North Sea may have played a role in the increase in growth rate observed in some bottom-dwelling flatfish (de Veen, 1976; Millner & Whiting, 1996; Rijnsdorp & van Leeuwen, 1996).

Fish play a central structuring role in the ecology of many aquatic systems (Hansson, 1985). Alterations in fish abundance, size distribution, or spatial or temporal distributions can induce changes in other aspects of the ecosystem through changes in the strength or direction of the ecological links. In many marine ecosystems, there is evidence of species other than the prey that are influenced by changes in fish predation (Parsons, 1991, 1992, 1996; Verity & Smetacek, 1996; ICES, 1998).

Fisheries impose size-selective predation on the stocks and as such exploited populations tend to have different size spectra when compared with unfished populations (e.g. Pope *et al.*, 1988). There is also empirical evidence that the heavier the exploitation of a stock, the steeper the slope of the log numbers per size class vs. size relationship (Pope *et al.*, 1988; Rice & Gislason, 1996). The ICES (1998) considers this to be a useful indicator of changes in fishing effort, but notes that this relationship is sensitive to environmental changes that alter growth rates or may lead to species replacements. The ecological consequences of these changes are difficult to predict, but include altered predation rates as small and large fish rarely feed on the same prey (see below).

Fishing also alters the species composition of the fish guild. In the North Sea, populations of the benthic feeding gadoids, cod, haddock and whiting, have declined over the last 30 years or so (Pope & Macer, 1996; Serchuk *et al.*, 1996; Greenstreet & Rogers, this volume, Chapter 14). During the same period, populations of non-target species such as long rough dab, common dab and lemon sole have increased (Heesen & Daan, 1996). Therefore, there is a need to provide an assessment the consequences of these changes for the level of predation pressure exerted on the benthos and changed strength of ecological interactions within the fish guild.

Other long-term factors influencing the benthos

Fishing is not the only factor to influence benthic ecosystems on decadal, or longer time scales and at the basin-wide scale. Other influences that effect large areas over extended periods include climatic/meteorological factors and anthropogenic changes in nutrient levels. Meteorological factors tend not to exert a strong direct impact on deeper water (> 100 m) sublittoral benthic communities, as the water column moderates these effects. Intertidal communities are obviously affected to a greater extent by climatic and meteorological changes. During severe winters, overwintering mortality of macrobenthic fauna on tidal flats in the Dutch Wadden Sea was observed to increase (Beukema, 1992b). During a period of eight successive mild winters, overwintering mortality was reduced (Beukema, 1992a), and resulted in a more stable biomass of intertidal fauna. Buchanan & Moore (1986a) noted that overwintering mortality of the soft-sediment community at a depth of 55 m off the Northumberland coast was affected by winter minimum sea surface temperatures. Fluctuations in the benthic community structure of Weymouth and Poole Bay were recorded following the severe winter of 1962–1963 (Holme, 1983). Increases in swell from storms have been shown to disturb shallow benthic communities (Turner *et al.*, 1995) and Drake & Cacchione (1985, 1986) have noted that waves associated with storms may influence the benthos at depths in excess of 50 m, although the maximum depth will depend mainly upon the hydrography of the region.

Assessing the influence of climate is difficult and a number of proxy variables have been used. Amongst these are the position of the North Wall of the Gulf Stream and the North Atlantic oscillation (NAO) index. Both have been shown to be good indicators of the climatic conditions over large areas of the northern hemisphere (Taylor, 1996; Taylor & Stepens, 1998) and are correlated with biological parameters such as zooplankton abundance (Taylor & Stephens, 1980; Frid & Huliselan, 1996; Taylor, 1996), benthic biomass (Wieking & Kröncke, 1999) and vegetation (Willis, 1995).

Indirect effects of climate on the benthos are more commonly observed. Benthic infauna that are associated with soft sediments consume organic matter that descends from the pelagic zone (Pearson & Rosenberg, 1978). Most of this food source consists of phytoplankton that have settled out of the water column. Changes in the abundance of plankton are linked to large-scale climatic variables and associated changes in meteorological variables, which indirectly affect the food availability for benthic organisms and subsequently changes in their abundance and community structure (Pearson & Rosenberg, 1986). Such 'benthic–pelagic coupling' has been observed at many sites where long-term trends or sudden changes in the pelagic community have manifested themselves in the benthic community (Austen *et al.*, 1991; Buchanan, 1993). This intimate relationship between the benthos and pelagic food supply was apparent off the Northumberland coast, where a good correlation was observed between phytoplankton index and benthic abundance when a 2-year time lag was used. Austen (1991) also noted a change in benthic communities across the entire North Sea area between 1980 and 1981, which was 2 years later than the change in plankton community structure. Austen (1991) concluded that the associated change in plankton community structure was due to enhanced organic input. Postma (1981) has suggested that if there was a switch from a benthic to a pelagic food web, organic material could be transported from the central North Sea to coastal areas. Lindeboom *et al.* (1994) has postulated that the observed increase in coastal benthic biomass has occurred as a result of the utilisation of the majority of primary production in the benthos rather than the pelagic system.

Separating out the ecosystem effects of coincidental changes in climate and nutrient input is difficult unless long-term data are available for both. Eutrophication of the pelagic environment also affects primary production and therefore the amount of organic matter reaching the benthos. Josefson *et al.* (1993) noted that benthic biomass and abundance had doubled between 1974 and 1988 in the Skagerrak–Kattegat area. This was attributed to eutrophication. Parallel to and preceding the faunal changes, nitrogen and chlorophyll-*a* concentrations increased in the surface layer and were correlated with increases in agricultural run-off. Similarly, increased eutrophication in the Baltic has been linked to changes in pelagic production, and the subsequent increase in benthic biomass (Cederwall & Elmgren,

1980). Long-term changes in the ecosystem of the Wadden Sea have also been observed over the 1970–1990 period when the biomass and abundance of benthic invertebrates doubled at 15 stations. Eutrophication is also considered responsible for these changes (Beukema, 1992c).

Medium-term changes in the benthos as revealed by time series

The longest continuous marine sampling programme in the world is probably that of the continuous plankton recorder (CPR) surveys conducted presently by the Sir Alistair Hardy Foundation for Ocean Science (SAHFOS) (Warner & Hays, 1994). Recent analyses of the meroplankton recorded in North Sea CPR surveys have shown an increase in the dominance of echinoderm larvae (Lindley *et al.*, 1995), which has been interpreted as due to an increase in their relative abundance in the benthos. It has then been suggested that this may be the result of increases in the population of scavenging starfishes and ophioroids that have occurred as a result of food subsidies from fishing disturbance (Kaiser, 1996; Ramsay *et al.*, this volume, Chapter 10). However, the ability of the CPR programme to quantify plankton and especially meroplankton is limited and the timing of the change, in 1978, is well after the major changes in fishing technology. These, and the widespread nature of the increase in echinoderms in the CPR data, indicate that environmental causes are responsible rather than fisheries alone.

   Effects of increased stress to the benthos attributed to fishing were observed in the central-western North Sea in the late 1980s (Frid *et al.*, in press (a)). Long-term monitoring of two benthic stations off the Northumberland coast has been carried out since 1971 (Buchanan, 1993) (Fig. 13.1). One station, located at a depth of 80 m (Station P), is located within a *Nephrops norvegicus* (Dublin Bay Prawn) fishing ground (subject to otter trawling), while the other station (Station M1) is located at a depth of 55 m on the edge of the fished area. Up to 1986, trends in the abundance of benthic infauna at both sites paralleled changes in phytoplankton index (with a 2-year lag to allow for growth to minimum sieve retention size (Buchanan, 1993), suggesting that benthic productivity was controlled by organic matter input (Fig. 13.2a). From 1986 until 1990 fishing effort increased within the *Nephrops* ground (Fig. 13.2b). This increase in fishing effort caused a decline in benthic abundance at Station P (Fig. 13.2b) and large-scale year-to-year changes in community structure indicative of a stressed community (Fig. 13.3) (Warwick & Clarke, 1993). These large year-to-year changes may be observed in Fig. 13.3, which shows year-to-year changes in Bray–Curtis similarity. Year-to-year changes at the less heavily fished Station M1, outside the *Nephrops* ground, continued closely to mirror changes in phytoplankton input throughout the time series (Fig. 13.2a). Furthermore, fluctuations in phytoplankton input had little effect on year-to-year changes in community structure at this site (Fig. 13.3), indicating a relatively unstressed community.

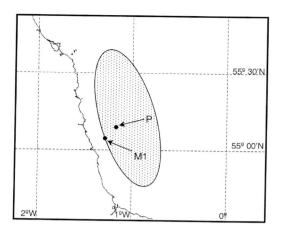

Fig. 13.1 Map showing locations of benthic sampling stations P and M1, off the north-east coast of England.

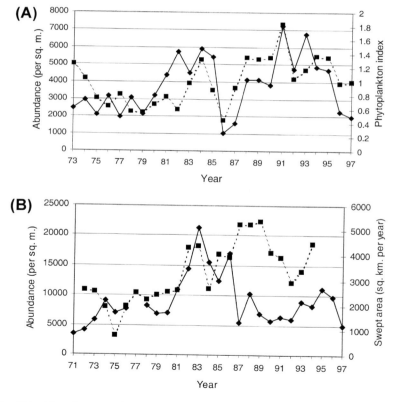

Fig. 13.2 (A) Abundances of individuals at Station M1 (continuous line) and CPR phytoplankton index (lagged by 2 years) (dashed line). (B) Abundances of individuals at Station P (continuous line) and fishing impact measured as swept area in the period 1972–1994 (dashed line).

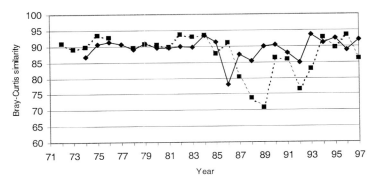

Fig. 13.3 Time series plot of year-to-year Bray–Curtis similarity for Station M1 (continuous line) and Station P (dashed line).

Long-term changes in the benthos as revealed by sampling

The North Sea has been one of the most intensively fished areas in the world over the last 100 years. It has also been the subject of early benthic ecological research. Some of the earliest quantitative benthic data were collected in 1920 (Davis, 1923, 1925). The *Dana* expedition in the 1950s provided another snapshot and, more recently, a series of research studies, e.g. ICES, the North Sea Benthos Survey (Kunitzer *et al.*, 1992), and pre-drilling studies for the oil and gas industries have provided greater coverage of the seabed fauna.

A number of authors have made historical comparisons based on these data. Rumohr & Kujawski (unpublished data) found that the German Bight macrobenthic community composition altered from the start of the 20th century compared with 1986 based on presence/absence records. Many of the taxa recorded in the early part of the century were bivalves, whereas, by 1986, scavenging crustaceans and seastars had increased in occurrence while bivalves were no longer prominent in the fauna. These patterns concur with the predictions of community change made on the basis of short-term trawl damage studies (Bergman & van Santbrink, this volume, Chapter 4). Similarly, Schroeder & Knust (1999) also examined long-term changes in the German Bight and concluded that fishing mortality had caused a decline in large, long-lived taxa, but that overall there had been an increase in total biomass as a result of increases in the abundance of opportunistic taxa.

Frid *et al.* (in press (c)) considered quantitative abundance in the macrofauna of five regions of both the central and southern North Sea (Fig. 13.4). They adopted a conservative approach to the 'quality control' of data and were still able to show definite changes in the macrofaunal communities in three of the five areas between the early 1920s and the late 1980s (Fig. 13.5). The lack of change in the remaining two areas was interpreted as evidence that the changes were not part of a broad-scale environmental change. One of the areas that showed no significant change included

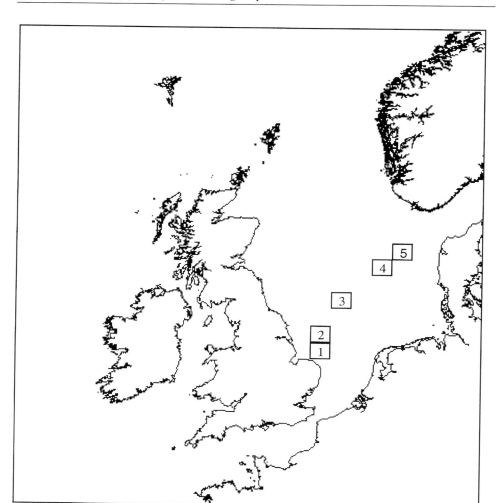

Fig. 13.4 Locations of the ICES statistical rectangles: (1) 35F1, Dowsing Shoal; (2) 36F1, Great Silver Pit; (3) 38F2, Dogger Bank; (4) Inner Shoal; and (5) 41F5, Fisher Bank.

much of the Dogger Bank, but this area may have already attained an alternative stable state as a result of fishing activities prior to 1920.

Kröncke and her co-workers (Kröncke, 1990, 1992; Kröncke & Rachor, 1992; Kröncke & Knust, 1995) have also studied the Dogger Bank benthos. They resampled stations sampled in the 1950s during the 1980s and found that many changes in faunal abundance occurred including the loss of *Spisula* spp. from certain areas. These changes were interpreted as responses to food availability due to eutrophication and/or climatic fluctuations but these authors were unable to make inferences about fishing effects.

In the German Bight, the taxa that were predicted to be most sensitive to fishing disturbance were changed most dramatically in response to fishing activity

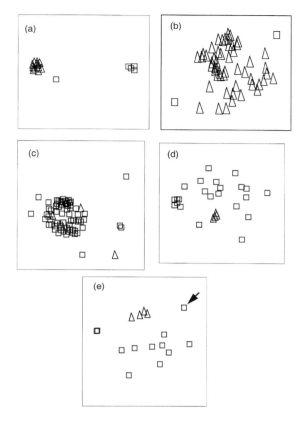

Fig. 13.5 MDS ordinations of the similarity in the composition of the macrobenthic community between the 1920s (△) and after 1986 (□) in five ICES statistical rectangles: (a) 35F1, Dowsing Shoal, stress = 0.01; (b) 36F1, Great Silver Pit, stress = 0.20; (c) 38F2, Dogger Bank, stress = 0.22; (d) 40F4, Inner Shoal, stress = 0.17; and (e) 41F5, Fisher Bank, arrowed sample had a similarity (Bray–Curtis) of <5% to the remaining samples, stress = 0.12.

(Schroeder & Knust, 1999). However, in the three regions in which community changes were apparent they were associated with changes in the abundance of many taxa and not just those sensitive to the direct effects of fishing. Hence, in some areas, the indirect effects of fishing (sediment changes, nutrient flux, predation pressure) may be at least as important as the direct effects.

Long-term changes in predation pressure

Fish predation is commonly seen to be the principal way that fish influence benthic communities (e.g. Whitman & Sebens, 1992; Sala & Zabala, 1996; Sala & Boudouresque, 1997). Benthic feeding fish do not take prey in proportion to their availability; rather they exhibit some degree of selection (e.g. Packer *et al.*, 1994), thereby altering relative abundances of benthic species. The removal of preferred prey may release resources for utilisation by other, less preferred, species while the act of predation may cause small-scale physical disturbance to the system and contribute to the spatial heterogeneity of the benthos (Hall, 1994). The most incontrovertible evidence of these effects comes from experimental studies that have shown that fish predation can act to control both the number of individuals in the system and the relative abundance of the species (see Wilson, 1990). Other controlling factors, such as physical disturbance, emigration, immigration or benthic predators, have been shown to be equally or more important than fish predation (e.g. Ambrose, 1984, 1991). Commercially important species can also structure the benthic community through indirect interactions, such as by predation on in-coming benthic larvae (Langton & Robinson, 1990).

By combining datasets on the abundance (Figs 13.6 and 13.7), size frequency and size-specific diet of North Sea fish, it has been possible to evaluate predation pressure over the period 1970–1993 for the eight most abundant demersal fish species (ICES, 1998; Frid *et al.*, in press (b)). Although target fish populations (gadoids and plaice) have declined, the overall estimated level of predation on the benthos has increased from around 23 Mt year^{-1} in 1970 to 29 Mt year^{-1} in 1993 (Fig. 13.8).

Frid *et al.* (in press (b)) have demonstrated that the consumption of North Sea benthos may have changed as stock sizes have changed. The principal factor influencing fish stock size of exploited species is fishing, and the expansion in the, non-target, dab population may be due to competitive or predatory release. Fishing has removed the larger gadoids that are principally piscivorous, and has allowed expansion of flatfish and young gadoids that prey upon benthos to a larger extent. However, the differences in diet of various species would also appear to have influenced the composition of the benthos consumed. Overall, crustaceans have declined in dietary importance, while echinoderms (predominantly ophioroids) have increased. The effects of fish predation on benthic communities are not well documented. The intensity of fish predation on North Sea benthos found by Frid *et al.* (in press (b)) (20–45% of the production is consumed by fish) is similar to the 39% of macrobenthic consumption consumed by fish estimated for this area by Greenstreet *et al.* (1997).

The challenge

Society expects to be given guidance on whether there is a need to provide management of the marine environment over and above that to manage target stocks, and yet scientists are reticent about making such statements, as the system is not readily amenable to experimental manipulation. The precautionary approach has been accepted as a guiding principal for fisheries management (Garcia, 1996; ICES, 1998) and carries with it an implicit integration of fisheries and ecosystem management. It also seeks to act when the available data indicate the possibility of

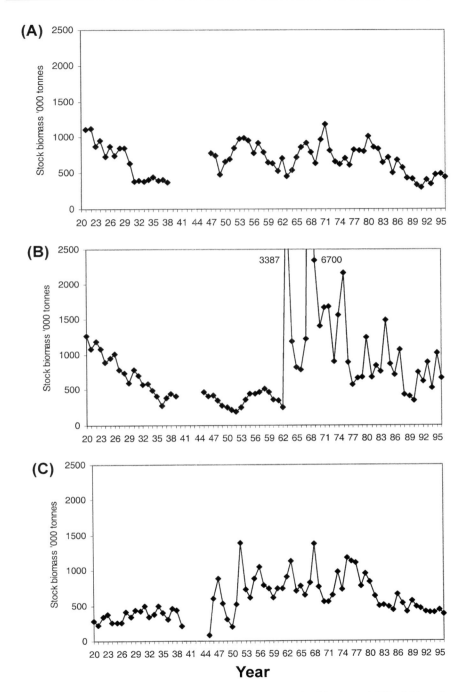

Fig. 13.6 North Sea stock biomass (kt) of: (A) cod; (B) haddock; and (C) whiting, from 1920 to 1996.

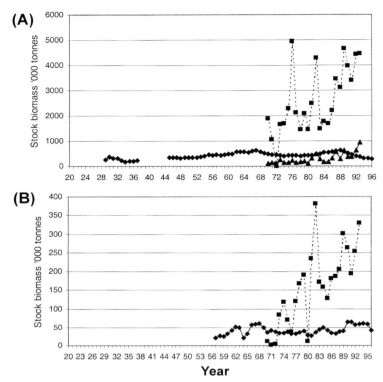

Fig. 13.7 North Sea stock biomass (kt) of: (A) plaice (continuous line), common dab (dashed line) and long rough dab (dotted line); (B) sole (continuous line) and lemon sole (dashed line), from 1920 to 1996.

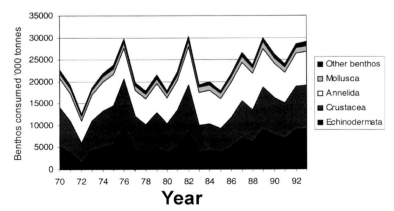

Fig. 13.8 Estimated total consumption of benthic food by eight species of demersal feeding fish in the North Sea, 1970–1993.

irreversible damage rather than needing the full demonstration of proof of a negative effect. Our inability to carry out scientifically rigorous experiments at the appropriate time and spatial scales should not hamper the incorporation of genuine concerns into policy. Fisheries and ecosystem managers will need to respond to a weight of evidence from studies that individually do not establish a cause and effect relationship, but that when taken together suggest a likely cause.

Discussion

A critical evaluation of the role of fishing in bringing about long-term changes in marine ecosystems is difficult to achieve. There are many studies that now show changes in the macrobenthos over the last 30 or so years. In many cases, these changes are linked to climatic influences on changing supplies of food (Buchanan & Moore, 1986a; Kröncke, 1990, 1992; Austen *et al.*, 1991; Buchanan, 1993; Kröncke & Knust, 1995). Such changes tend to occur over decadal time scales.

Extending such investigations back in time is hampered by the quality of the data available. Stretching throughout the history of mechanised fishing, there is now a body of data that suggests community changes arising from both direct fishing mortality and indirect effects of fishing (Frid *et al.* in press (a, c)). The latter includes changes in benthic water column fluxes of nutrients. The importance of benthic processes to phytoplankton productivity is well established (Rowe *et al.*, 1975; Prins & Smaal, 1990), as is the direct link of phytoplankton production to benthic ecology (Pearson & Rosenberg, 1986; Buchanan, 1993). Fishing perturbs both sides of this equation, physically altering sediments and flux rates, thereby causing direct and indirect changes in benthic abundances.

To date, in spite of sound science, we have failed to prevent overexploitation of just about every fish stock, the challenge now is to manage the ecosystem sustainably, based on incomplete knowledge of its functional processes and response to fishing. How can such management be achieved? Clearly, while marine reserves have an important role in protecting key sites, they are not feasible on the scales required definitely to preserve ecosystem function. Reductions in effort required to meet sustainability targets on exploited species will contribute, but we believe that, ultimately, specific measures will be needed to address ecosystem concerns. Firstly, changes in gear design to increase selectivity and reduce incidental mortality/ damage, and secondly, a more diverse fishery – less heavy exploitation of a key stocks, and more equitable use of the available productivity of the marine ecosystem.

Acknowledgements

This synthesis draws upon many pieces of work, and we gratefully acknowledge the efforts of the original authors. In particular, we have drawn extensively on work carried out under European Commission, study contract number 94/77 to C.L.J.F.

and S.J. Hall and the work of the *ICES Working Group on the Ecosystem Effects of Fishing*. Development of the ideas presented here has benefited from discussions with many colleagues and in particular we would like to thank: Steve Hall, Jack Buchanan, Jane Lancaster and all the members of the Marine Ecology Research Group at the Dove Marine Laboratory. We gratefully acknowledge the assistance of the anonymous reviewers and the editors in production of this chapter.

References

Aebischer, N., Coulson, J. & Colebrook, J. (1990) Parallel long-term trends across four marine trophic levels and weather. *Nature*, **347**, 753–5.

Ambrose, W.G. (1984) Role of predatory infauna in structuring marine soft-bottom communities. *Marine Ecology Progress Series*, **17**, 109–15.

Ambrose, W.G. (1991) are infaunal predators important in structuring marine soft-bottom communities. *American Zoologist*, **31**, 849–60.

Arrhenius, F. & Hansson, S. (1993) Food consumption of larval, young and adult herring and sprat in the Baltic Sea. *Marine Ecology Progress Series*, **96**, 125–37.

Austen, M.C., Buchanan, J.B., Hunt, H.G., Josefson, A.B. & Kendall, M.A. (1991) Comparison of long-term trends in benthic and pelagic communities of the North-Sea. *Journal of the Marine Biological Association of the United Kingdom*, **71**, 179–90.

Beukema, J.J. (1992a) Expected changes in the Wadden Sea benthos in a warmer world: Lessons from periods with milder winters. *Netherlands Journal of Sea Research*, **30**, 73–9.

Beukema, J.J. (1992b) Expected effects of changes in winter temperatures on benthic animals living in soft sediments in coastal North Sea areas. In *Expected Effects of Climatic Change of Marine Coastal Ecosystems* (eds J.J. Beukema, W.J. Wolff & J.W.M. Brouns), pp. 83–92. Kluwer, Dordrecht.

Beukema, J.J. (1992c) Long-term and recent changes in the benthic macrofauna living on tidal flats in the western part of the Wadden Sea. *Netherlands Institute for Sea Research*, **20**, 135–41.

Buchanan, J.B. (1993) Evidence of Benthic pelagic coupling at a station off the Northumberland coast. *Journal of Experimental Marine Biology and Ecology*, **172**, 1–10.

Buchanan, J.B. & Moore, J.J. (1986a) A broad review of variability and persistence in the Northumberland benthic fauna – 1971–85. *Journal of the Marine Biological Association of the United Kingdom*, **66**, 641–57.

Buchanan, J.B. & Moore, J.J. (1986b) Long-term studies at a benthic station off the coast of Northumberland. *Hydrobiologia*, **142**, 121–7.

Camphuysen, C.J., Ensor, K.W.F.R., Garthe, S., Hüppop, O.G., L., H., O. & L., T.M. (1993) *Seabirds feeding on discards in winter in the North Sea. Final Report to the European Commission 92/3505.* Netherlands Institute for Sea Research, Den Burg, Texel, The Netherlands.

Camphuysen, C.J., Calvo, B., Durinck, J., Ensor, K., Follestad, A.W.F.R., Garthe, S., Leaper, G., Skov, H., Tasker, M.L. & N., W.C.J. (1995) *Consumption of discards in the North Sea. Final Report EC DG XIV research contract BIOECO/93/10,* 202 pp. Netherlands Institute for Sea Research, Den Burg, Texel, The Netherlands.

Carpenter, S.R. (1988) *Complex Interactions in Lake Communities.* Springer, New York.

Cederwall, H. & Elmgren, R. (1980) Biomass increase of benthic macrofauna demonstrates eutrophication of the Baltic Sea. *Ophelia*, Suppl. 1, 287–304.

Cryer, M. G., Pierson, G. & Townsend, C.R. (1986) Reciprocal interactions between roach (Rutilus *rutilus* L.) and zooplankton in a small lake: prey dynamics and fish growth and recruitment. *Limnology and Oceanography*, **31**, 1022–38.

Davis, F.M. (1923) No. 1. Preliminary investigations of the Dogger Bank. Fishery Investigations. In: *Quantitative Studies on the Fauna of the Sea Bottom*, Vol. 2, pp. 1–54. Ministry of Agriculture, Fisheries and Food Series, London, UK.

Davis, F.M. (1925) No. 2. Results of the investigations into the southern North Sea 1921–1924. Fishery Investigations. In: *Quantitative Studies on the Fauna of the Sea Bottom*, Vol. 2, pp. 1–50. Ministry of Agriculture, Fisheries and Food Series, London, UK.

Dayton, P.K., Thrush, S.F., Agardy, M.T. & Hofman, R.J. (1995) Environmental-effects of marine fishing. *Aquatic Conservation: Marine and Freshwater Ecosystems*, **5**, 205–32.

De Angelis, D.L. & Waterhouse, J.C. (1987) Equilibrium and non-equilibrium concepts in ecological models. *Ecological Monographs*, **57**, 1–21.

Drake, D.E. & Cacchione, D.A. (1985) Seasonal-variation in sediment transport on the Russian River shelf, California. *Continental Shelf Research*, **4**, 495–514.

Drake, D.E. & Cacchione, D.A. (1986) Field observations of bed shear-stress and sediment resuspension on continental shelves, Alaska and California. *Continental Shelf Research*, **6**, 415–29.

Dunnet, G.M., Furness, R.W., Tasker, M.L. & Becker, P.H. (1990) Seabird ecology in the North Sea. *Netherlands Journal of Sea Research*, **26**, 387–425.

Estes, J.A. (1996) The influence of large, mobile predators in aquatic food webs: examples from sea otters and kelp forests. In: *Aquatic Predators and their Prey* (eds S.P.R Greenstreet. & M.L. Tasker), pp. 65–72. Blackwell Science, Oxford.

Frid, C.L.J. & Huliselan, N.V. (1996) Far-field control of long-term changes in Northumberland (NW North Sea) coastal zooplankton. *ICES Journal of Marine Science*, **53**, 972–7.

Frid, C.L.J., Buchanan, J.B. & Garwood, P.R. (1996) Variability and stability in benthos: twenty-two years of monitoring off Northumberland. *ICES Journal of Marine Science*, **53**, 978–80.

Frid, C.L.J., Clark, R.A., Hall, J.A. & Hall, S.J. (in press (a)) Long term changes in the benthos on a heavily fished ground off the NE coast of England. *Journal of the Marine Biological Association of the United Kingdom*.

Frid, C.L.J., Hansson, S., Ragnarsson, S.A., Rijnsdorp, A. & Steingrimsson, S.A. (in press (b)) Changes in levels of predation on benthos as a result of exploitation induced changes in fish populations. *Ambio*.

Frid, C.L.J., Harwood, K.G., Hall, S.J. & Hall, J.A. (in press (c)) Long term trends in benthic communities on North Sea fishing grounds. *ICES Journal of Marine Science*.

Garcia, S.M. (1996) The precautionary approach and its implications for fisheries research, technology and management: An updated review. In: *Precautionary Approach to Fisheries. Part 2: Scientific Papers Prepared for Technical Consultation on the Precautionary Approach to capture fisheries (including species introductions)*, FAO Fisheries Technical Paper No. 350, 210 pp. Fisheries and Agriculture Organisation, Rome.

Greenstreet, S.P.R., Bryant, A.D., Broekhuizen, N., Hall, S.J. & Heath, M.R. (1997) Seasonal variation in the consumption of food by fish in the North Sea and implications for food web dynamics. *ICES Journal of Marine Science*, **54**, 243–66.

De Groot, S.J. & Lindeboom, H.J. (1994) *Environmental impact of bottom gears on benthic fauna in relation to natural resources management and protection of the North Sea*. NIOZ Report 1994-11/RIVO-DLO Report C026/94, 257 pp. Netherlands Institute for Sea Research, Den Burg, Texel, The Netherlands.

Hall, S.J. (1994) Physical disturbance and marine benthic communities: life in unconsolidated sediments. *Oceanography and Marine Biology: An Annual Review*, **32**, 179–239.

Hansson, S. (1985) *Effects of Eutrophication on Fish Communities, with Special Reference to the Baltic Sea – A Literature Review*, pp. 36–56. Institute of Freshwater Research, Drottningholm.

Heesen, H.J.L. & Daan, N. (1996) Long term trends in non-target North Sea fish species. *ICES Journal of Marine Science*, **53**, 1063–78.

Hill, B.J. & Wassenberg, T.J. (1990) Fate of discards from prawn trawlers in Tores Straight. *Australian Journal of Marine & Freshwater Research*, **41**, 53–64.

Holme, N.A. (1983) Fluctuations in the benthos of the western English Channel. In: *Proceedings of the 17th European Marine Biology Symposium*, Brest, France, pp. 121–4.

ICES (1998) Report of the Working Group on the Ecosystem Effects of Fishing Activities, *International Council for the Exploration of the Sea*, C.M. 1998/ACFM/ACME:1, 263 pp.

Josefson, A.B., Jensen, J.N. & Aertebjerg, G. (1993) The benthos community structure in the late 1970s and early 1980s – a result of a major food pulse. *Journal of Experimental Marine Biology and Ecology*, **172**, 31–45.

Kaiser, M.J. (1996) Starfish damage as an indicator of trawling intensity. *Marine Ecology-Progress Series*, **134**, 303–7.

Kaiser, M.J. & Spencer, B.E. (1994) Fish scavenging behavior in recently trawled areas. *Marine Ecology Progress Series*, **112**, 41–9.

Kneib, R.T. (1991) Indirect effects in experimental studies of marine soft-sediment communities. *American Zoologist*, **31**, 874–85.

Kröncke, I. (1990) Macrofauna standing stock of the Dogger Bank – a comparison. 2. 1951–1952 versus 1985–1987 are changes in the community of the northeastern part of the Dogger Bank due to environmental changes. *Netherlands Journal of Sea Research*, **25**, 189–98.

Kröncke, I. (1992) Macrofauna standing stock of the Dogger Bank – a comparison .3. 1950–54 versus 1985–87 – a final summary. *Helgolander Meeresuntersuchungen*, **46**, 137–69.

Kröncke, I. & Knust, R. (1995) The Dogger-Bank – a special ecological region in the central North Sea. *Helgolander Meeresuntersuchungen*, **49**, 335–53.

Kröncke, I. & Rachor, E. (1992) Macrofauna investigations along a transect from the inner German Bight towards the Dogger Bank. *Marine Ecology Progress Series*, **91**, 269–76.

Kunitzer, A., Basford, D., Craeymeersch, J.A., Dewarumez, J.M., Dorjes, J., Duineveld, G.C.A., Eleftheriou, A., Heip, C., Herman, P., Kingston, P., Niermann, U., Rachor, E., Rumohr, H. & Dewilde, P.A.J. (1992) The benthic infauna of the North-Sea – species distribution and assemblages. *ICES Journal of Marine Science*, **49**, 127–43.

Langton, R.W. & Robinson, W.E. (1990) Faunal associations on scallop grounds in the western Gulf of Maine. *Journal of Experimental Marine Biology and Ecology*, **144**, 157–71.

Lindeboom, H.J., van Raaphorst, W., Beukema, J.J., Cadee, G.C. & Swennen, C. (1994) (Sudden) changes in the biota of the North Sea: oceanic influences underestimated? In: *Actual Problems of the Marine Environment. Lectures of the 4th International Scientific Symposium*, Vol. 2, pp. 87–100. Hamburg, Germany.

Lindley, J.A., Gamble, J.C. & Hunt, H.G. (1995) A change in the zooplankton of the central North-Sea (55° to 58° N) – a possible consequence of changes in the benthos. *Marine Ecology Progress Series*, **119**, 299–303.

Mann, K.H. & Lazier, J.R.N. (1991) *Dynamics of Marine Ecosystems: Biological – Physical Interactions in the Ocean*. Blackwell Science, Oxford.

Millner, R.S. & Whiting, C.L. (1996) Long-term changes in growth and population abundance of sole in the North Sea from 1940 to the present. *ICES Journal of Marine Science*, **53**, 1185–95.

Moreno, C.A., Lunecke, K.M. & Lopez, M.I. (1986) The response of an intertidal *Concholepas concholepas* (Gastropoda) population to protection from man in southern Chile and the effects on benthic sessile assemblages. *Oikos*, **43**, 359–64.

Packer, D.B., Watling, L. & Langton, R.W. (1994) The population structure of the brittle star *Ohiura sarsi* Lutken in the Gulf of Maine and its trophic relationship to American plaice (*Hippoglossoides platessoides* Fabricus). *Journal of Experimental Marine Biology and Ecology*, **179**, 207–22.

Parsons, T.R. (1991) The impact of fish harvesting on ocean ecology. *Marine Pollution Bulletin*, **22**, 217.

Parsons, T.R. (1992) The removal of marine predators by fisheries and the impact on trophic structure. *Marine Pollution Bulletin*, **25**, 51–3.

Parsons, T.R. (1996) The impact of industrial fisheries on the trophic structure of marine ecosystems. In: *Food Webs: Integration of Patterns and Dynamics* (eds G.A. Polis & K.O. Winemiller), pp. 352–7. Chapman & Hall, New York.

Pauly, D. (1988) Fisheries research and demersal fisheries of southeast Asia. In: *Fish Population Dynamics* (ed. J.A. Gulland), pp. 329–48. John Wiley, Chichester.

Pearson, T.H. & Rosenberg, R. (1978) Macrobenthic succession in relation to organic enrichment and pollution of the marine environment. *Oceanography and Marine Biology: An Annual Review*, **16**, 229–311.

Pearson, T.H. & Rosenberg, R. (1986) Feast and famine: structuring factors in marine benthic communities. In: *Organization of Communities* (eds J.H.R. Gee & P.S. Gill), pp. 373–98. Blackwell Science, Oxford.

Pope, J.G. & Macer, C.T. (1996) An evaluation of the stock structure of North sea cod, haddock and whiting since 1920, together with a consideration of the impacts of fisheries and predation effects on their biomass and recruitment. *ICES Journal of Marine Science*, **53**, 1157–69.

Pope, J.G., Stokes, T.K., Murawski, S.A. & Idoine, S.I. (1988) A comparison of fish size composition in the North Sea and on Georges Bank. In *Ecodynamics: Contributions to Theoretical Ecology* (eds W. Wolff, C.-J. Soeder & F.R. Drepper), pp. 146–52. Springer, Berlin.

Postma, H. (1981) Exchange of materials between the North Sea and the Wadden Sea. *Marine Geology*, **40**, 199–213.

Prins, T.C. & Smaal, A.C. (1990) Benthic–pelagic coupling: the release of inorganic nutrients by an intertidal bed of *Mytilus edulis*. In: *Trophic Interactions in the Marine Environment* (eds M. Barnes & R.N. Gibson), pp. 89–103. Aberdeen University Press, Aberdeen.

Ramsay, K., Kaiser, M.J. & Hughes, R.N. (1996) Changes in hermit crab feeding patterns in response to trawling disturbance. *Marine Ecology Progress Series*, **144**, 63–72.

Ramsay, K., Kaiser, M.J. & Hughes, R.N. (1998) Responses of benthic scavengers to fishing disturbance by towed gears in different habitats. *Journal of Experimental Marine Biology and Ecology*, **224**, 73–89.

Rice, J. & Gislason, H. (1996) Changes in abundance and diversity size spectra of the North sea, as reflected in surveys and models. *ICES Journal of Marine Science*, **53**, 1214–225.

Rijnsdorp, A.D. & van Leeuwen, P.I. (1996) Changes in growth of North Sea plaice since 1950 in relation to density, eutrophication, beam-trawl effort, and temperature. *ICES Journal of Marine Science*, **53**, 1199–213.

Rijnsdorp, A.D. & Millner, R.S. (1996) Trends in population dynamics and exploitation of North Sea plaice (*Pleuronectes platessa* L) since the late 1800s. *ICES Journal of Marine Science*, **53**, 1170–84.

Rijnsdorp, A.D., Buys, A.M., Storbeck, F. & Visser, E.G. (1998) Micro-scale distribution of beam trawl effort in the southern North Sea between 1993 and 1996 in relation to the trawling frequency of the sea bed and the impact on benthic organisms. *ICES Journal of Marine Science*, **55**, 403–19.

Rowe, G.T., Clifford, C.H., Smith, K.L. Jr & Hamilton, P.L. (1975) Benthic nutrient regeneration and its coupling to primary productivity in coastal waters. *Nature*, **225**, 215–17.

Rudstam, L.G., Hansson, S., Johansson, S. & Larsson, U. (1992) Dynamics of planktivory in a coastal area of the northern Baltic Sea. *Marine Ecology Progress Series*, **80**, 159–73.

Sala, E. & Boudouresque, C.F. (1997) The role of fishes in the organization of a Mediterranean sublittoral. *Journal of Experimental Marine Biology and Ecology*, **212**, 25–44.

Sala, E. & Zabala, M. (1996) Fish predation and the structure of the sea-urchin Paracentrotus. *Marine Ecology Progress Series*, **140**, 71–81.

Schroeder, A. & Knust, R. (1999) Long term changes in the benthos of the German Bight (North Sea) – possible influences of fisheries? *ICES Journal of Marine Science*.

Serchuk, F.M., Kirkegaard, E. & Daan, N. (1996) Status and trends of the major roundfish, flatfish and pelagic fish stocks in the North Sea: Thirty-year overview. *ICES Journal of Marine Sciences*, **53**, 1130–45.

Simenstad, C.A., Estes, J.A. & Kenyon, K.W. (1978) Aleuts, sea otters and alternate stable state communities. *Science*, **200**, 403–11.

Sørensen, J. (1978) Capacity for denitrification and reduction of nitrate to ammonia in a coastal marine environment. *Applied Environmental Microbiology*, **35**, 301–5.

Taylor, A.H. (1996) North–south shifts of the Gulf-Stream – ocean–atmosphere interactions in the North-Atlantic. *International Journal of Climatology*, **16**, 559–83.

Taylor, A.H. & Stepens, J.A. (1998) The North Atlantic oscillation and the latitude of the Gulf Stream. *Tellus Series a – Dynamic Meteorology & Oceanography*, **50**, 134–42.

Taylor, A.H. & Stephens, J.A. (1980) Latitudinal displacements of the Gulf Stream (1966 to 1977) and their relation to changes in temperature and zooplankton abundance in the NE Atlantic. *Oceanologia Acta*, **3**, 145–9.

Taylor, A.H., Colebrook, J.M., Stephens, J.A. & Baker, N.G. (1992) Latitudinal displacements of the gulf-stream and the abundance of plankton in the north-east Atlantic. *Journal of the Marine Biological Association of the United Kingdom*, **72**, 919–21.

Turner, S.J., Thrush, S.F., Pridmore, R.D., Hewitt, J.E., Cummings, V.J. & Maskery, M. (1995) Are soft-sediment communities stable – an example from a windy harbor. *Marine Ecology Progress Series*, **120**, 219–30.

Underwood, A.J. (1990) Experiments in ecology and management – their logics, functions and interpretations. *Australian Journal of Ecology*, **15**, 365–89.

Underwood, A.J. (1992) Beyond baci – the detection of environmental impacts on populations in the real, but variable, world. *Journal of Experimental Marine Biology and Ecology*, **161**, 145–78.

Underwood, A.J. (1996) Detection, interpretation, prediction and management of environmental disturbances – some roles for experimental marine ecology. *Journal of Experimental Marine Biology and Ecology*, **200**, 1–27.

De Veen, J.F. (1976) On changes in some biological parameters in the North Sea sole. *Journal du Conseil Permanent International pour L'exploration de la Mer*, **37**, 60–90.

Verity, M. & Smetacek, V. (1996) Organism life cycles, predation and the structure of marine pelagic ecosystems. *Marine Ecology Progress Series*, **130**, 277–93.

Warner, A.J. & Hays, G.C. (1994) Sampling by the continuous plankton recorder survey. *Progress in Oceanography*, **34**, 237–56.

Warwick, R.M. & Clarke, K.R. (1993) Increased variability as a symptom of stress in marine communities. *Journal of Experimental Marine Biology and Ecology*, **172**, 215–26.

Wassenberg, T.J. & Hill, B.J. (1990) Partitioning of material discarded from prawn trawlers in Moreton Bay. *Australian Journal of Marine and Freshwater Research*, **41**, 27–36.

Whitman, J.D. & Sebens, K.P. (1992) Regional variation in fish predation intensity: a historical perspective in the Gulf of Maine. *Oecologia*, **90**, 305–15.

Wieking, G. & Kröncke, I. (1999) Long term comparison of macrofaunal communities on the Dogger Bank between 1996 and 1985–1987. *ICES Journal of Marine Science*.

Willis, A.J. (1995) Does Gulf Stream position affect vegetation dynamics in western Europe? *Oikos*, **73**, 408–10.

Wilson, W.H. (1990) Competition and predation in marine soft-sediment communities. *Annual Review of Ecology and Systematics*, **21**, 221–41.

Chapter 14
Effects of fishing on non-target fish species

S.P.R. GREENSTREET[1] and S.I. ROGERS[2]

[1]FRS, Marine Laboratory, PO Box 101, Victoria Road, Aberdeen, AB11 9DB, UK
[2]Centre for Environment, Fisheries and Aquaculture Science, Lowestoft Laboratory, Pakefield Road, Lowestoft, NR33 0HT, UK

Summary

1. Although some of the studies reviewed examine datasets extending back over seven decades or more, none of the time series is sufficiently long to allow us to compare the current fished situation with the original unfished ecosystem.
2. Attempts to correlate time-series trends in the abundance of non-target species with fishing-disturbance trends have to date proved inconclusive.
3. A better approach towards determining fishing effects on non-target species lies through the development of underlying theory, which would enable the establishment of specific testable hypotheses. In particular, the development of hypotheses to predict the effects of fishing on species diversity are required. This would enable diversity changes to be more strongly related to changes in fisheries exploitation patterns, allowing changes in the abundance of different non-target species in time and space to be linked more directly to fishing.
4. Current applied theory has identified specific life-history characteristics likely to make a species vulnerable to fishing disturbance. These life-history characteristics include large ultimate size, slow growth rate, and large size and higher age at maturity.
5. The elasmobranchs have been identified as a group of species that have such life-history characteristics likely to render them susceptible to fishing disturbance. In general, trends in the abundance of the different shark, skate and ray species in the North Sea can be attributed to fishing mortality, since they follow predictions based on the life-history characteristics of each species. On the Georges Bank, skate and dogfish abundance actually increased, probably because they were always discarded, and likely to have a high survival rate following discarding.
6. An increased ability to predict the likely consequences of continued high levels of fishing disturbance would be a likely further benefit to be gained through the development of underlying theory.
7. Whilst fishing undoubtedly causes increased mortality for many non-target species, in some cases, it may also allow increased scope for population growth through scavenging and reduced predation and competition.

Keywords: non-target species, species diversity, elasmobranchs, fishing disturbance, life history.

Introduction

Demersal fisheries in northern continental shelf seas tend to be mixed-species fisheries. Fishing vessels exploiting these stocks generally direct their activities

---

towards a specific target species, e.g. haddock *Melanogrammus aeglefinus* in the north-western North Sea or plaice *Pleuronectes platessa* or sole *Solea solea* in the southern North Sea. However, although the different gears employed (i.e. seine net, otter trawl, pair trawl and beam trawl) are designed to catch certain types of fish (e.g. the heavy chains of beam trawls make them particularly good at catching flatfish), they can also catch fish of any species present in the path of the gear. Few, if any, demersal species are completely invulnerable. Consequently, catches made using these gears in the North Sea and other north-east Atlantic shelf seas may contain up to 20 fish species. In many tropical multispecies fisheries, the entire catch is generally considered to have value as a food source. However, approximately half the fish species in a typical North Sea catch will have little, or no, commercial value and are rarely, if ever, landed. These non-target, non-commercial species are therefore discarded soon after being brought aboard the vessel. Should we be concerned about the mortality inflicted by fishing activities on this commercially valueless component of the catch, and the consequences of this mortality on their population dynamics?

While we may exploit only a relatively small number of species in any one marine region, it is a mistake to assume that just the consequences of fishing on this limited number of species are all that is important. The situation on the Georges Bank in the north-west Atlantic during the 1970s and 1980s clearly illustrates this point. The Georges Bank supported a highly productive mixed-species demersal fishery. Exploitation levels increased rapidly during the 1960s and, by the 1970s, catches of the economically desirable gadoid species, principally cod *Gadus morhua* and haddock, had declined markedly (Brown *et al.*, 1976; Overholtz & Tyler, 1985). Traditional fisheries' science and management practices were applied to the fishery, and strict quotas were set for each of the major commercial species. Despite the stringent nature of these controls, the fishery failed to recover and, ultimately, the fishing industry that had built up along the east coast of America and Canada went into decline. What happened, and why didn't the fisheries management regime imposed have the anticipated effect? One plausible explanation is that the traditional fisheries' science and management ethos prevalent at this time, with its single-species focus directed more or less entirely on the targeted commercial species, failed to take account of the non-target species. Non-target species made up a substantial fraction of the Georges Bank groundfish assemblage and, with hindsight, it would seem that these interacted strongly with the target species.

Pimm & Hymen (1987) have suggested that exploited species may be replaced by ecologically similar species. The gadoid species, which were the mainstay of the fishery on the Georges Bank, tended to be relatively large-bodied piscivorous or benthivorous fish. Another group of fish, the dogfish, skates and rays, also large-bodied piscivorous and benthivorous fish, were not targeted by the fishing fleets, as there was no market for them on the western seaboard of the north Atlantic. Although undoubtedly taken in the by-catch, elasmobranch fish were generally discarded. The absence of swim bladders, combined with presence of tough scaleless skins, would tend to ensure that they had a high survival rate following capture and discarding. In the early days of the fishery, these elasmobranch species made up less than 25% of the groundfish assemblage biomass. By the 1980s, this fraction had tripled and the assemblage biomass was now dominated by dogfish, skates and rays; the fraction of the assemblage biomass consisting of gadoid species had fallen from about 55% to less than 15% (Sherman, 1991; Murawski & Idoine, 1992). A real increase in the combined biomass of the elasmobranch species, as much as any decrease in gadoid biomass, was responsible for these changes in the proportional composition of the groundfish assemblage (Murawski & Idoine, 1992). One group of large benthivorous or piscivorous species had been replaced by another.

In a review of theoretical studies, Beddington (1984) suggested that the community structure of exploited ecosystems may be characterised by several alternative stable states. The Georges Bank may be an example of this (Fogarty & Murawski, 1998). The initial gadoid-dominated assemblage constituted one stable state, while the elasmobranch-dominated assemblage, present after a decade or more of heavy fishing, seems to have been equally stable. Despite greatly reducing the fishing pressure on the system, particularly on the principal gadoid species, the elasmobranch-dominated system has persisted and the gadoid species have yet to recover their former status in the assemblage. The precise mechanism underlying such a stable state 'flip' is not clear. One may, however, speculate that the formerly large populations of gadoid fish, perhaps through competition for some limiting resource (Grosslein *et al.*, 1980), may have limited the scope for growth in the populations of dogfish, skates and rays. The high reproductive potential of the gadoid species would have helped them to maintain this dominant position, hence making this a stable state. The reduction in the abundance of the large gadoid fish may have released the elasmobranch species from competitive pressure, allowing their populations to expand to the point where they dominated the assemblage. Despite the reduction in fishing pressure directed at the gadoid species, dominance of the assemblage by the elasmobranch species has persisted. Elasmobranch populations may then have been sufficiently large that the predation pressure inflicted by them on young gadoid fish kept gadoid recruitment levels low (Rothschild, 1991), or perhaps competition from such large numbers of elasmobranchs prevented the recovery of the gadoids (Grosslein *et al.*, 1980; Fogarty & Murawski, 1998). Alternatively, changes in seabed habitat structure, caused by the passage of trawls over the seabed, could have resulted in the loss of refuges for juvenile gadoids, making them more vulnerable to predation (Auster *et al.*, 1996).

The lesson here is that there may be undesirable consequences to ignoring the effects of fishing on non-target species. If multiple stable states exist, the Georges Bank scenario suggests that, under certain circumstances, fishing may provide the impetus to flip the system from one stable state to another. Not only may this have profound knock-on effects on other components of marine ecosystems, but, if the alternative stable state is dominated by species of little or no market value, it could be of considerable economic consequence. The Georges Bank example illustrates how difficult it might be to restore the original state. Monitoring changes in the target species populations alone is not enough to provide sufficient warning of impending catastrophic changes in the species composition of marine fish species assemblages; contemporaneous changes in the populations of non-target species must also be followed. The Georges Bank is an example of an extreme change in the species composition of a groundfish assemblage. Is there any evidence to suggest that the populations of non-target species on this, the eastern, side of the Atlantic may be going through similar changes? If so, is it possible to relate these changes directly to fishing?

Trends in the abundance of non-target fish species in the North Sea

Closer to home, the North Sea has received the greatest attention with regard to the amount of scientific survey work undertaken. More datasets are available, have been analysed and the results published for this region than anywhere else in northern European waters. The marine research institutes of countries bordering the North Sea have been involved in groundfish surveys for many years. Some of these surveys, particularly in recent decades, have been carried out as joint international collaborative exercises covering the entire North Sea, under the auspices of the International Council for the Exploration of the Sea (ICES). The time span of these various ICES-coordinated surveys is variable, but the longest time series, for the winter International Bottom Trawl Survey (IBTS, initially the International Young Fish Survey, IYFS), extends back to 1970. In other instances, individual countries have routinely carried out their own surveys to meet their own specific requirements. These national surveys have tended to be more restricted geographically and, again, time-series duration is extremely variable. In one case, the time series extends back to the 1920s (e.g. Greenstreet *et al.*, 1999a), and some groundfish survey data from the early 1900s are also available (e.g. Rijnsdorp *et al.*, 1996). Groundfish survey data therefore tend to fall into two types; data covering a long time period collected within a relatively confined area, or data collected over a large area, but over a relatively short period.

Predictably, early analyses of these datasets generally considered only the commercially important species (e.g. Jones & Hislop, 1978), and just a few of the more abundant non-target species (Richards *et al.*, 1978; Hempel, 1978). Of some 175 groundfish species recorded in the North Sea (Yang, 1982), fewer than a dozen were included in these initial analyses; trends in the remaining 160 or so, mainly non-target, species were ignored. Only in the last few years have trends in the abundance of 44 of the more commonly caught non-target species been studied (Table 14.1). At first glance this table suggests little consistency between studies. Table 14.2 helps to explain this confusion, indicating clear differences in the types of dataset analysed, the extent of geographic coverage and the time span of the different studies involved. Species trends over a particular time period apparent on the North Sea-wide scale, e.g. those for bib *Trisopterus luscus* and John Dory *Zeus faber*, may not have been observed in studies concentrating on more restricted regions within the North Sea. For some species, e.g. red gurnard *Aspitrigla cuculus* and grey gurnard *Eutrigla gurnardus*, trends apparent on the North Sea-wide scale were in the opposite direction to trends observed on a regional scale. There are several examples of species, e.g. poor cod *Trisopterus minutus* and long rough dab *Hippoglossoides platessoides*, which appear to have increased in abundance in one part of the North

Table 14.1 Direction of trends in the abundance of non-target species published in recent studies

Species	Study	Trend in abundance
Lesser spotted dogfish *Scyliorhinus canicula*	Heessen (1996)	Not determinable
Tope *Galeorhinus galeus*	Heessen (1996)	Possible increase
Smooth hound *Mustelus mustelus*	Heessen (1996)	Possible increase
Blond ray *Raja brachyura*	Walker & Heessen (1996)	No trend
Spotted ray *Raja montagui*	Rogers & Millner (1996)	Decrease
	Walker & Heessen (1996)	Not determinable
Cuckoo ray *Raja naevus*	Walker & Heessen (1996)	Possible decrease
Starry ray *Raja radiata*	Heessen & Daan (1996)	Increase
	Walker & Heessen (1996)	Increase
Torsk *Brosme brosme*	Heessen (1996)	Possible increase
5-bearded rockling *Ciliata mustela*	Rijnsdorp *et al.* (1996)	Decrease
3-bearded rockling *Gaidropsarus vulgaris*	Rijnsdorp *et al.* (1996)	Decrease
4-bearded rockling *Enchelyopus cimbrius*	Heessen & Daan (1996)	Possible increase
Ling *Molva molva*	Heessen (1996)	Not determinable
Pollack *Pollachius pollachius*	Heessen (1996)	Not determinable
Bib *Tricopterus luscus*	Heessen & Daan (1996)	Increase
	Corten & van den Kamp (1996)	No trend
Poor cod *Trisopterus minutus*	Rijnsdorp *et al.* (1996)	Decrease
	Heessen & Daan (1996)	Increase
	Greenstreet & Hall (1996)	No trend
	Corten & van den Kamp (1996)	Increase
Hake *Merluccius merluccius*[a]	Heessen (1996)	No trend
John Dory *Zeus faber*	Heessen (1996)	Possible increase
	Corten & van den Kamp (1996)	Not determinable
Boarfish *Capros aper*	Heessen (1996)	Increase
Bluemouth *Helicolenus dactylopterus*	Heessen (1996)	Increase
Norway haddock *Sebastes viviparus*	Greenstreet & Hall (1996)	Increase
Red gurnard *Aspitrigla cuculus*	Rijnsdorp *et al.* (1996)	Decrease
	Heessen (1996)	Possible increase
Grey gurnard *Eutrigla gurnardus*	Rijnsdorp *et al.* (1996)	Decrease
	Heessen & Daan (1996)	Increase
	Greenstreet & Hall (1996)	Decrease
Tub gurnard *Trigla lucerna*	Rijnsdorp *et al.* (1996)	Possible decrease
	Heessen (1996)	No trend

Table 14.1 *continued*

Species	Study	Trend in abundance
	Corten & van den Kamp (1996)	Not determinable
Bull rout *Myxocephalus scorpius*	Rijnsdorp *et al.* (1996)	Possible decrease
	Heessen & Daan (1996)	Increase
Hooknose *Agonus cataphractus*	Rijnsdorp *et al.* (1996)	Decrease
	Rogers & Millner (1996)	No trend
Lumpsucker *Cyclopterus lumpus*	Heessen (1996)	Possible increase
Sea snail *Liparis liparis*	Rogers & Millner (1996)	No trend
Red mullet *Mullus surmuletus*	Rijnsdorp *et al.* (1996)	Increase
	Heessen (1996)	Not determinable
	Corten & van den Kamp (1996)	Not determinable
Ballan wrasse *Labrus bergylta*	Rogers & Millner (1996)	Not determinable
Eelpout *Zoarces viviparus*	Rogers & Millner (1996)	Decrease
Butterfish *Phollis gunnellus*	Rogers & Millner (1996)	Decrease
Catfish *Anarhichas lupus*	Heessen (1996)	Increase
Lesser weever *Echiichthys vipera*	Rijnsdorp *et al.* (1996)	Decrease
	Heessen (1996)	Increase
	Rogers & Millner (1996)	No trend
Greater weever *Trachinus draco*	Rijnsdorp *et al.* (1996)	Decrease
Dragonet *Callionymus lyra*	Rijnsdorp *et al.* (1996)	Decrease
	Heessen (1996)	Decrease
Spotted dragonet *Callionymus maculatus*	Heessen (1996)	Not determinable
Norwegian topknot *Phrynorhombus norvegicus*	Heessen (1996)	Possible increase
Scaldfish *Arnoglossus laterna*	Rijnsdorp *et al.* (1996)	Decrease
	Heessen (1996)	Possible increase
Witch *Glyptocephalus cynoglossus*	Heessen (1996)	No trend
	Greenstreet *et al.* (1999a)	Decrease[b]
Long rough dab *Hippoglossoides platessoides*	Rijnsdorp *et al.* (1996)	Decrease
	Heessen & Daan (1996)	Increase
	Greenstreet & Hall (1996)	No trend
Halibut *Hippoglossus hippoglossus*[a]	Heessen (1996)	Possible increase
Common dab *Limanda limanda*	Rijnsdorp *et al.* (1996)	Decrease
	Heessen & Daan (1996)	Possible increase
	Greenstreet & Hall (1996)	No trend
Lemon sole *Microstomus kitt*[a]	Rijnsdorp *et al.* (1996)	Possible decrease
	Heessen & Daan (1996)	Increase
	Greenstreet & Hall (1996)	Increase

Table 14.1 *continued*

Species	Study	Trend in abundance
Solenette *Buglossidium luteum*	Rijnsdorp *et al.* (1996)	Decrease
	Rogers & Millner (1996)	No trend
	Heessen (1996)	Not determinable

[a]This study dealt exclusively with the North Sea, where hake, halibut and lemon sole are not targeted specifically, although they are usually landed if taken in the by-catch. In other parts of the continental shelf waters of Northern Europe, i.e. in western and southern areas, hake are specifically targeted by fishing operations.
[b]A decrease is suggested in all four of the regions studied.

Sea, but to have declined in other areas. The length of the time series is also an important factor in determining the direction of any trend. Thus, poor cod in the southern North Sea appear to have increased in abundance over the period 1970–1993. However, data spanning a longer period suggest that they have decreased in abundance between 1906 and 1909 and 1990 and 1995. Distinguishing short-term fluctuations from long-term trends becomes increasingly more problematic as the

Table 14.2 Type of study, spatial and temporal coverage, and gears used in the studies cited in Table 14.1

Study	Study type	Period(s)	Gears	Spatial coverage
Rijnsdorp *et al.* (1996)	Two period comparison	1906—1909, 1990—1995	OT20[a] vs. GOV[b]	Southern North Sea
Heessen & Daan (1996)	Time series	1970—1993	GOV[b,c]	North Sea
Heessen (1996)	Time series	1970—1993	GOV[b,c]	North Sea
Walker & Heessen (1996)	Time series	1970—1993	GOV[b,c]	North Sea
Rogers & Millner (1996)	Time series	1973—1995	BT2 & PN1.5[d]	SE English coast
Corten & van de Kamp (1996)	Time series	1970—1993	GOV[b,c]	Southern North Sea
Greenstreet & Hall (1996)	Two period comparison	1929—1953, 1980—1993	OT48 vs. AOT[e]	North-western North Sea
Greenstreet *et al.* (1999a)	Time series	1925—1996	OT48 vs. AOT[e]	North-western North Sea

[a]20-ft (6-m) otter trawl with 40-mm codend mesh with a swept area rate of 15 000 m^2 h^{-1}.
[b]Grande ouverture verticale otter trawl with 20-mm codend mesh with a swept area rate of 530 000 m^2 h^{-1}.
[c]In the early 1980s countries participating in this ICES-coordinated survey switched to using the GOV, prior to this other gears were used, e.g. the herring bottom trawl with 20-mm mesh codend on Scottish vessels.
[d]2-m beam trawl and 1.5-m push net designed to have similar efficiency and selectivity, enabling catches per unit swept area to be compared directly.
[e]These two gears are essentially the same: 48-ft otter trawls with 35-mm codend mesh.

time span of any dataset decreases, a point illustrated by Corten & van de Kamp's (1996) study.

In explaining such trends, correlations with some causative factor are often implicitly assumed, if not explicitly investigated. It is generally assumed that fishing disturbance in the North Sea has increased over the course of the 20th century and so it is easy to suggest that straightforward increases or decreases in the abundance of particular species might be indications of a fishing effect. Is it possible that the complex pattern of trends in species abundance apparent in Table 14.1, in different regions and at differing spatial and temporal scales, can still be attributable to changes in the pattern of fishing exploitation? In fact, the development of the fishing industry in the North Sea has been far from simple (ICES, 1995). Fishing effort has never been evenly distributed across the North Sea. Different gears, directed at different target species, with differing levels of impact on the seabed and varying catchability rates for different by-catch species, have been used in different regions of the North Sea, and at varying intensities. Some gears, such as the Danish seine net, have gone out of fashion, declines in their use being balanced by increased use of otter and pair trawls (Greenstreet *et al.*, 1999b). The designs of some gears, such as the beam trawl, have been improved and their use has increased markedly, albeit in restricted regions of the North Sea (Jennings *et al.*, 1999a). Improvements in vessel design and technology have enabled fishing boats to tow larger and heavier gears, to travel further and faster, and to stay at sea for longer. As a result of all these developments, long-term changes in fishing effort reveal a complex pattern of spatial and temporal interactions (Greenstreet *et al.*, 1999b; Jennings *et al.*, 1999a). Thus, the changes in the pattern of fishing effort distribution over the last two to three decades may perhaps be potentially complicated enough to have caused the variety of trends in species abundance evident in Table 14.1. However, the problem now is one of designing an experiment capable of teasing out these relationships!

A further problem with any correlative approach lies in excluding potentially confounding effects and linking actual cause and effect. Changes in fishing practices are far from being the only changes to have occurred in the North Sea during the last few decades. Variation in surface sea temperature and salinity, the strength of Atlantic inflow in the north and dissolved inorganic phosphate are just some of the physical factors for which decadal-scale changes have been documented (Becker & Pauly, 1996; Danielssen *et al.*, 1996; Laane *et al.*, 1996; van Leussen *et al.*, 1996; Turrell *et al.*, 1996). Similarly, chlorophyll concentration and primary productivity, phytoplankton, zooplankton and benthos species composition and abundance are just a few components of the biota for which long-term variation has been described (Reid *et al.*, 1990; Beukema *et al.*, 1996; Bot & Colijn, 1996; Frid & Huliselan, 1996; Greve *et al*, 1996). All these variables, and probably many more, could affect the population dynamics of fish (e.g. Nielsen & Richardson, 1996). Indeed, significant correlations have been identified between mean annual water temperature and the catch rates of at least two non-target fish species, the hooknose *Agonus cataphractus* and gunnel *Pholis gunnellus* (Rogers & Millner, 1996). Separating fishing effects from these other potentially influential factors as a definitive cause of change in the abundance of particular non-target fish species should pose considerable problems.

Corten & van de Kamp (1996) conclude that variation in environmental conditions explains much of the changes in abundance of several fish species in the southern North Sea. Heessen & Daan (1996) also conclude that their 'results certainly do not single out a particular factor as the causal explanation for any of the observed trends'.

Finally, the relatively short duration of all the datasets cited above makes the identification of fishing as a causative factor determining the trends in any species' abundance a difficult task. Even the longest time series started well after the initiation of fishing activities in the North Sea (Cushing, 1988; Smith, 1994). For example, by the 1930s, the total UK landings of demersal species from the North Sea had exceeded 200 000 t year^{-1} (Wood, 1956). Thus, since none of the datasets captures the situation prior to the onset of fishing, the change between an unfished groundfish assemblage and a fished one has passed unobserved.

Evidence that fishing has affected non-target fish species

Greenstreet & Hall (1996), looking at long-term changes in the structure and species composition of the groundfish assemblage, also explicitly examined trends in the non-target species component of the assemblage. This initial study was carried out on a limited dataset consisting of 9 years within the period 1929–1953 and the 13 years between 1980 and 1993. They examined data for three contiguous areas in the north-western North Sea for which they had reasonably reliable estimates of trends in fishing effort from 1960 onwards (see also Greenstreet *et al.*, 1999b). In the two offshore areas, fishing effort had increased steadily, while in the third, inshore, area effort had actually declined. However, despite this reduction in effort, current effort levels were similar to those recorded in the two offshore regions. The data further indicated that fishing disturbance in the inshore area had been considerably higher than ever observed in either of the two offshore regions, and had been at high levels for a prolonged period of time. Significant changes in species composition, in both the entire groundfish species assemblage and the non-target species component only, were apparent in all three areas. These changes in species relative abundance resulted in a reduction in species diversity (and corresponding increases in species dominance) between the two time periods in all three areas when data for all species were analysed. However, only in the heavily disturbed inshore area was any change, a significant increase in species dominance, apparent in the structure of the non-target species assemblage. Greenstreet & Hall (1996) concluded that the changes in species relative abundance, giving rise to this alteration in the structure of the non-target species component of the groundfish assemblage, may well therefore constitute a fishing effect.

The dataset analysed by Greenstreet & Hall (1996) has recently been considerably expanded. Geographic coverage was increased to include a fourth region in the centre of the North Sea, and data for many more years were added so that, rather than comparing two separate periods, the data could now be treated as a continuous time series (Greenstreet *et al.*, 1999a). No trend in species diversity/dominance was detected in the new central North Sea region studied, for either the entire species assemblage or the non-target species subset. Since estimated fishing effort was least in this area, the lack of any trend fitted the anticipated pattern. Trends in species diversity and dominance in the three areas originally covered were similar to those noted in the earlier study. It was now possible, however, to determine that much of the increase in species dominance (decrease in species diversity); that took place among the non-target species in the historically heavily fished inshore area, occurred during the late 1970s or early 1980s. Considering how fishing effort has actually changed in this area, the timing of the dominance/diversity shift among the non-target species raises some doubt as to whether it really was the result of sustained fishing disturbance over a prolonged period.

In another recent study, fish catch-rate data collected from extensive research vessel surveys in three mainly western coastal areas of the British Isles during the period 1901–1907 (Garstang, 1903, 1905; Holt, 1910) were compared with the equivalent data from recent groundfish surveys collected between 1989 and 1997 (Rogers & Ellis, in press). Records of mesh size and other gear characteristics suggested that these data were broadly comparable and could be used to identify changes that had occurred in these demersal fish assemblages. Results of the analysis showed that in the English coastal regions of the southern North Sea, during an interval of over 80 years, fish assemblages became more diverse; several non-target species such as the dragonets Callionymiidae, bib and bull-rout *Myoxocephalus scorpius* were now more abundant, while the previous levels of dominance by plaice and whiting *Merlangius merlangus* had decreased. In the Irish Sea and on the south coast of England, however, the distribution of species' relative abundance remained constant, although the identity of the highest-ranking species had changed. The relative proportion of small fish (maximum body length <30 cm) in the catches, which were all non-target species and the least vulnerable to commercial trawling because of mesh size limitations, increased in all regions between the two survey periods.

These trends towards the dominance of groundfish assemblages by an increasing diversity of small-bodied non-target species should not be considered as an inevitable consequence of commercial fishing activity. For example, they conflict with observations from similar historic data made by Rijnsdorp *et al.* (1996), in which a recent increase in dominance (decrease in diversity) resulted from greatly increased population sizes of just two species, dab *Limanda limanda* and whiting. In the north-western North Sea, similar long-term increases in dominance were thought to be caused by a recent increase in abundance of Norway pout *Trisopterus esmarki* (Greenstreet & Hall, 1996; Greenstreet *et al.*, 1999a). Trends in dominance and diversity are thus not consistent across these studies; both increases and decreases in diversity have been attributed to fishing disturbance. Theoretical considerations suggest that assemblage responses to variation in disturbance can involve either an increase or a decrease in diversity, and the direction of any response may depend on system productivity (Huston, 1994; Rosenzweig, 1995).

Shifts in species aggregated length frequency distributions (often referred to as size spectra) towards assemblages dominated by smaller fish have been demonstrated in a number of studies (Pope & Knights, 1982; Pope *et al.*, 1988; Murawski & Idoine, 1992; Rice & Gislason, 1996; Gislason & Rice, 1998). However, while this appears to be a feature of the whole groundfish assemblage, when only the non-target component of the assemblage is considered, such shifts in the assemblage size spectrum are not apparent (Greenstreet & Hall, 1996; Greenstreet *et al.*, 1999a). These results suggest that shifts in the species aggregated length frequency distribution of the assemblage, thought to be primarily caused by fishing activities, arise mainly through changes in the size structure of the target species' populations. By virtue of the small size of the majority of non-target species, the species aggregated size frequency distribution of the non-target component of the assemblage appears to be relatively resilient to fishing.

Theoretical population dynamics suggests that populations of slow-growing animals, which grow to a large ultimate size, and which mature at a late age and large size, should decrease faster when subject to a given mortality rate compared with populations consisting of fast-growing, ultimately relatively small, and early maturing animals (Southwood, 1976; Hoenig & Gruber, 1990). If this is the case, it should be possible to predict which species of fish are likely to be most susceptible to fishing disturbance (Adams, 1980; Beddington & Cooke, 1983; Roff, 1984; Kirkwood *et al.*, 1994). Jennings *et al.* (1998) examined this idea using phylogenetic comparisons to identify life-history correlates of abundance trends in 18 intensively exploited fish stocks in the north-east Atlantic. After allowing for differences in the rates of fishing mortality, they demonstrated that those stocks that had decreased in abundance, compared with their nearest phylogenetic relatives, did indeed mature later and at a larger size, attained a larger maximum size, and had slower daily specific growth rates. Jennings *et al.* (1999b) then went on to examine long-term trends in the average ultimate size, mean age and length at maturity, and mean growth rate in the whole ground fish assemblage in a region of the North Sea, east of Shetland, where fishing effort (hours fishing) had more than doubled since 1960. Greenstreet *et al.* (1999a) had previously demonstrated significant changes in species composition in this area over the period 1925–1996, both within the whole groundfish assemblage and for a subset of the assemblage consisting only of non-target species. These changes in species composition resulted in a significant increase in the average growth rate among individual fish making up the groundfish assemblage, and significant decreases in the maximum attainable length, age at maturity and length at maturity. Individual comparisons between nine pairs of closely related species confirmed that, in eight instances, the species that had declined in abundance, compared with its nearest relative, had a greater maximum size, slower growth rate, and matured at greater length and age. Here, we repeat Jennings *et al.*'s (1999b) analysis, but this time only for the non-target species in the assemblage. With the exclusion of those species specifically targeted by fishermen, which dominate the groundfish assemblage, the data are, not surprisingly, much more variable. Nevertheless, the same trends in growth rate, maximum attainable length and length at maturity are still apparent (Fig. 14.1). This is strong evidence

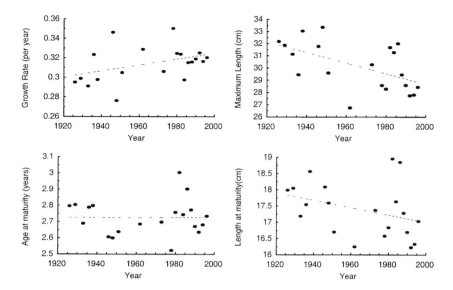

Fig. 14.1 Trends in the mean growth rate, maximum attainable length, age at maturity and length at maturity in the non-target component of the groundfish species assemblage in the region of the North Sea to the east and south-east of Shetland (see Jennings *et al.*, 1999a, for further details).

that the increase in fishing disturbance in this part of the North Sea has, to some extent at least, contributed to the observed changes in species composition within the groundfish assemblage, and that the changes observed in the non-target component of the assemblage may also be attributable to fishing activity.

  With this theoretical support in place, we can now be more confident in attributing the decline in populations of large, slow-growing species to the effect of consistent high levels of fishing mortality. Thus, the elasmobranchs, with their slow growth rates, large sizes and high ages at maturation, and lower potential rates of population increase (Hoenig & Gruber, 1990), are a potentially valuable group for monitoring fishing effects. Examples of population declines in these species are becoming more common in the literature. For example, in the Bay of Biscay, the long-term decline in abundance of the blonde ray *Raja brachyura* and other species of large-bodied carcharhiniform sharks (*Galeorhinus galeus, Mustelus asterias*) was attributed to the long-term effects of bottom trawling throughout the area (Quéro & Cendrero, 1996). The angel shark *S. squatina* was relatively abundant in the North Sea and all along the south coast, including Devon and Cornwall in the late 19th century (Day, 1880). Data from the west coast of France showed that catch rates of 0.5–1.0 angel shark per 24 h fishing were common during the period 1869–1891 (Quéro & Cendrero, 1996), but recent surveys by European research laboratories throughout the continental shelf rarely encounter this species now. Two skate species, the common skate *R. batis*, and the barndoor skate *R. laevis*, are thought to be vulnerable to localised extinction in regions of the North Atlantic (Brander,

1981). In the North Sea, while several ray species have shown constant or declining catch rates, only the starry ray *R. radiata*, which is discarded by English trawl fisheries, appears to have increased in abundance. Of all the skates and rays in the North Sea, the starry ray has the smallest ultimate size, the fastest growth rate, and the shortest length and youngest age at maturity (Walker & Hislop, 1998). The spotted ray *R. montagui* is considered vulnerable to fishing activity in the southern North Sea, and overexploitation is thought to have caused an absence of this species from the German Bight in the southern North Sea (Walker & Heessen, 1996). In theory, the changes in the relative abundance of sharks and rays in Northern European waters should correspond to their perceived level of vulnerability to fishing mortality, which can be predicted from their life-history characteristics. The rank order of a selection of rays, from most to least vulnerable, has been suggested as *R. batis* > *R. clavata* > *R. montagui* > *R. naevus* > *R. radiata* (Walker & Hislop, 1998). In practice, records from contemporary surveys tend to support this theoretical vulnerability approach, both by proposing reasons for species declines and in helping to explain the apparent dominance of the elasmobranchs in some groundfish assemblages by the starry ray, cuckoo ray *R. naevus* and lesser spotted dogfish *Scyliorhinus canicula* (Walker & Hislop, 1998). The percentage composition of the elasmobranch fauna collected during research surveys between 1901 and 1907 were dominated by the thornback ray *R. clavata*, and a number of other larger- and smaller-bodied rays and small sharks (Fig. 14.2). Contemporary data from the same regions reveal a decline in the relative abundance of thornback ray, and a corresponding increase in lesser spotted dogfish as a proportion of the entire

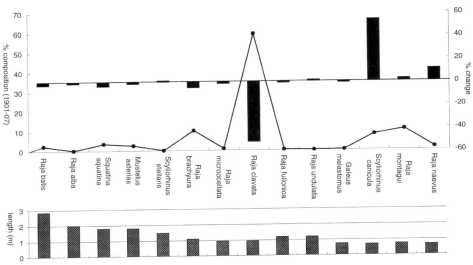

Fig. 14.2 Percentage composition (by number) of the elasmobranch fauna of the British Isles determined from 1901–1907 survey data (line), and the change in percentage composition when compared with contemporary samples (% in contemporary data minus % in historic data) (bars). The bars in the lower panel indicate the maximum length normally reached by each species.

elasmobranch fraction of the assemblage. Species in this figure are ranked by perceived level of vulnerability, from the larger-bodied skates and sharks on the left, towards the smaller and potentially less vulnerable species on the right (Rogers & Ellis, unpublished data).

Conclusions

There is little evidence to suggest that major stable state flips, similar to the sort of fluctuation apparent on the Georges Bank, have occurred in the North Sea. Paradoxically, the high intensity of fishing activity in the North Sea may itself have been a factor in preventing such a change; no major group of species having been exempt from fishing pressure and thus able to gain a selective advantage and dominate the assemblage. In the Georges Bank region, elasmobranch populations were not subjected to fishing mortality in the same manner as in the North Sea, there being no market for these species on the western seaboard of the USA and Canada. All sharks, skates and rays taken in the by-catch were discarded and, since they are likely to have a high probability of surviving capture and subsequent discarding, their populations were able to expand, to the point where they dominated the groundfish assemblage. Although not specifically targeted in the North Sea many elasmobranchs taken in the by-catch were nevertheless landed. Landings of sharks and rays from the North Sea peaked at nearly 50 000 t in the early 1960s, since when they have declined to less than 20 000 t (ICES, 1995). Thus, in the North Sea many elasmobranch species were subject to fishing mortality, to which, because of their life-history characteristics, most species were particularly vulnerable. Consequently, declines in their populations have been considered indicative of a fishing effect on the non-target component of the groundfish assemblage. On the Georges Bank, however, because this same group of fish were not subject to appreciable fishing mortality, they appear to have been able to exploit a 'niche' made available through the removal of the other major group of occupants of the same niche, the commercially valuable gadoid species. Consequently, the effect of fishing in this region has been to cause an increase in their population sizes.

Although no major species composition 'stable state flip' has been observed, nevertheless, the demersal fish assemblage in many parts of the European continental shelf seas has undergone marked changes in species composition. These have also often been associated with changes in species diversity, or other aspects of assemblage structure. However, the direction of the change in these measures of assemblage structure have not always been consistent. Both increases and decreases in species diversity have, for example, been associated with an increase in fishing disturbance, and there are sound theoretical reasons why this should be the case. We are not yet in a position, with respect to our understanding of how marine ecosystems function, to be able to predict in any particular situation how increasing fishing disturbance is likely to affect structural aspects of the groundfish assemblage, such as species diversity or species richness.

We do, however, have a much better understanding of the sort of life-history characteristics likely to render a species vulnerable to increasing fishing activity. Thus, species that are slow growing, reach a large ultimate size and mature at a larger size and older age are likely to be much less able to sustain high levels of fishing mortality. Changes in the species composition of the groundfish assemblage observed in an area where fishing effort has doubled in just over three decades have been associated with an increase in the average growth rate and decreases in the average ultimate size, and size and age at maturity in the assemblage as a whole. Similar trends were also apparent in the non-target species component of the assemblage, suggesting that, although not the specific target of fishing activities, these species were still affected by the increase in fishing disturbance.

Such studies lend credence to the idea that certain non-target species, on the basis of their life-history characteristics, such as some of the sharks, dogfish, skates and rays, could play a useful role as indicators of the impact of fishing disturbance on the wider groundfish assemblage. Indeed, many elasmobranch species in the heavily fished North Sea have undergone considerable population declines.

Acknowledgements

We thank John Hislop, Nick Bailey and two anonymous referees for helpful comments and suggestions. We are also indebted to various colleagues who have contributed to this work through the numerous chats and discussions we have had with them over the years. This work was funded by the Scottish Office, Ministry of Agriculture, Fisheries and Food, and by the European Commission as part of the project 'Monitoring Biodiversity in the North Sea Using Groundfish Surveys' (FAIR CT95-0817).

References

Adams, P.B. (1980) Life history patterns in marine fishes and their consequences for management. *Fisheries Bulletin*, **78**, 1–12.

Auster, P.J., Malatesta, R.J., Langton, R.W., Watling, L., Valentine, P.C., Donaldson, C.L.S., Langton, E.W., Shepard, A.N. & Babb, I.G. (1996) The impacts of mobile fishing gear on seafloor habitats in the gulf of Maine (northwest Atlantic): implications for conservation of fish populations. *Reviews in Fisheries Science*, **4**, 185–202.

Becker, G.A. & Pauly, M. (1996) Sea surface temperature changes in the North Sea and their causes. *ICES Journal of Marine Science*, **53**, 887–98.

Beddington, J.R. (1984) The responses of multispecies systems to perturbations. *Exploitation of Marine Communities* (ed. R.M. May). *Life Sciences Research Report* 32, pp. 209–25. Springer, New York

Beddington, J.R. & Cooke, J.G. (1983) The potential yield of fish stocks. *FAO Fisheries Technical Paper*, 242.

Beukema, J.J., Essink, K. & Michaelis, H. (1996) The geographical scale of synchronized fluctuation patterns in zoobenthos populations as a key to underlying factors: climatic or man-induced. *ICES Journal of Marine Science*, **53**, 964–71.

Bot, P.V.M. & Colijn F. (1996) A method for estimating primary production from chlorophyll concentrations with results showing trends in the Irish Sea and Dutch coastal zone. *ICES Journal of Marine Science*, **53**, 945–50.

Brander, K. (1981) Disappearance of common skate *Raja batis* from the Irish Sea. *Nature*, **290**, 48–9.

Brown, B.E., Brennan, J.A., Grosslein, M.D., Heyerdahl, E.G. & Hennemuth, R.C. (1976) The effect of fishing on the marine finfish biomass in the Northwest Atlantic from the Gulf of Maine to Cape Hatteras. *International Commission for the Northwest Atlantic Fisheries Research Bulletin*, **12**, 49–68.

Corten, A. & van den Kamp, G. (1996) Variation in the abundance of southern fish species in the southern North sea in relation to hydrography and wind. *ICES Journal of Marine Science*, **53**, 1113–19.

Cushing, D.H. (1988) *The Provident Sea.* Cambridge University Press, Cambridge.

Danielssen, D.S., Svendsen, E. & Ostrowski, M. (1996) Long-term hydrographic variation in the Skagerrak based on the section Torungen-Hirtshals. *ICES Journal of Marine Science*, **53**, 917–25.

Day, F. (1880) *The Fishes of Great Britain and Ireland*, 2 vols, 388 pp. Williams & Norgate, London.

Fogarty, M.J. & Murawski, S.A. (1998) Large scale disturbance and the structure of marine systems: fishery impacts on Georges Bank. *Ecological Applications*, **8** (Suppl.), 6–22.

Frid, C.L.J. & Huliselan, N.V. (1996) Far-field control of long-term changes in Northumberland (NW North Sea) coastal zooplankton. *ICES Journal of Marine Science*, **53**, 972–7.

Garstang, W. (1903) Report on trawling and other investigations carried out in the bays on the south-east coast of Devon during 1901 and 1902. *Journal of the Marine Biological Association of the United Kingdom*, **6**, 435–527.

Garstang, W. (1905) Report on the trawling investigations, 1902–3, with especial reference to the distribution of the plaice. *First Report on Fishery and hydrographic investigations in the North Sea and adjacent waters (southern area)*, pp. 67–198. International Fisheries Investigations, Marine Biological Association. UK.

Gislason, H. & Rice, J. (1998) Modelling the response of size and diversity spectra of fish assemblages to changes in exploitation. *ICES Journal of Marine Science*, **55**, 362–70.

Greenstreet, S.P.R. & Hall, S.J. (1996) Fishing and the ground-fish assemblage structure in the north-western North Sea: an analysis of long-term and spatial trends. *Journal of Animal Ecology*, **65**, 577–98.

Greenstreet, S.P.R., Spence, F.E. & McMillan, J.A. (1999a) Changes in the groundfish species assemblage of the northwestern North Sea between 1925 and 1996. *Fisheries Research*, **40**, 153–83.

Greenstreet, S.P.R., Spence, F.E., Shanks, A.M. & McMillan, J.A. (1999b) Trends in fishing effort by U.K. registered vessels landing in Scotland: Implications for Marine ecosystem perturbation. *Fisheries Research*, **40**, 107–24.

Greve, W., Reiners, F. & Nast, J. (1996) Biocoenotic changes of the zooplankton in the German Bight: possible effects of eutrophication and climate. *ICES Journal of Marine Science*, **53**, 951–6.

Grosslein, M.D., Langton, R.W. & Sissenwine, M.P. (1980). Recent fluctuations in pelagic fish stocks of the Northwest Atlantic, Georges Bank region, in relation to species interactions. *Rapports et Proces-verbaux des Reunions Conseil International pour l'Exploration de la Mer*, **177**, 374–404.

Heessen, H.J.L. (1996) Time series data for a selection of forty fish species caught during the International Beam Trawl Survey. *ICES Journal of Marine Science*, **53**, 1079–84.

Heessen, H.J.L. & Daan, N. (1996) Long-term trends in ten non-target North Sea fish species. *ICES Journal of Marine Science*, **53**, 1063–78.

Hempel, G. (1978) North Sea fisheries and fish stocks – a review of recent changes. *Rapports et Proces-verbaux des Reunions Conseil International pour l'Exploration de la Mer*, **173**, 145–67.

Hoenig, J. M. & Gruber, S. H. (1990) Life history patterns in the elasmobranchs: Implications for fisheries management. In: *Advances in the Biology, Ecology, Systematics, and the Status of the Fisheries* (eds H.L. Pratt, S.H. Gruber & T. Taniuchi). *NOAA Technical Report*, **90**, 1–16

Holt, E.W.L. (1910) Report of a survey of trawling grounds on the coasts of Counties Down, Louth, Meath and Dublin. Part I: Record of fishing operations. *Fisheries, Ireland Scientific Investigations*, **1901**, 1–538.

Huston, A.H. (1994). *Biological Diversity: The Coexistence of Species on Changing Landscapes*, 681 pp. Cambridge University Press, Cambridge.

ICES (1995) Report of the study group on ecosystem effects of fishing activities. *ICES Co-operative Research Report*, **200**, 120pp.

Jennings, S., Alvsvåg, J., Cotter, A.J., Ehrich, S., Greenstreet, S.P.R. Jarre-Teichmann, A., Mergardt, N., Rijnsdorp A.D., & Smedstad, O. (1999a) Fishing effects in northeast Atlantic shelf seas: patterns in fishing effort, diversity and community structure. III. International fishing effort in the North Sea: an analysis of spatial and temporal trends. *Fisheries Research*, **40**, 125–34.

Jennings, S., Greenstreet, S.P.R. & Reynolds, J. (1999b) Structural change in an exploited fish community: a consequence of differential fishing effects on species with contrasting life histories. *Journal of Animal Ecology*, **68**, 617–27.

Jennings, S., Reynolds, J.D. & Mills, S.C. (1998) Life history correlates of responses to fisheries exploitation. *Proceedings of the Royal Society of London*, **265**, 1–7

Jones, R. & Hislop, J.R.G. (1978) Changes in North Sea haddock and whiting. *Rapports et Proces-verbaux des Reunions Conseil International pour l'Exploration de la Mer*, **172**, 58–71.

Kirkwood, G.P., Beddington, J.R. & Rossouw, J.A. (1994) Harvesting species of different life-spans. In: *Large Scale Ecology and Conservation Biology* (eds. P.J. Edwards, R.M. May & N.R. Webb), pp. 199–227. Blackwell Science, Oxford.

Laane, R.W.P.M., Southward, A.J., Slinn, D.J., Allen, J., Groeneveld, G. & de Vries, A. (1996) Changes and causes of variability in salinity and dissolved inorganic phosphate in the Irish Sea, English Channel and Dutch coastal zone. *ICES Journal of Marine Science*, **53**, 933–44.

Leussen, W. van, Radach, G., van Raaphorst, W., Colijn, F. & Laane, R. (1996) The North-West European Shelf Programme (NOWESP): integrated analysis of shelf processes based on existing data sets and models. *ICES Journal of Marine Science*, **53**, 926–32.

Murawski, S.A. & Idoine, J.S. (1992) Multispecies size composition: a conservative property of exploited fishery systems. *Journal of Northwest Atlantic Fisheries Science*, **14**, 79–85.

Nielsen, E. & Richardson, K. (1996) Can changes in the fisheries yield in the Kattegat (1950–1992) be linked to changes in primary production. *ICES Journal of Marine Science*, **53**, 988–94.

Overholtz, W.J. & Tyler, A.V. (1985) Long-term responses of the demersal fish assemblages of Georges Bank. *Fishery Bulletin*, **83**, 507–20.

Pimm, S.L. & Hyman, J.B. (1987) Ecological stability in the context of multispecies fisheries. *Canadian Journal of Fisheries and Aquatic Science*, **44** (Suppl. 2), 84–94.

Pope, J.G. & Knights, B.J. (1982) Comparison of length distributions of combined catches of all demersal fishes in surveys in the North Sea and at Faroe Bank. In: *Multispecies Approaches to Fisheries Management Advice* (ed. M.C. Mercer), *Canadian Special Publication in Fisheries and Aquatic Science*, **59**, 116–18.

Pope, J.G., Stokes, T.K., Murawski, S.A. & Idoine, S.I. (1988) A comparison of fish size-composition in the North Sea and on Georges Bank. In: *Ecodynamics; Contributions to Theoretical Ecology* (eds W. Wolff, C.-J.Soeder & F.R. Drepper), pp. 146–52. Springer, Berlin.

Quéro, J.-C. & Cendrero, O. (1996) Incidence de la pêche sur la biodiversité ichtyologique marine: le bassin d'Arcachon et le plateau continental sud Gascogne. *Cybium*, **20**, 323–56.

Reid, P.C., Lancelot, C., Gieskes, W.W.C., Hagmeier, E. & Weichart, G. (1990) Phytoplankton of the North Sea and its dynamics: a review. *Netherlands Journal of Sea Research*, **26**, 295–331.

Rice J. & Gislason, H. (1996) Patterns of change in the size spectra of numbers and diversity of the North Sea fish assemblage, as reflected in surveys and models. *ICES Journal of Marine Science*, **53**, 1214–25.

Richards, J., Armstrong, D.W., Hislop, J.R.G., Jermyn, A.S. & Nicholson, M.D. (1978) Trends in Scottish research-vessel catches of various fish species in the North Sea. *Rapports et Procesverbaux des Reunions Conseil International pour l'Exploration de la Mer*, **172**, 211–24.

Rijnsdorp, A.D., Leeuwen, P.I. van, Daan, N. & Heessen, H.J.L (1996) Changes in the abundance of demersal fish species in the North Sea between 1906–1909 and 1990–1995. *ICES Journal of Marine Science*, **53**, 1054–62.

Roff, D.A. (1984) The evolution of life history parameters in teleosts. *Canadian Journal of Fisheries and Aquatic Sciences*, **41**, 989–1000.

Rogers, S.I. & Ellis, J.R. (in press) Changes in the demersal fish assemblages of British coastal waters in the 20th century. *ICES Journal of Marine Science*.

Rogers, S.I. & Millner, R.S. (1996) Factors affecting the annual abundance and regional distribution of English inshore demersal fish populations: 1973 to 1995. *ICES Journal of Marine Science*, **53**, 1094–112.

Rosenzweig, M.L. (1995) *Species Diversity in Time and Space*, 436 pp. Cambridge University Press, Cambridge.

Rothschild, B.J. (1991) Multispecies interactions on Georges Bank. *ICES Marine Science Symposia*, **193**, 86–92.

Sherman, K. (1991) The large marine ecosystem concept: research and management strategy for living marine resources. *Ecological Applications*, **1**, 349–60.

Smith, T.D. (1994) *Scaling Fisheries: The Science of Measuring the Effects of Fishing, 1855–1955*. Cambridge University Press, Cambridge.

Southwood, T.R.E. (1976) Bionomic strategies and population parameters. In: *Theoretical Ecology: Principles and Applications* (ed. R.M. May), pp. 26–48. Blackwell Science, Oxford.

Turrell, W.R., Slesser, G., Payne, R., Adams, R.D. & Gillibrand, P.A. (1996) Hydrography of the East Shetland Basin in relation to decadal North Sea variability. *ICES Journal of Marine Science*, **53**, 899–916.

De Vooys, C.G.N. & van der Meer, J. (1998) Changes between 1931 and 1990 in by-catches of 27 animal species from the souther North Sea. *Netherlands Journal of Sea Research*, **39**, 291–8.

Walker, P.A. & Heessen, H.J.L. (1996) Long-term changes in ray populations in the North Sea. *ICES Journal of Marine Science*, **53**, 1085–93.

Walker, P.A. & Hislop J.R.G. (1998) Sensitive skates or resilient rays? Spatial and temporal shifts in ray species composition in the central and northwestern North Sea between 1930 and the present day. *ICES Journal of Marine Science*, **55**, 392–402.

Wood, H. (1956) Fisheries of the United Kingdom. In: *Sea Fisheries: Their Investigation in the United Kingdom* (ed. M. Graham). Arnold, London.

Yang, J. (1982) An estimate of the fish biomass in the North Sea. *Journal du Conseil International pour l'Exploration de la Mer*, **40**, 161–72.

Chapter 15
Impacts of fishing on diversity: from pattern to process

S. JENNINGS[1,2] and J.D. REYNOLDS[1]

[1]*School of Biological Sciences, University of East Anglia, Norwich, NR4 7TJ, UK*
[2]*Centre for Environment, Fisheries and Aquaculture Science, Lowestoft Laboratory, Pakefield Road, Lowestoft, NR33 0HT, UK*

Summary

1. Fishing has led to reductions in the diversity of fish and invertebrate communities in the north-east Atlantic.
2. Diversity can be measured in many ways. Some approaches emphasise species richness, while others emphasise the distribution of individuals among species and the evolutionary relatedness among species.
3. Reductions in fish diversity result from the direct mortality of target species rather than the indirect effects of fishing on trophic relationships.
4. Reductions in invertebrate diversity result from the effects of towed gears on the seabed. This effect is particularly apparent on stable substrates, but may not be detectable where mobile sediments are continually resuspended by waves and tides.
5. It is difficult to separate biogeographical patterns in diversity from patterns induced by fishing. Large-scale studies of fishing effects on invertebrate diversity can only proceed if the spatial resolution of fishing effort data is reduced to metres rather than tens of kilometres.
6. Links between fish or invertebrate diversity and the stability or productivity of marine communities are not known. We should aim to start describing both pattern and process in order that we can describe the effects of fishing on ecosystem function.
7. Considerable resources are required for diversity studies and diversity is not a particularly sensitive measure of fishing effects. An alternative is to use multivariate methods to identify indicator species that are vulnerable to fishing. Studies of the abundance and distribution of these species would provide a cost-effective approach for identifying areas impacted by fishing.

Keywords: fishing, diversity, fish, scale, multivariate analysis.

Introduction

Biodiversity, a contraction of biological diversity, can be defined as the variability among living organisms from all sources, including diversity within species, between species and of ecosystems (Gaston, 1996). Biodiversity loss is often viewed as the most undesirable consequence of human activities on earth. In this chapter, we consider the effects of fishing on species diversity in the north-east Atlantic and discuss ways in which studies of diversity could be improved to detect fishing effects and to determine

---

relationships between diversity and ecosystem processes. We focus on species diversity because this is the focus of existing research and not because it is necessarily of greater significance than intraspecific diversity or diversity among ecosystems.

We begin by reviewing methods for measuring species diversity. Historically, there have been many attempts to identify the 'best' diversity measures, but the measures describe different aspects of diversity from species richness to the distribution of individuals among species and evolutionary relatedness. Since no single measure encompasses all aspects of diversity, the favoured approach is to report a range of measures. Many diversity measures have been used in studies of the effects of fishing. We describe these measures, their advantages and disadvantages.

We proceed to describe studies of the effects of fishing on fish and invertebrate diversity. The studies of fish diversity are based on long time-scales, over 50 years in some cases, and spatial scales of hundreds of square kilometres. Invertebrate studies, conversely, are based on simultaneous or short-term comparisons between fished and unfished areas. The studies demonstrate that fishing often leads to reductions in diversity but, apart from the loss of species or habitats that may be of conservation concern, it is not clear whether diversity loss should be viewed as 'good' or 'bad'. We propose that patterns in diversity should be linked to patterns in community stability or productivity and that vulnerable species may be better indicators of fishing effects than changes in diversity. We conclude the chapter by suggesting areas for future research.

Measurement of diversity

The following review is not comprehensive, but covers the techniques that have been used in studies of fishing effects. There have been many attempts to describe diversity using a single statistic that is amenable to statistical analysis (Table 15.1). These statistics usually encapsulate species richness or the distribution of individuals among species. They range from counts of the total number of species recorded (species richness) to statistics that indicate both richness and the way in which the total number of individuals is divided amongst the total number of species (equitability). When using equitability indices, a community of high evenness and low dominance is generally considered more diverse than one with low evenness and high dominance (where few species account for most of the total abundance).

It is clear from Table 15.1 that different indices describe different aspects of diversity and that any one index cannot encapsulate all of these. Species richness is highly dependent on sampling effort and can be used only for comparing samples of the same size; as such, species richness is often used as a measure of species density per unit area. The Margalef and Menhinick's indices combine the number of species recorded (S) and the total number of individuals of all species (D). They are effectively measures of the number of species present for a given number of individuals. The Shannon–Wiener and Brillouin indices indicate species richness and the distribution of individuals among species. These indices do not make assumptions about the underlying species abundance distribution. Both indices

Table 15.1 Some widely used indices of diversity. Further details in Hill (1973), Magurran (1988) and Clarke & Warwick (1994)

Index	Measures	Formula	Notes
Species richness (S)	Number of species	S	
Margalef (D)	Number of species for given number of individuals	$(S - 1)/\ln N$	1
Menhinick's (D)	Number of species for given number of individuals	$S/\sqrt{N}$	
Shannon–Wiener (H')	Richness and equitability	$-\sum p_i \ln p_i$	2
Evenness (for H')	Evenness	$H'/H_{\max}$ or $H'/\ln S$	
Brillouin (HB)	Richness and equitability	$(\ln N! - \sum \ln n_i)/N$	3
Pielou evenness (for HB)	Evenness	$HB/HB_{\max}$	
Simpson's (D)	Dominance	$\sum (n_i(n_i - 1)/N(N - 1))$	4
Hill N_0	Number of species	S	5
Hill N_1	Number of 'abundant' species	$\exp H'$	
Hill N_2	Number of 'very abundant' species	$1/D$	6
Taxonomic diversity (Δ)	Species diversity with taxonomic separation	$\left[\sum\sum_{i<j} \omega_{ij} x_i x_j\right]/[n(n-1)/2]$	7
Taxonomic distinctness (Δ^*)	Taxonomic distinctness without species diversity	$\left[\sum\sum_{i<j} \omega_{ij} x_i x_j\right]/\left[\sum\sum_{i<j} x_i x_j\right]$	7
Taxonomic distinctness (Δ^+)	Taxonomic distinctness for presence/absences data	$\left[\sum\sum_{i<j} \omega_{ij}\right]/[s(s-1)/2]$	7

Notes: 1. Where S = number of species and N = number of individuals. 2. Where p_i is the proportion of individuals of the ith species. Shannon–Weiner index assumes random sampling from an infinitely large population and that all species present are represented in the sample. In reality, the true value of p_i is unknown and is estimated from $p_i = n_i/N$. 3. Alternative to Shannon–Wiener used to describe a known collection. The form appropriate to an infinitely large community would be $\sum p_i^2$, where p_i is the proportion of individuals in the ith species. n is the number of individuals in the ith species. 4. Gives probability that any two species drawn at random from an infinitely large community belong to different species, since D increases with decreasing diversity this index is often expressed as $1-D$ or $1/D$. 5. Hill proposed a family of diversity measures ranging from those that emphasise uncommon species (richness) to those that emphasise dominance. N_0, N_1 and N_2 cover most aspects of diversity and are usually reported together. Note relationships between Hill numbers and the Shannon–Wiener and Simpson indices. 6. Where D is Simpson's index. 7. x_i is abundance of ith of s species observed, $n = \sum_i x_i$ is total number of individuals in a sample, and ω_{ij} weights the path length linking species i and j through the taxonomy.

have related measures of evenness that describe how evenly individuals are distributed amongst species. Simpson's index measures dominance and is weighted towards the abundance of the commonest species rather than the total number of species. Dominance can be regarded as the opposite of evenness. The differences between indices indicate that several measures are needed to describe patterns of species richness, evenness and dominance in a community. Hill (1973) proposed that a series of indices should be reported to describe aspects of diversity ranging from species richness (N_0) through evenness (N_1, effectively the number of abundant species) to dominance (N_2, effectively the number of very abundant species).

Diversity ordering procedures are used to overcome some of the inconsistencies that arise when different diversity indices are used to compare communities. For example, a comparison of Shannon–Wiener indices may suggest that one community is more diverse than another, while a comparison of Simpson indices suggests that it is not (Tóthmérész, 1995). As such, the communities cannot be ordered on the basis of their diversity. Diversity ordering procedures involve calculating diversity indices for a range of α values, where α determines the relative weighting towards species richness or dominance in the diversity measure. The Réyni (1961) index has been used in studies of fishing effects on diversity where

$$H_\alpha = \left(\ln \sum p_i^\alpha \right) / (1 - \alpha)$$

where p_i is the proportional abundance of the ith species. Once H_α has been calculated for a range of α values, H_α is plotted against α. If the trajectories for two communities cross, they are non-comparable, but when one trajectory falls consistently below the other, that community is said to be less diverse. Hill (1973) and others (Tóthmérész, 1995) have defined similar families of diversity indices. Thus, the Hill numbers N_0, N_1 and N_2 (Table 15.1) are indices for three specific values of α (0, 1 and 2) in the Hill (1973) family.

The indices described above capture only a very specific aspect of 'biodiversity', namely patterns in species abundance. Yet most people intuitively include some notion of genetic or phylogenetic hierarchy into their concept of 'biodiversity' (May, 1990). The last three indices in Table 15.1 attempt to capture this formally (Warwick & Clark, 1995; Clarke & Warwick, 1999). For example, consider the species in Fig. 15.1a, which are more distantly related to each other, and therefore more diverse taxonomically (and genetically) than those in Fig. 15.1b. One could focus solely on taxonomic depths and ignore taxonomic evenness (Faith, 1994), but this would not distinguish between Fig. 15.1c and d. Arguably, Fig. 15.1c is more diverse, when richness is combined with taxonomic diversity (Clarke & Warwick 1999).

To quantify these aspects of evolutionary diversity, Warwick & Clarke (1995) and Clarke & Warwick (1999) introduced three indices. Taxonomic diversity (Δ) can be thought of as Simpson diversity, with an additional component of taxonomic separation. It is the mean path length along a taxonomic hierarchy between any two randomly chosen individuals. Δ^* is similar, but with a reduced role of species abundance, such that it measures the mean path length between any two randomly chosen individuals, conditional on them being from different species. The simplest variant on this approach ignores abundance information altogether, such that Δ^+ measures taxonomic path lengths in presence/absence data. Clarke & Warwick

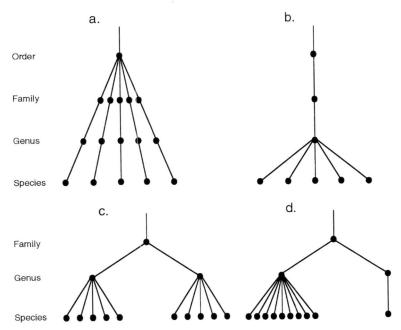

Fig. 15.1 Examples of taxonomic hierarchies for species in samples. Mean evolutionary path lengths between individuals are greater in a than in b. A balanced tree is shown in c, which is arguably more taxonomically diverse than d. (After Clarke & Warwick, 1999.)

(1999) have shown that these taxonomic indices have the very desirable property of being independent of sample size, a considerable advance over traditional indices such as Shannon richness. Furthermore, Warwick & Clark (1995) showed that the indices were able to pick up reductions in taxonomic distinctness of benthic organisms near an oilfield in the North Sea which could not be seen with three traditional indices: Shannon species diversity, richness (Margalef's *D*), and evenness (Pielou's *J*).

While diversity indices offer some statistical convenience, they have several undesirable properties. In particular, the reduction of information provides little information on the overall distribution of individuals between species. An alternative is to look at the actual relationships between species and abundances using graphical techniques. Changes in the relationships over time or comparisons among fished and unfished areas can indicate the effects of fishing or other stresses on community structure. Plots of the number of individuals (*y*) against the number of species (*x*) within a sample are commonly known as rarefraction curves, and communities of high diversity produce curves of steeper gradient and greater elevation (Fig. 15.2a). Various models have been proposed to fit the relationships between numbers of individuals and numbers of species. The most widely used methods for reporting species abundance relationships are based on the assumption that they loosely conform to log normal distributions, geometric series or log series (e.g. May, 1975; Magurran, 1988). The geometric series generally applies when few species are

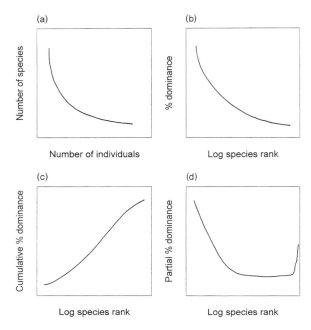

Fig. 15.2   Species abundance plots that have been used to describe community diversity: (a) rarefaction curve; (b) dominance curve; (c) k dominance curve; (d) partial dominance curve. See Platt *et al.* (1984), Magurran (1988) and Clarke (1990).

dominant and the log series or log normal when species of intermediate abundance are more common.

Four methods of reporting species abundance relationships have been used for the study of human impacts on fish and invertebrate communities. Dominance curves are the abundance of each species, after ranking by abundance, expressed as a percentage of the total abundance of all species and plotted against log rank (Fig. 15.2b). k dominance curves are the cumulative ranked abundance against log species rank, such that the most elevated curves show the lowest diversity (Fig. 15.2c; Platt *et al.*, 1984). The cumulative abundance scale is often log or modified log transformed (Clarke, 1990). Since dominance curves can be dramatically affected by individual species of high abundance, Clarke (1990) proposed the use of partial dominance curves where the dominance of a species at a given rank is compared only with species at lower ranks. With partial dominance curves (Fig. 15.2d), preceding ranks do not affect later points on curve. The most elevated curves show the lowest diversity.

Fishing effects on diversity

While many aspects of diversity have ecological significance, studies of the effects of fishing have largely focused on species diversity. Fishing may affect species diversity by killing target and non-target species of fish or invertebrates, by changing habitat structure and by altering the ecological relationships between species. Beam and otter trawls are used widely in the north-east Atlantic, and have effects on the diversity of benthic communities. In stable sediments, such as those found in deeper areas not affected by wave stress, trawling has been shown to reduce the species diversity of the benthic invertebrate fauna, to reduce evenness and increase dominance. Kaiser & Spencer (1996), for example, looked at the effects of beam trawling on benthic communities at an experimental study site in the Irish Sea. In fished areas, species richness, Shannon–Wiener and Simpson indices were lower. The Shannon–Wiener index showed the greatest reductions in diversity in fished areas, suggesting that a decrease in the abundance of rare species contributed most to the difference between fished and unfished systems. This was supported by *k* dominance plots which indicated a shift towards a few dominant species in fished areas.

Epifaunal invertebrates that structure marine habitats may be particularly vulnerable to disturbance by towed fishing gears. Trawling is well known to reduce the density of tube-building worms such as *Sabellaria* and cerianthid anemones (Riesen & Reise, 1982; Langton & Robinson, 1990). In a study of fishing effects on Georges Bank, Collie *et al.* (1997) demonstrated the consequences of reductions in habitat complexity that follow trawling. In fished areas, species richness and dominance were lower. The reduction in dominance was attributed to the loss of structuring species such as bryozoans and hydroids that were very abundant in unfished areas. The reduction in diversity was due to the loss of many small species such as polychaetes that formerly lived in association with the biogenic fauna. Biogenic structures that are not attached to the seabed, such as serpulid tubeheads, may be pushed aside by the pressure wave that precedes the passage of a trawl. Tubeheads are formed by the overlapping calcareous tubes of serpulid worms, and harbour a diverse community of small invertebrates. Kaiser *et al.* (1999) compared tubehead density and the diversity of associated fauna at fished and unfished sites in the Irish Sea. There was no evidence to suggest that trawling had adverse effects on these biogenic structures and their fauna.

Studies of the effects of fishing on invertebrate diversity have tended to focus on comparisons between fished and unfished areas. There are few long-term and large-scale studies of fishing effects on invertebrate communities other than those based on the comparisons of surveys repeated in the same areas after intervals of many years (e.g. Riesen & Reise, 1982; Kröncke, 1990). Some of these studies have shown reductions in abundance of species that are assumed to be vulnerable to fishing. However, the replication or level of taxonomic identification is usually inadequate to look in detail at diversity and it is virtually impossible to disentangle the effects of fishing and other influences such as climate change and eutrophication.

Fish communities have been sampled on larger spatial scales and over longer periods than invertebrate communities and the effects of fishing on diversity are increasingly well known. In particular, the long-term trawl surveys of the North Sea demersal fish community have provided evidence for the changes in diversity that are associated with intensive size-selective fishing. Rice & Gislason (1996) examined changes in Shannon–Wiener indices, by length classes, for the North Sea demersal fish assemblage from 1977 to 1993 (Fig. 15.3): a period of increasing fishing intensity (Jennings *et al.*, 1999a). The slope of the diversity spectra did not change consistently with time (Fig. 15.4). Greenstreet & Hall (1996) looked at longer-term changes by comparing the North Sea groundfish assemblage, as sampled by Scottish trawl surveys, in the periods 1929–1953 and 1980–1993. They calculated Hill's N_1 (exponential of Shannon–Wiener index, Table 15.1) and Hill's N_2 (reciprocal of Simpson's index) for the whole groundfish assemblage and for non-target species in three areas: to the east of the Shetland Islands, in the central north-western North Sea and off the north-east coasts of the UK. In general, Hill's N_1 and N_2 for the whole assemblage declined between 1929 and 1953 and 1980 and 1993, while there was little change in the diversity of the non-target assemblage. Dominance was investigated using k dominance curves, and there were significant temporal differences between the entire assemblages in all three areas (Fig. 15.5). In every case, the assemblage in recent years is dominated by fewer species (the curves are steeper and more elevated). When species targeted by the fishery are excluded from the analysis, the difference is significant only on the east coast of the UK. The results suggest that direct effects of fishing on target species have an important impact on diversity but that there have been limited indirect effects on the non-target assemblage. In a subsequent study, Greenstreet *et al.* (1999) extended the data series

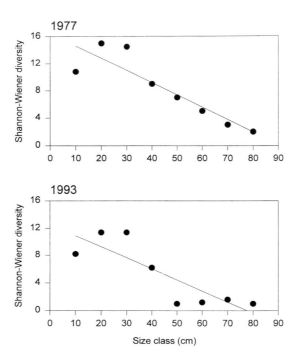

Fig. 15.3 Relationship between Shannon–Wiener diversity and 10-cm size class (diversity plotted at upper limit of class) for the North Sea demersal fish community in 1977 and 1993. Data from Rice & Gislason (1996).

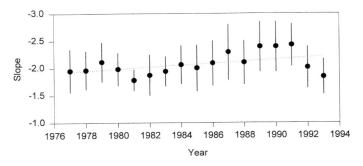

Fig. 15.4 Relationships between the slopes of annual Shannon–Wiener diversity spectra (Fig. 15.2 shows the slopes for 1977 and 1993 only) and time for the North Sea demersal fish community in the period 1977–1993. Data from Rice & Gislason (1996).

and described diversity changes in the period 1925–1996. Species diversity declined in the areas where fishing intensity was highest but fishing effects were largely confined to decreases in the abundance of those few species, such as skates and rays, which have high ages at maturity and are particularly vulnerable to exploitation (Jennings *et al.*, 1998). This suggests that changes in the abundance of vulnerable indicator species, rather than broad measures of community diversity, may provide a better measure of the effects of fishing on the North Sea fish community.

Rogers *et al.* (1999a) tried to link diversity changes with fishing effort by relating spatial variations in the diversity of target and non-target species in the North Sea, English Channel and Irish Sea to current patterns of fisheries exploitation. They used Rényi's diversity index family to rank the diversity of fish faunas throughout the region. The North Sea fauna was the least diverse, partly as a result of the uniform nature of the seabed. West of the Dover Strait, the more heterogeneous substrate supported a more diverse fauna of smaller-sized fish, with the occurrence of southern species such as red gurnard and thickback sole, and an increasing abundance of elasmobranchs. Rogers *et al.* (1998) looked at the same assemblage using species abundance methods such as *k* dominance plots, and these also showed that the fish communities were less diverse in the east (North Sea) than the west (English Channel and Irish Sea). Their analyses suggested that diversity patterns were largely explained by biogeographic factors, seabed structure and regional hydrography. They also confirmed the Greenstreet *et al.* (1999) hypothesis that the abundance of vulnerable species would provide a better indicator of fishing effects than diversity measures. Vulnerable species were those that contributed most to the dissimilarity between the structure of fished and unfished communities and they were identified using multivariate techniques such as similarity percentages analysis (Clarke & Warwick, 1994).

To date only two studies of fishes have used the new taxonomic indices proposed by Warwick & Clark (1995). Hall & Greenstreet (1998) examined a total dataset of 76 species from two time series in the northern North Sea: 1929–1953 and 1980–1993. They showed a very strong congruence between two measures of distinctness

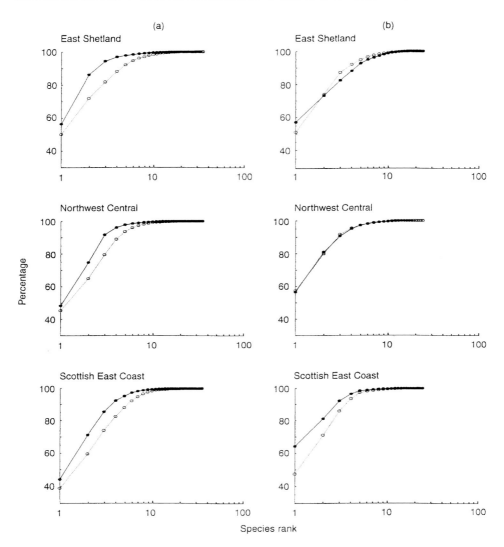

Fig. 15.5 *k* dominance curves for (a) all species and (b) non-target species in the North Sea groundfish assemblage during 1929–1953 (○) and 1980–1993 (●). East Shetland, Northwest Central and Scottish East Coast refer to the three sea areas for which analyses are presented. From Greenstreet *et al.* (1996).

(Δ and Δ^*), and two traditional diversity indices (Hill's N_1 and Hill's N_2). There were both increases and decreases over time. Although Hall & Greenstreet (1998) did not relate their results directly to fishing, they suggested that fisheries may have been a cause of the fluctuations documented, and they speculated that the indices could have changed together, because any perturbations that were large enough to have been picked up by traditional indices may have already affected the taxonomic indices.

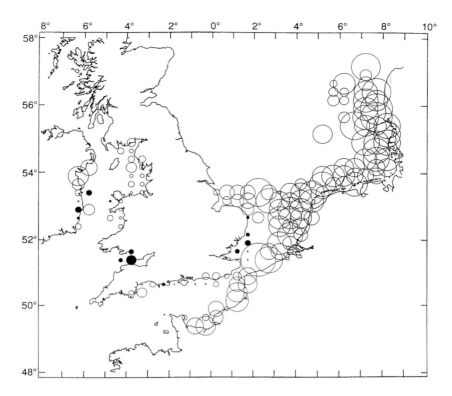

Fig. 15.6 Taxonomic distinctness (Δ^+) of benthic samples of fishes in the north-east Atlantic: ● indicate areas with samples that are more taxonomically distinct than the theoretical mean, while ○ indicate areas with fishes that are less taxonomically distinct than the theoretical mean. Larger circles represent greater differences from the mean. (From Rogers *et al.*, 1999a.)

In a wider study of coastal benthic fishes in the north-east Atlantic, Rogers *et al.* (1999a) found a strong trend toward greater taxonomic distinctness to the south and west of the British Isles than in the north-east (Fig. 15.6). As with Hall & Greenstreet (1998), no direct comparison was made with fishing effort, but it was noted that most of the spatial differences disappeared when data from elasmobranchs were removed from the analyses. Since the 'slow' life histories of sharks and rays is expected to render them particularly vulnerable to fisheries (Brander, 1981; Casey & Myers, 1998; Jennings *et al.*, 1998), it is possible that the spatial pattern could in part be due to differences in fishing mortality on these species.

Detecting changes in diversity

In general, fishing reduces the diversity of fish and invertebrate communities, but large-scale environmental and biogeographical factors have a greater influence on spatial patterns of diversity than fishing. In shallow waters affected by waves and tides, fishing effects on invertebrate diversity are difficult to detect because benthic communities are well adapted to the continual movement and redistribution of the sediment. Thus, trawling did not appear to affect invertebrate diversity in mobile sediments in the Irish Sea (Kaiser & Spencer, 1996). At present, the overriding role of biogeographical factors in determining invertebrate diversity in regions such as the North Sea (Dyer *et al.*, 1983; Heip *et al.*, 1992; Heip & Craeymeersch, 1995; Jennings *et al.*, 1999), coupled with a lack of accurate information on the spatial and temporal distribution of fishing effort, has hampered the study of fishing effects on large-scale patterns in diversity. Moreover, areas may be trawled or not trawled because they had different benthic communities to begin with. An experimental design based on comparing replicated fished–unfished 'treatments' within many biogeographic regions is not possible, because the spatial resolution of fishing effort data are too poor to identify areas that have, or have not, been trawled (Chapter 1). Most fishing effort data are currently collected within ICES rectangles (boxes of 0.5° latitude by 1° longitude), while replicate benthos samples can only be collected from areas of a few metres to a few hundred square metres. As such, it is never clear whether the area where the benthos is sampled has or has not been trawled. Satellite tracking devices will be added to EC fishing vessels in the future, and this will allow ecologists to design experiments in areas where they know the history of fishing activity.

There has been a tendency for studies of invertebrate diversity to focus on a restricted range of larger infaunal and epifaunal species. These species are often significant because they structure habitats and provide food for fishes of commercial significance. They are relatively easy to study because there are good taxonomic guides to assist with their identification, and they can often be sorted and identified without the aid of microscopes. Moreover, they can be sampled using gears such as small beam trawls, which integrate much of the small-scale patchiness in their distribution and reduce the levels of replication required to obtain acceptable statistical power to detect differences between control and treatment sites. The macroinvertebrates do, however, only account for a small proportion of the total species richness and biomass of the benthic invertebrate community, and changes in their diversity are not necessarily indicative of changes in the rest of the community. The first results of studies with taxonomic distinctness indices for invertebrates are promising (Warwick & Clark, 1995), since these proved more sensitive as environmental indicators than did traditional diversity indices. It would be well worthwhile to explore them further in the context of fishing effects.

There is little doubt that the study of invertebrate diversity at the species level requires high levels of sampling effort and taxonomic expertise. Indeed, in many parts of the north-east Atlantic, species identification still poses serious problems and only a limited number of specialists have the ability to consistently and accurately identify marine invertebrates. For the purposes of describing general regional patterns in diversity, there is much to be gained using higher taxonomic richness as a surrogate for species richness, an approach increasingly used in terrestrial environments (Balmford *et al.*, 1996). Moreover, if the aim is to use diversity indices for detecting the effects of fishing, in much the same way that diversity is used as a measure of pollution, there is little doubt that the resources required will limit such research to small-scale studies in a few locations. A better alternative would be to use full multivariate analyses (Clarke, 1993; Clarke & Warwick, 1994) to identify species that discriminate between fished and unfished communities and to use these species, and others with similar life histories, as indicators of fishing effects. The results of studies with both invertebrate and fish communities suggest that these species are better indicators of fishing effects than gross measures of community diversity (Greenstreet & Hall, 1996; Greenstreet *et al.*, 1999; Kaiser & Spencer, 1996: Rogers *et al.*, 1999a, b) and it is easier and more cost-effective to study fewer species. It would, of course, be necessary to use a selection of species with a sufficiently wide range of environmental preferences to ensure that changes in abundance and distribution due to fishing could be reliably separated from those due to environmental change.

Although diversity measures may not be the best indicators of fishing effects, they may be correlates of ecosystem stability or productivity. Relationships between diversity and stability have been widely explored by theorists, but there remain few comprehensive and realistic empirical tests of their hypotheses. Studies on terrestrial grassland suggest that biodiversity begets stability (Tilman & Downing, 1994; Tilman *et al.*, 1996) because diverse systems are more likely to contain some species that can thrive during a given environmental perturbation and thus compensate for the loss of species resulting from the disturbance. In the marine environment, we need to know whether changes in diversity have effects on ecosystem processes. Existing studies have consistently described pattern rather than process.

Conclusions

Fishing often leads to reductions in diversity but more small-scale experiments will do little to improve our understanding of fishing effects. We need to study invertebrate diversity on much larger scales, equivalent to those that have been used for fish, and to separate fishing and biogeographical effects. If these studies are to be successful, we will need more accurate information on the distribution of fishing effort, such as would be obtained from satellite tracking of vessel movements.

Diversity studies are not a particularly powerful means of detecting fishing effects on fish or invertebrate communities. This is because they are costly in time and expertise for the level of information they provide. It would be preferable to use full multivariate community analyses to identify indicator species that are particularly sensitive to fishing and then to monitor their distribution and abundance. This circumvents the problems of having to take representative samples of entire communities and lessens the breadth of taxonomic expertise required by those conducting the monitoring.

At present, there is no information on relationships between patterns in diversity and community process. The diversity measures we have described allow us to detect change, but do not allow us to state that the change is good or bad in relation to system processes. While convention dictates that low diversity is bad, we need empirical studies that link changes in diversity with changes in the stability or productivity of marine communities.

Acknowledgements

We wish to thank Simon Greenstreet and Stuart Rogers for permission to reproduce some of the figures in this article and John Cotter for his support. The senior author was funded by the European Commission.

References

Balmford, A., Jayasuriya, A.H.M. & Green, M.J.B. (1996) Using higher-taxon richness as a surrogate for species richness. II. Local applications. *Proceedings of the Royal Society*, **B263**, 1571–5.

Clarke, K.R. (1990) Comparisons of dominance curves. *Journal of Experimental Marine Biology and Ecology*, **138**, 143–57.

Clarke, K.R. (1993) Non-parametric multivariate analysis of changes in community structure. *Australian Journal of Ecology*, **18**, 117–43.

Clarke, K.R. & Warwick, R.M. (1994) *Change in Marine Communities: An Approach to Statistical Analysis and Interpretation*, 144 pp. Natural Environment Research Council, Plymouth, UK.

Clarke, K.R. & Warwick, R.M. (1999) A taxonomic distinctness index and its statistical properties. *Journal of Applied Ecology*, **35**, 523–31.

Collie, J.S., Escanero, G.A. & Valentine, P.C. (1997) Effects of bottom fishing on the benthic megafauna of Georges Bank. *Marine Ecology Progress Series*, **155**, 159–72.

Dyer, M.F., Fry, W.G., Fry, P.D. & Cranmer, G.J. (1983) Benthic regions within the North Sea. *Journal of the Marine Biological Association of the United Kingdom*, **63**, 683–93.

Faith, D.P. (1994) Phylogenetic pattern and the quantification of organismal biodiversity. *Philosophical Transactions of the Royal Society of London Series* B, **345**, 45–58.

Gaston, K.J. (1996) What is biodiversity. In: *Biodiversity: A Biology of Numbers and Difference* (ed. K.J. Gaston), pp. 1–9. Blackwell Science, Oxford.

Greenstreet, S.P.R. & Hall, S.J. (1996) Fishing and ground-fish assemblage structure in the north-western North Sea: an analysis of long-term and spatial trends. *Journal of Animal Ecology*, **65**, 577–98.

Greenstreet, S.P.R., Spence, F.E. & McMillan, J.A. (1999) Fishing effects in northeast Atlantic shelf seas: patterns in fishing effort, diversity and community structure. V. Changes in the groundfish species assemblage of the northwestern North Sea between 1925 and 1996. *Fisheries Research*, **40**, 153–84.

Hall, S.J. & Greenstreet, S.P.R. (1998) Taxonomic distinctness and diversity measures: responses in marine fish communities. *Marine Ecology Progress Series*, **166**, 227–9.

Heip, C., Basford, D., Craeymeersch, J.A., Dewarumez, J.-M., Dorjes, J., de Wilde, P., Duineveld, G., Eleftheriou, A., Herman, P.M.J., Niermann, U., Kingston, P., Künitzer, A., Rachor, E., Rumohr, H., Soetaert, K. & Soltwedel, T. (1992) Trends in biomass, density and diversity of North Sea macrofauna. *ICES Journal of Marine Science*, **49**, 13–22.

Heip, C. & Craeymeersch, J.A. (1995) Benthic community structures in the North Sea. *Helgoländer Meeresuntersuchungen*, **49**, 313–28.

Hill, M.O. (1973) Diversity and evenness: a unifying notation and its consequences. *Ecology*, **54**, 427–32.

Jennings, S., Reynolds, J.D. & Mills, S.C. (1998). Life history correlates of responses to fisheries exploitation. *Proceedings of the Royal Society of London*, **B265**, 333–9.

Jennings, S., Alvsvåg, J., Cotter, A.J., Ehrich, S., Greenstreet, S.P.R., Jarre-Teichmann, A., Mergardt, N., Rijnsdorp, A.D. & Smedstad, O. (1999a) Fishing effects in northeast Atlantic shelf seas: patterns in fishing effort, diversity and community structure. III. International fishing effort in the North Sea: an analysis of temporal and spatial trends. *Fisheries Research*, **40**, 125–34.

Jennings, S., Reynolds, J.D. & Greenstreet, S.P.R. (1999b) Structural change in an exploited fish community: a consequence of differential fishing effects on species with contrasting life histories. *Journal of Animal Ecology*, **68**, 617–27.

Jennings, S., Lancaster, J.E., Woolmer, A. & Cotter, A.J. (1999c) Distribution, diversity and abundance of epibenthic fauna in the North Sea. *Journal of the Marine Biological Association of the United Kingdom*, **79**, 385–99.

Kaiser, M.J. & Spencer, B.E. (1996) The effects of beam-trawl disturbance on infaunal communities in different habitats. *Journal of Animal Ecology*, **65**, 348–58.

Kaiser, M.J., Cheney, K., Spence, F.E., Edwards, D.B. & Radford, K. (1999) Fishing effects in northeast Atlantic shelf seas: patterns in fishing effort, diversity and community structure. VII. The effects of trawling disturbance on the fauna associated with the tubeheads of serpulid worms. *Fisheries Research*, **40**, 195–205.

Kröncke, I. (1990) Macrofauna standing stock of the Dogger Bank, a comparison: 1950–1954 versus 1985–1987. *International Council for the Exploration of the Seas*, C.M. 1990/Mini 3, 22 pp.

Langton, R.W. & Robinson, W.E. (1990) Faunal associations on scallop grounds in the western Gulf of Maine. *Journal of Experimental Marine Biology and Ecology*, **144**, 157–71.

Magurran, A.E. (1988) *Ecological Diversity and its Measurement*. Croom Helm, London.

May, R.M. (1975) Patterns of species abundance and diversity. In: *Ecology and Evolution of Communities* (eds M.L. Cody & J.M. Diamond), pp. 81–120. Harvard University Press, Cambridge, MA.

May, R.M. (1990) Taxonomy as destiny. *Nature*, **347**, 129–30.

McCann, K., Hastings, A. & Huxel, G.R. (1998) Weak trophic interactions and the balance of nature. *Nature*, **395**, 794–8.

Platt, H.M., Shaw, K.M. & Lambshead, P.J.D. (1984) Nematode species abundance patterns and their use in detection of environmental perturbations. *Hydrobiologia*, **118**, 59–66.

Réyni, A. (1961) On measures of entropy and information. In: *Proceedings of the 4th Berkeley Symposium on Mathematical Statistics and Probability* (ed. J. Neyman), pp. 547–61. University of California Press, Berkeley.

Rice, J. & Gislason, H. (1996) Patterns of change in the size spectra of numbers and diversity of the North Sea fish assemblage, as reflected in surveys and models. *ICES Journal of Marine Science*, **53**, 1214–25.

Riesen, W. & Reise, K. (1982) Macrobenthos of the subtidal Wadden Sea: revisited after 55 years. *Helgolander Meeresuntersuchungen*, **35**, 409–23.

Rogers, S.I., Rijnsdorp, A.D., Damm, U. & Vanhee, W. (1998) Demersal fish populations in the coastal waters of the UK and continental NW Europe from beam trawl survey data collected from 1990–1995. *Journal of Sea Research*, **39**, 79–102.

Rogers, S.I., Clarke, K.R. & Reynolds, J.D. (1999a) The taxonomic distinctness of coastal bottom-dwelling fish communities of the Northeast Atlantic. *Journal of Animal Ecology*, **68**, 769–82.

Rogers, S.I., Maxwell, D., Rijnsdorp, A.D., Damm, U. & Vanhee, W. (1999b) Fishing effects in northeast Atlantic shelf seas: patterns in fishing effort, diversity and community structure. IV.

Can comparisons of species diversity be used to assess human impacts on demersal fish faunas. *Fisheries Research*, **40**, 135–52.

Tilman, D. & Downing, J.A. (1994) Biodiversity and stability in grasslands. *Nature*, **367**, 363–5.

Tilman, D., Wedin, D. & Knops, J. (1996) Productivity and sustainability influenced by biodiversity in grassland ecosystems. *Nature*, **379**, 718–20.

Tóthmérész, B. (1995) Comparison of different methods for diversity ordering. *Journal of Vegetation Science*, **6**, 283–90.

Warwick, R.M. & Clarke, K.R. (1995) New 'biodiversity' measures reveal a decrease in taxonomic distinctness with increasing stress. *Marine Ecology Progress Series*, **129**, 301–5.

Part 5
Conservation methods, issues and implications for biodiversity

A typical catch from a commercial beam trawl of which a large proportion is composed of non-target fishes and invertebrates. Areas closed to fishing activities and no-take zones may protect benthic communities that are characterised by long-lived and emergent species, but their potential as fish stock management tools in temperate waters is probably limited to a few sedentary species such as scallops.

Chapter 16
Technical modifications to reduce the by-catches and impacts of bottom-fishing gears

B. VAN MARLEN

Netherlands Institute for Fisheries Research (RIVO-DLO), Haringkade 1, PO Box 68, NL-1970 AB IJmuiden, The Netherlands

Summary

1. Many techniques have been developed to improve the species and size selectivity of fishing gears and to reduce discards.
2. Given a proper and clear incentive, fishermen use techniques to improve gear selectivity (e.g. sorting grids, square mesh windows).
3. Application of these techniques contributes to stock conservation.
4. It is much more difficult to reduce mortality of benthic organisms due to demersal trawling, as these gears need bottom contact to achieve their required catch efficiency.
5. Possibly alternative stimulation techniques (electrical fields, water injection) could be applied, but these still require further research and development.
6. Techniques to release benthic animals from nets at sea may have a smaller, but nevertheless worthy, contribution to the conservation of benthic fauna.

Keywords: technical modification, conservation measures, bottom-fishing gear, by-catches, impact.

Introduction

The management problem of avoiding the capture of juvenile and undersized fish is very old, and mesh-size restrictions can be traced back to the Middle Ages (Anon., 1921). Mesh-size regulations are imposed on many fleets, on specific gears and are currently applied to fisheries for a wide range of species. Fish can only pass through meshes of a certain size (and shape) depending on their body dimensions, such as girth, and their morphology. If size selection were 'knife-edged', it would imply that above a certain length all fish are retained and below this length none is retained. However, size selection is rarely this well defined.

The term by-catch, 'unwanted' or 'wanted' by-catch are terms that are ill defined. Regulations imply that there are differences in the desire to catch certain species, depending on the fishing rights or quota allocated to fishermen. Unwanted by-catch is usually discarded, i.e. thrown overboard. Most of these animals have very low chances of survival, which is another issue that plays an important role in assessing the overall effects of by-catch on fish stocks.

Unwanted by-catch can be divided into four classes:

1. Species of commercial value
    ● Undersized species
    ● Fish for which there is no more quota
    ● Fish of lower value others of the same species
2. Species with low commercial value
    ● No market value
    ● Rare or protected species
3. Some invertebrates
4. Inert material and debris.

As selection is not precise, there is always a fraction of undersized fishes in the catch that have to be discarded. In mixed fisheries, the choice for a minimum mesh size for one species will lead to a mismatch in the percentage of undersized fish caught of another species, depending on differences in their size and behaviour. Apart from discarding these undersized fish, fishermen may also choose to discard fish of lower value to enable them to land fish of higher value within the constraints of their quota allocations. This behaviour is known as 'high-grading' (see Pascoe, this volume, Chapter 21).

In general, technical measures are aimed at striking a balance between fishing effort and the available resources. In particular, the aim is to limit the capture of undersized fish and species that have endangered stocks.

A large range of design features can be incorporated into fishing gears in order to improve species and size selectivity. The ability to separate species depends on mechanical sorting where a distinct difference in size exists and/or on behavioural differences between various species (see Wardle, 1995 for review). The ability of fish to respond to certain stimuli is limited by thresholds that are species and water temperature dependent. Light and sound play a major role, as well as the swimming ability of fish. Wardle (1995) advocated the use of colour and contrast in netting to stimulate escape behaviour by non-target individuals and species.

Technical modification of fishing gears to reduce capture of undersized fish and discards

Mesh size

The scientific reasoning behind mesh-size regulation is given in many studies aimed at quantifying selection parameters for particular gears. The use of selectivity and effort data in stock assessment is explained in a recent International Council for the Exploration of the Sea (ICES) study group report (Anon., 1998a), and a review of gear selectivity data obtained for many gears, areas and species is given by Wileman (1991). Generally, a larger mesh size will increase the selection factor and the 50% retention length, resulting in escape by a greater proportion of juvenile fish (Wileman *et al.*, 1996). The importance of the mesh size in determining the 50%

retention length of fish has been demonstrated clearly by Ferro & Graham (1998) and is justification for the legislative control of codend mesh size.

Separator panels

Observations that different species behave differently when swimming within towed nets led to research to examine the possibility of exploiting this behaviour to release non-target species. For instance, haddock (*Melanogrammus aeglefinus*) and whiting (*Merlangius merlangus*) rise and even pass over the headline of demersal trawls, whereas cod (*Gadus morhua*) tend to dive down and appear in the lower parts of the net (Main & Sangster, 1982). Studies in Norway confirmed some of the basic differences in the behaviour of different species in a demersal trawl fitted with a horizontal separator panel (Fig. 16.1). Flatfish and cod appeared in the lower parts of the net while saithe (*Pollachius virens*) were caught in the higher parts (Valdemarsen *et al.*, 1985; Engås & West, 1995). Horizontal panels were also introduced in the *Nephrops* fishery in the Irish Sea to select out the unwanted fish such as juvenile haddock and whiting (Hillis, 1989). The majority of *Nephrops* can be guided into a codend attached to the lower half of the net, whilst whiting and haddock are diverted to a codend attached to the top half. Such devices have the advantage that they reduce the sorting time on deck and the amount of discarded material caught by the gear. Eventually, the use of these horizontal separator panels was made compulsory through legislation. Separator panels with diamond and square mesh were also tried out in whiting trawls. Cod once again orientated towards the lower parts of the net while whiting were caught in the top section (Moth Poulsen, 1994; Wileman, 1995). More recently, behavioural studies in US waters

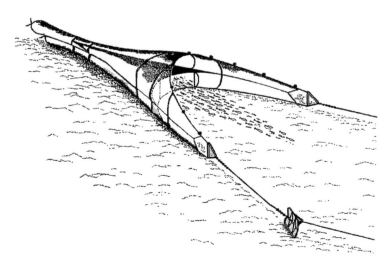

Fig. 16.1   Horizontal separator panel in a towed trawl net.

have demonstrated the benefits of inserting a horizontal separator panel to reduce fish by-catches in the squid (*Loligo paelei*) fishery (Glass *et al.*, 1998).

Sorting grids or grates

Rigid sorting grids first originated in the Norwegian prawn (*Pandalus borealis*) fishery in which there are large differences in the size and morphology of fish and prawns. These grids usually consist of a rigid metal frame placed inside a section of the trawl that allows prawns to pass through, while larger white fish are guided towards an escape hole. Various designs were tested, e.g. the 'Nordmøre Grid' and the 'Sort-X', and recently the 'Sort-V', which is simpler in design. The latter is lighter, which may improve safety when this design is used in heavy sea conditions (Fig. 16.2). At present these grids are used in many fleets throughout the world either on a voluntary basis or through mandatory enforcement (Anon., 1998b). Most designs consist of grids with parallel bars running lengthwise in the direction of the line of tow. Meillat *et al.* (1994) recommended a combination grid with vertical and horizontal bars in a mixed fishery for monkfish (*Lophius piscatorius*), megrim (*Lepidorhombus whiffiagonis*), rays (*Raja* spp.) and hake (*Merluccius vulgaris*) in the Celtic Sea and the Bay of Biscay. Smaller versions were designed and tested successfully for the brown shrimp (*Crangon crangon*) fishery (Graham & Radcliffe, 1998; Polet, 1998) and, as a result, the European Commission has suggested that these grids should become mandatory in the shrimp fishery, starting in the year 2000. Suuronen (1995) investigated the possibility of improving selectivity in the Baltic herring fishery by using various configurations of a sorting panel in the top of the codend. They also attempted to manipulate the water flow through the net to lead the fish towards the grids and enhance the selection mechanism by so-called flow boosters mounted underneath the codend (Fig. 16.3). Similar attempts were made to find a suitable grid configuration that would separate herring (*Clupea harengus* L.), mackerel (*Scomber scombrus* L.) and horse mackerel (*Trachurus trachurus* L.) from mixed catches (van Marlen *et al.*, 1994; van Marlen, 1995). Catch size affected the efficiency with which fish were sorted and this was further complicated by the similar behaviour of these species in the trawl. The use of a grid can sort 19% or more of small mackerel from catches (Beltestad & Misund, 1993).

A grid inserted in the net of a beam trawl used to catch brown shrimps was shown to improve the size selectivity of shrimps because the cleaner (i.e. less by-catch) catches meant that the net meshes remained relatively unobstructed. However, this grid had no effect on the selectivity of flatfish and also resulted in a 10–15% loss of marketable shrimp. In addition, operational problems may occur due to the grid's clogging with seaweed or starfish (Anon., 1998b; Graham & Radcliffe, 1998; Polet, 1998). A considerable improvement in size selectivity of finfish was found in the Barents Sea when using the 'Sort-X' grid in shrimp trawls. However, extremely high catch rates led to a decrease in sorting efficiency.

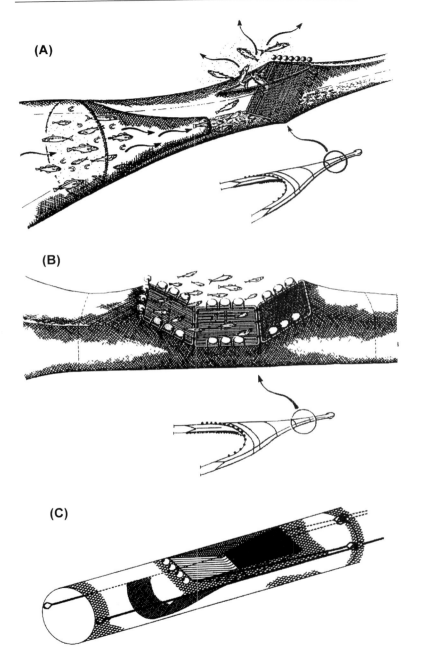

Fig. 16.2 Various grids designs to separate non-target species from target species: (A) 'Nordmøre' grid; (B) 'Sort-X' grid; (C) 'Sort-V' grid.

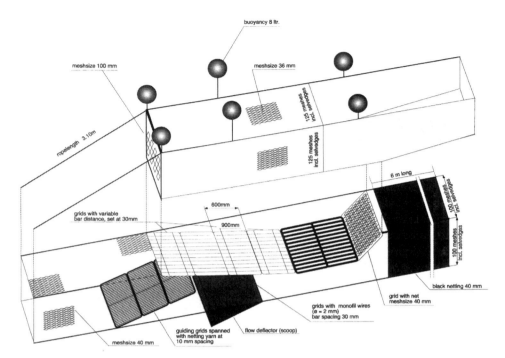

Fig. 16.3   Flow boosters in grid configuration.

The impact of introducing grids in various fisheries can be marked, e.g. in the Canadian *Pandalus borealis* fishery, the use of grids resulted in a decrease in the by-catch of juvenile fish of between 10 000 and 15 000 t year^{-1} since its adoption by the industry in 1994. The percentage of by-catch in catches dropped from about 15% to 2%. It is estimated that the use of grids in the Faroese lemon sole and plaice fishery would avoid the annual by-catch of 330 t of cod and 350 t of haddock. Similarly, a 50–70% reduction in the by-catch of undersized cod and haddock could be achieved in the Barents Sea groundtrawl fishery.

Square-meshed codends

Mesh shape in relation to the cross-sectional dimensions of a fish determines its chances of escape. In conventional codends, the meshes tend to close owing to the load of accumulating catch. Hanging meshes on the square (i.e. with one of the mesh bars in the transverse direction and the other in the lengthwise direction) produces a more cylindrical codend shape and thus the meshes open more effectively (Robertson & Stewart, 1988). For demersal species such as haddock and whiting, this technique was found to release a large number of juveniles. The potential to release Baltic cod was demonstrated by Larsson *et al.* (1988), who found that the use of a square-mesh codend in demersal trawls could significantly reduce the proportion of the juvenile cod in catches. However, square meshes do not improve the escape of flatfish species (Fonteyne & M'Rabet, 1992). There are practical problems, however, when applying square hanging using conventional netting. Isaksen & Valdemarsen (1986) reported that net distortion occurred frequently in Norwegian experiments on demersal trawls and seines. Distortion can be avoided by the use of knotless netting, but this is not easily purchased and is more complicated to repair. While better selectivity was found for cod and haddock using square-mesh codends, redfish (*Sebastes marines* L.) were enmeshed in greater numbers. Canadian experiments also indicated the potential of square-mesh codend to release juvenile fish (Cooper & Hickey, 1989, 1994), although again they were less effective for flatfish (Walsh *et al.*, 1989).

Square-meshed panels or windows

Placing the square meshes in a panel or window in front of the codend seems to avoid many practical problems and opposition from fishermen experienced with square-mesh codends (Hillis *et al.*, 1991; Briggs & Robertson, 1993). These windows were introduced in Ireland, the UK and Iceland in demersal trawls in the *Nephrops* fishery and have been mandatory since 1992 in the UK and 1994 in Ireland (Fig. 16.4). The optimum position of these panels remains a topic of debate. Irish researchers advocated a more forward position (Briggs *et al.*, 1996). The insertion of a square-mesh panel did not affect the catches of *Nephrops* in Danish experiments, but some losses of marketable round fish such as haddock and whiting occurred (Lowry *et al.*, 1995). In the Baltic cod fishery, several designs of square-mesh windows were tried without significant improvement in selectivity for this species (Madsen *et al.*, 1998). An interesting variant called the Swedish exit window is placed in the lower sides of the codend (Anon., 1995b; Suuronen *et al.*, 1995; Tschernij *et al*, 1996). In the Portuguese crustacean fishery sorting panels in combination with square-mesh windows were tried out in 1993 and 1994 (Campos *et al.*, 1996). The authors reported some success, but also losses of marketable hake, which will prevent commercial introduction. The work done by Brewer *et al.* (1998) also incorporated investigations of square-mesh windows in the Australian prawn fishery. Such windows were found to reduce the level of by-catch of undersized fish significantly, especially in combination with grids.

Codend modification

Design characteristics other than mesh size have proven to affect selectivity, these include the number of meshes in the circumference of the codend and its length (Robertson & Ferro, 1988; Reeves *et al.*, 1992). Adding more meshes to the circumference of the codend tends to close down the individual meshes as does increasing the length of the extension piece. Fishermen can utilise these design

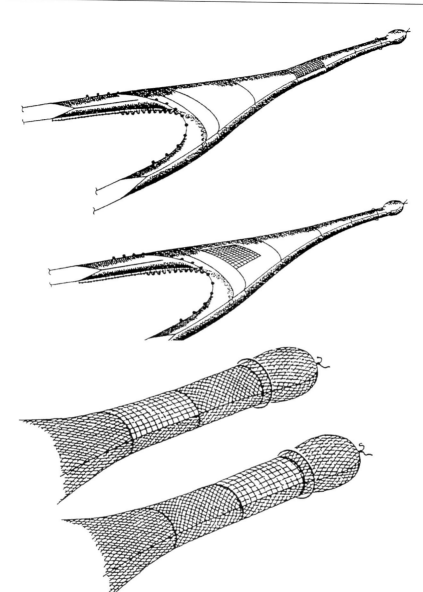

Fig. 16.4 Square-mesh windows as used in trawl nets.

characteristics to manipulate the selectivity of their gears, i.e. to catch more fish and still adhere to the regulations relating to mesh and codend characteristics. This research lead to additional technical constraints in the design of nets, such as the 100-mesh rule for demersal trawls.

Experiments in the cod fishery using shortened lastridge ropes aimed to avoid closure of the meshes under strain in the codend (Isaksen & Valdemarsen, 1990).

These ropes are shortened by 12–15% and, while selectivity can be enhanced, the method is relatively simple and yet successful in a number of fisheries (Anon., 1995a).

Increasing twine thickness has a negative effect on escape of (juvenile) fish (Lowry & Robertson, 1994; Lowry, 1995; Ferro & Graham, 1998) and *Nephrops* (Polet & Redant, 1994). As fish have to wriggle through a mesh, the force that has to be applied to open it determines whether such escape attempts will be successful. Thicker twines are usually also stiffer. The European Commission has suggested therefore a limit of 8 mm for codend twine material as a new technical measure to improve the survival of escaping fish.

Escape openings

Many techniques introduced successfully in otter-trawl fisheries did not seem to work for flatfish beam trawls. The reasons might be differences in fish morphology, behavioural characteristics and the higher towing speeds used. Several new design features have been developed recently, including a top panel made of very large diamond or hexagonal meshes that appeared to be effective in releasing (juvenile) whiting and cod from V-nets (Fig. 16.5a) (van Marlen, 1993; Fonteyne, 1997). Similarly, cutting a large hole in the top panel seemed an easy and relatively effective way of releasing juvenile roundfish, although the effect might be better in longer nets (Fig. 16.5b). An alternative possibility is the use of a top panel of square meshes that also increases the escape of juvenile roundfish (Fonteyne, 1997). Distortion of the square meshes might be limited by lacing two sections of net to a centre line (Fig. 16.5c).

Technical modifications to reduce mortality and capture of benthic organisms

Alternative stimulation

Earlier chapters in this book demonstrate clearly the effects of tickler chains on benthic fauna. If these chains could be replaced by some other means of stimulating fish in the seabed, the adverse effects on benthic fauna might be averted. Pumping water into the seabed through a system of hoses as an alternative method of stimulation has been investigated recently, but currently catches fall short in comparison with a conventional tickler chain gear of similar dimensions. Preliminary results show a tendency of decreasing catches of benthos, but the catches of target flatfish such as sole, plaice and dab remain unsatisfactory (K. Lange, personal communication).

Electrical stimulation has been investigated since the early 1970s in various countries as a means of catching brown shrimps and sole (de Groot & Boonstra,

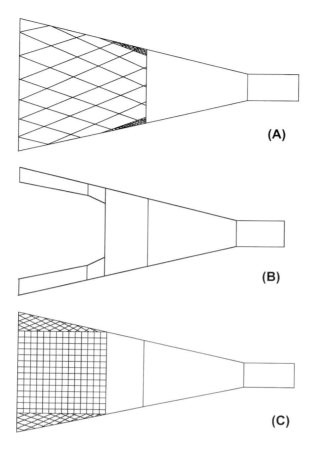

Fig. 16.5 Various net modifications designed to reduce catches of roundfish in beam trawls: (A) large mesh top panel in a beam trawl; (B) reduced or cut-away top panel in a beam trawl; (C) square-mesh top panel in a beam trawl.

1970; vanden Broucke, 1973; de Groot & Boonstra, 1974; Horn, 1976; Stewart, 1979; Horton, 1984; Agricola, 1985; van Marlen, 1985). Although earlier research was motivated through possible reductions in gear drag and fuel consumption, the potential to reduce impacts on the seabed has become a new incentive for this research. This technique is effective for catching sole (van Marlen & de Haan, 1988). A key problem was the robustness of an electrified beam trawl. Most systems were complex and vulnerable to damage with considerable downtime in the harsh conditions of fishing at sea. Other problems are the large investment needed and the prolonged payback time with current relatively low fuel prices (van Marlen, 1988). Recent research in collaboration with the fishing industry in The Netherlands indicates the potential to reduce the catches of some benthic species while maintaining sole catches. Further research is needed to improve this technique and to evaluate the effects of electrical stimuli on marine life.

Release holes

Beam-trawl fishermen use several types of release holes in their nets that reduce by-catches of invertebrates, but these designs are poorly documented. Increasing mesh size from 80 to 90 mm in the lower panel of a beam-trawl net did not significantly reduce catches of benthos, but did lower catches of sole and undersized flatfish (Fonds & Blom, 1996). In a similar experiment, large meshes were cut out behind the footrope of a beam trawl in sequential order. The catches of the modified trawl were compared to those of a standard beam trawl fished simultaneously, resulting in a series of pair-wise experiments (unpublished data). Cutting away nine meshes did not seem to affect the catches of target species, nor did it reduce catches of benthos significantly. Apparently, the large mesh section needs to be extended further aft, which was tested in a series of experiments (Fig. 16.6). Configurations b and c in Fig. 16.6 yielded significantly reduced catches of sole. Nevertheless, benthic species such as hermit crabs (*Pagurus* sp.), prickly cockle (*Acanthocardia echinata*) and quahogs (*Artica islandica*) were released in greater proportions from the nets with large meshes. Although it was not observed directly, a possible explanation is that these species pass over the footrope and sink more quickly in the water flow owing to their weight and, consequently, pass through the large holes cut in the net. Lighter species, such as the sea mouse (*Aphrodita aculeata*), pass over these holes and are still captured in the net.

Conclusions

Many techniques can be used to reduce by-catches and the impacts of fishing gears operating on the seabed. However, it is difficult to generalise solutions and make them applicable to all metiers in the fishery. As gears and local circumstances differ greatly, solutions should be aimed at specific sectors in the fishing industry. Legislation should allow for these differences to be effective and acceptable, and be kept as simple as possible. Co-operation in the industry can be expected if clear incentives are given. A good example is the threat of closure of fishing grounds when by-catches exceed a predefined level. Research can help the fishing industry to design more efficient and environmentally effective gears. It should be emphasized that researchers have tools and methods often not available to fishermen (e.g. flume tanks, direct observation techniques), which can considerably speed up the process of development compared with conventional trial and error methods.

The aim of reducing the impact on benthic organisms is far more difficult to achieve than releasing undersized fish. Many fishing gears depend on good contact with the seabed, and the catches of many target species drop when bottom contact is reduced. Replacing tickler chains in beam trawls by the use of alternative stimuli is possible, and research and development along these lines should be encouraged. Changes in the design of conventional nets, such as drop-out zones in the lower net panel, offer limited benefits for the conservation of benthic invertebrates. Most

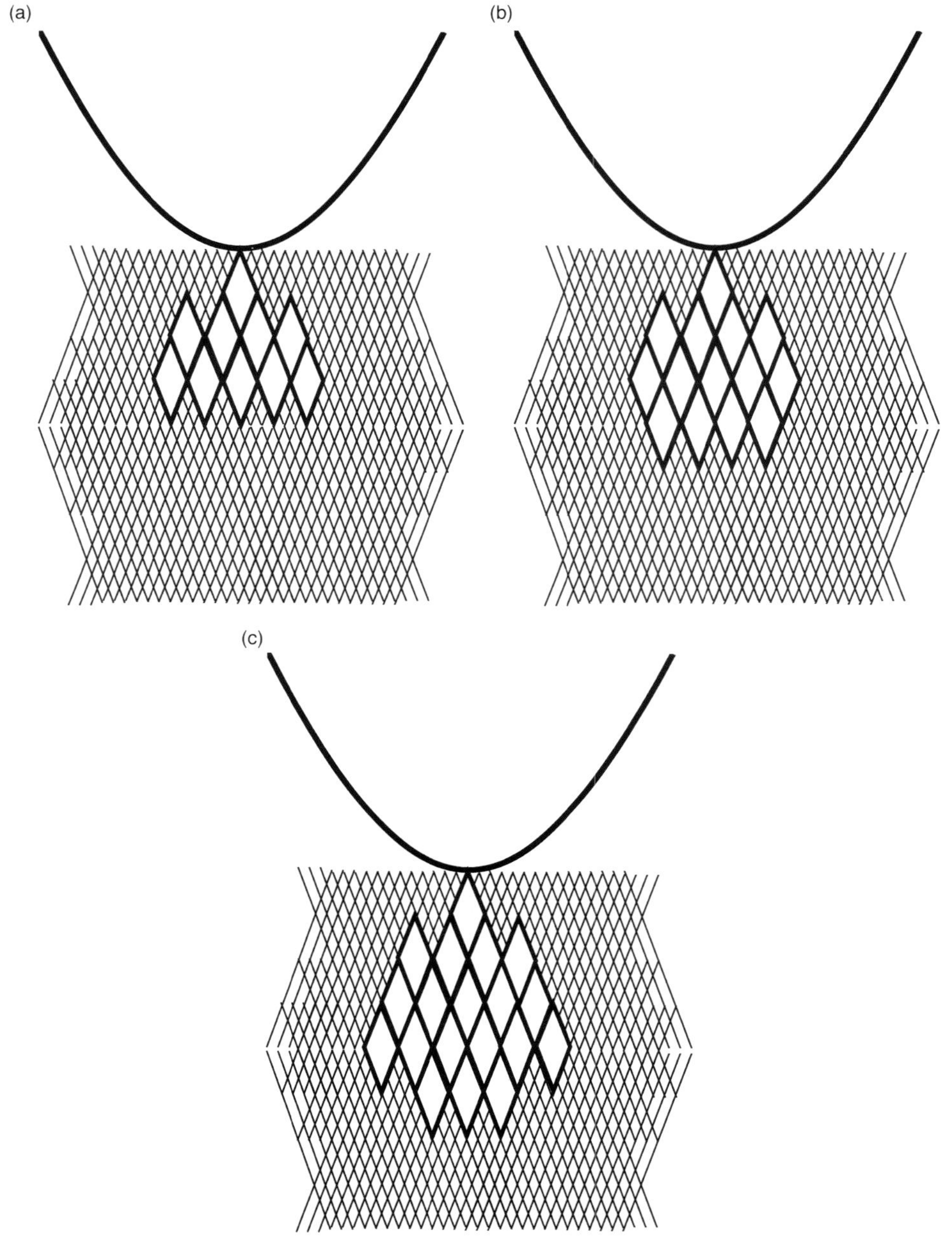

Fig. 16.6 Various designs of large meshes in the lower panel of beam trawls, which are designed to lower catches of benthos. Only the design shown in (c) had a significant effect by lowering by-catches.

mortality occurs in the trawl path and does not result from capture in the net (Bergman & van Santbrink, this volume, Chapter 4). Nevertheless, one could also argue that providing the design changes are simple and easy to implement, there is no harm in legislating them, even if the effect may be relatively small.

References

Agricola, J.B. (1985) Experiments on electrical stimulation of flatfish in beamtrawling during 1984. *International Council for the Exploration of the Sea*, C.M. 1985/B:36, 22 pp.

Anon. (1921) The history of trawling: its rise and development from the earliest times to the present day. *Fish Trade Gazette*, **38** (1974), 21–71.

Anon. (1995a) *Consideration of recently tested methods for changing the selectivity of towed fishing gears. Final Report of EU Concerted Action* AIR2-CT93-1656, March 1995, 120 pp.

Anon. (1995b) Report of the working group on fishing technology and fish behaviour. *International Council for the Exploration of the Sea*, C.M. 1995/B:2, 75 pp.

Anon. (1998a) Report of the study group on the use of selectivity and effort measurements in stock assessment. *International Council for the Exploration of the Sea*, C.M. 1998/B:6, 77 pp.

Anon. (1998b) Report of the study group on grid (grate) sorting systems in trawls, beam trawls and seine nets. *International Council for the Exploration of the Sea*, C.M. 1998/B:2, 62 pp.

Van Beek, F.A. (1982) On the effects of mesh enlargement in the North Sea sole fishery. *International Council for the Exploration of the Sea*, C.M. 1982/B:39, 11 pp.

Beltestad, A. & Misund, O.A. (1993) Sorting of mackerel by grids in purse seines and trawls. *Fisken og havet*, No. 8-1993, 10 pp.

Briggs, R.P., Armstrong, M.J. & Rihan, D. (1996) The use of parallel haul and twin-trawl techniques to study optimum position of square mesh escape panels in *Nephrops* trawls. *International Council for the Exploration of the Sea*, C.M. 1996/B:39, 15 pp.

Briggs, R.P., & Robertson, R.H.B. (1993) Square mesh panel studies in the Irish Sea Nephrops fishery. *International Council for the Exploration of the Sea*, C.M. 1993/B:20, 10 pp.

vanden Broucke, G. (1973) Further investigations on electrical fishing. *International Council for the Exploration of the Sea*, C.M. 1973/B:14, 6 pp.

Campos, A., Fonseca, P. & Wileman, D.A. (1996) Experiments with sorting panels and square mesh windows in the Portuguese crustacean fishery. *International Council for the Exploration of the Sea*, C.M. 1996/B:15, 16 pp.

Cooper C.G. & Hickey W.M. (1989) Selectivity experiments with square mesh codends of 130, 140, and 155mm. In: *Proceedings of World Symposium of Fishing Gear and Fishing Vessel Design*, Maritime Institute, St. John's, Newfoundland, Canada , 1988, pp. 52–9.

Cooper C.G. & Hickey W.M. (1994) *Selectivity Experiments with Square Mesh Codends on Haddock and Cod*, 10 pp. Department of Fisheries and Oceans, Canada.

Creutzberg, F., Duineveld, G.C.A. & van Noort, G.J. (1987) The effect of different numbers of tickler chains on beam trawl catches. *Journal du Conseil International pour l'Exploration de la Mer*, **43**, 159–68.

Engås, A. & West, C.W. (1995) Development of a species-selective trawl for demersal gadoid fisheries. *International Council for the Exploration of the Sea*, C.M. 1995/B + G + H + J + K, 20 pp.

Ferro, R.S.T. & Graham, G.N. (1998) Recent Scottish data on demersal fish selectivity. *International Council for the Exploration of the Sea*, C.M. 1998/OPEN:3.

Fonds, M. & Blom, W.C. (1996) *Onderzoek naar mogelijkheden tot vermindering van discardproductie door technische aanpassing van boomkornetten. BEON Report* 96-15, 42 pp.

Fonteyne, R. (1997) *Optimization of a species selective beam trawl. Final Report to the European Commission*, Project AIR2-CT93-1015, SOBETRA, 250 pp.

Fonteyne, R. & M'Rabet, R. (1992) Selectivity experiments on sole with diamond and square mesh codends in the Belgian coastal beam trawl fishery. *Fisheries Research*, **13**, 221–33.

Glass, C.W., Sarno, B., Milliken, H.O., Morris, G.D. & Carr, H.A. (1998) Squid (*Loligo paelei*) reactions to towed fishing gears; the role of behaviour in bycatch reduction, *International Council for the Exploration of the Sea*, C.M. 1998/M:04, 15 pp.

Graham, N. & Radcliffe, C. (1998) By-catch reduction in the brown shrimp, *Crangon crangon*, fishery using a rigid separation grid. Paper presented at the *International Council for the Exploration of the Sea*, FTFB-WG meeting, La Coruña, 14 pp.

De Groot, S.J. & Boonstra, G.P. (1970) Report on the development of an electrified shrimp-trawl in The Netherlands. *International Council for the Exploration of the Sea*, C.M. 1970/B:5, 6 pp.

De Groot, S.J. & Apeldoorn, J. (1971) Some experiments on the influence of the beam trawl on the bottomfauna. *International Council for the Exploration of the Sea*, C.M. 1971/B:2, 5pp.

De Groot, S.J. & Boonstra, G.P. (1974) The development of an electrified shrimp-trawl in The Netherlands. *Journal du Conseil International pour l'Exploration de la Mer*, **35**, 165–70.

Hillis, J.P. (1989) Further separator trawl experiments on Nephrops and whiting. *International Council for the Exploration of the Sea*, C.M. 1989/**B:46,** 6 pp.

Hillis, J.P., McCormick, R., Rihan, D. & Geary, M. (1991) Square mesh experiments in the Irish Sea. *International Council for the Exploration of the Sea*, C.M. 1991/B:58, 21 pp.

Horn, W. (1976) Rationalization of sole fisheries by means of electrified beam trawls. Report of the working group on research on engineering aspects of fishing gear, vessels and equipment. International Council for the Exploration of the Sea, C.M. 1976/B:7, 63–6.

Horton, R.S. (1984) *Trials of the electric beam trawling system on MFV Zuiderkruis, summer 1983. Seafish Report* IR 1180, 25 pp. Hull, UK.

Isaksen, B. & Valdemarsen, J.W. (1986) Selectivity experiments with square mesh codends in bottom trawl. *International Council for the Exploration of the Sea*, C.M. 1986/B:28, 18 pp.

Isaksen, B. & Valdemarsen, J.W. (1990) Codend with short lastridge ropes to improve size selectivity in fish trawls. *International Council for the Exploration of the Sea*, C.M. 1990/B:46, 8 pp.

Kvalsvik, K., Misund, O.A., Gamst, K. Skeide, R. Svellingen, I. & Vetrhus, H. (1998) Size selectivity experiments using sorting grid in pelagic mackerel (*Scomber scombrus*) trawl. Paper presented at the *International Council for the Exploration of the Sea*, FTFB-WG meeting, La Coruña, 7 pp.

Larsson, P.O., Claesson, B. & Nyberg, L. (1988) Catches of undersized cod in codends with square and diamond meshes. *International Council for the Exploration of the Sea*, C.M. 1988/B:57, 8 pp.

Lowry, N. (1995) The effect of twine size on bottom trawl codend selectivity. *International Council for the Exploration of the Sea*, C.M. 1995/B:6, 15 pp.

Lowry, N. & Robertson, J.H.B. (1994) The effect of twine thickness on codend selectivity of trawls for haddock in the North Sea. *International Council for the Exploration of the Sea*, C.M. 1994/B:34, 13 pp.

Lowry, N., Knudsen, L.H. & Wileman, D.A. (1995) Selectivity in Baltic cod trawls with square mesh codend windows. *International Council for the Exploration of the Sea*, C.M. 1995/B:5, 18 pp.

Madsen, N. Moth-Poulsen, T. & Lowry, N. (1998) Selectivity experiments with window codends fished in the Baltic Sea cod (*Gadus morhua*) fishery. *Fisheries Research*, **36**, 1–14.

Main, J. & Sangster, G.I. (1982) *A study of a multi-level bottom trawl for species separation using direct observation techniques. Scottish Fisheries Research Report* No. 26, 17 pp.

Van Marlen, B. (1985) Report of a seminar on electro-fishing at RIVO-IJmuiden on 24 January 1985. *International Council for the Exploration of the Sea*, C.M. 1985/B:37, 6 pp.

Van Marlen, B. (1988) A note on the investment appraisal of new fishing techniques. *International Council for the Exploration of the Sea*, C.M. 1988/B:16, 24 pp.

Van Marlen, B. (1993) EC project FAR TE-2–554 'Improved selectivity of fishing gears in the North Sea fishery – beam trawling'. *International Council for the Exploration of the Sea*, C.M. 1993/B:13, 26 pp.

Van Marlen, B. (1995) Recent developments in selective mid-water trawling. EC-project TE-3–613 (SELMITRA). *International Council for the Exploration of the Sea*, C.M. 1995/B+G+H+J+K, 32 pp.

Van Marlen, B. & de Haan, D. (1988) *Elektrische stimulering van platvis (verleden, heden en toekomst)*. *RIVO Report* TO 88–06, 97 pp.

Van Marlen, B., Lange, K., Wardle, C.S., Glass, C.W. & Ashcroft, B. (1994) Intermediate results in EC-project TE-3–613 'Improved species and size selectivity of midwater trawls (SELMITRA.)' *International Council for the Exploration of the Sea*, C.M. 1994/B:13, 14 pp.

Meillat, M., Dupouy, H., Bavouzet, G., Kergoat, B., Morandeau, F., Gaudou, O. & Vacherot, J.P. (1994) Preliminary results of a trawl fitted with a selective grid for the fishery of benthic species from Celtic Sea and Bay of Biscay. *International Council for the Exploration of the Sea*, C.M. 1994/B:23, 15 pp.

Moth Poulsen, T. (1994) Development of a species selective whiting trawl. *International Council for the Exploration of the Sea*, C.M. 1994/B:22, 12 pp.

Polet, H. (1998) Experiments with sorting grids in the Belgian brown shrimp (*Crangon crangon*) fishery. Paper presented at the *International Council for the Exploration of the Sea*, FTFB-WG meeting, La Coruña.

Polet, H. & Redant, F. (1994) Selectivity experiments in the Belgian Norway lobster (Nephrops norvegicus) fishery. *International Council for the Exploration of the Sea*, C.M. 1994/B:39, 22 pp.

Reeves, S.A., Armstrong, D.W., Fryer, R.J., & Coull, K.A. (1992) The effects of mesh size, codend extension length and codend diameter on the selectivity of Scottish trawls and seines. *ICES Journal of Marine Science*, **49**, 279–88.

Robertson, J.H.B. & Ferro, R.S.T. (1988) *Mesh selection within the codends of trawls. The effects of narrowing the codend and shortening the extension. Scottish Fisheries Research Report*, No. 39, 11 pp.

Robertson, J.H.B. & Stewart, P.A.M. (1988) A comparison of size selection of haddock and whiting by square and diamond mesh codends. *Journal du Conseil International pour l'Exploration de la Mer*, **44**, 148–61.

Stewart, P.A.M. (1979) A study of the response of flatfish (Pleuronectidae) to electrical stimulation. *Journal du Conseil International pour l'Exploration de la Mer*, **37**, 123–9.

Suuronen, P. (1995) Conservation of young fish by management of trawl selectivity. Ph.D. thesis, University of Helsinki, Finland, 130 pp.

Suuronen, P. Lehtonen, E., Tschernij, V. & Larsson, P. (1995) Skin injury and mortality of Baltic cod escaping from trawl codends equipped with exit windows. *International Council for the Exploration of the Sea*, C.M. 1995/B:8, 13 pp.

Tschernij, V., Larsson, P-O, Suuronen, P. & Holst, R. (1996) Swedish trials in the Baltic Sea to improve selectivity in demersal trawls. *International Council for the Exploration of the Sea*, C.M. 1996/B:25, 22 pp.

Valdemarsen, J.W., Engås, A. & Isaksen, B. (1985) Vertical entrance into a trawl of Barents Sea Gadoids as studied with a two level fish trawl. *International Council for the Exploration of the Sea*, C.M. 1985/B:46, 16 pp.

Walsh, S.J., Cooper, C. & Hickey, W. (1989) Size selection of plaice by square and diamond mesh codends. *International Council for the Exploration of the Sea*, C.M. 1989/B:22, 13 pp.

Wardle, C.S. (1995. A review of fish behaviour in relation to species separation and selectivity in mixed fisheries. *International Council for the Exploration of the Sea*, C.M. 1995/B+G+H+J+K, 11 pp.

Wileman D.A. (1991) *Codend selectivity: updated review of available data. Final Report to the European Commission* 1991/15, , 200 pp. Directorate General for Fisheries. DIFTA, North Sea Centre, Hirtshals, Denmark.

Wileman, D.A. (1995) Review of the progress made in Denmark in the development of species selective gears. *International Council for the Exploration of the Sea*, C.M. 1995/ B + G + H + J + K, 19 pp.

Wileman, D.A., Ferro, R.S.T., Fonteyne, R. & Millar, R.B. (1996) Manual of methods of measuring the selectivity of towed fishing gears. *International Council for the Exploration of the Sea Cooperative Research Report*, No. 215, 126 pp.

Chapter 17
Fishing and cetacean by-catches

N.J.C. TREGENZA

Institute of Marine Studies, University of Plymouth, Drake Circus, Plymouth, PL4 8AA, UK

Summary

1. Incidental by-catch of cetaceans occurs mainly in gill-net and midwater trawl fisheries in Europe. Harbour porpoise by-catch in bottom-set gill-nets appears to be the most serious and is at levels that may greatly reduce some populations of this species, which has already shown major declines around most European coasts. Common dolphins are incidentally caught by several fisheries, of which midwater trawls take the largest number. Most strandings of this species are a consequence of fishery interactions.
2. In some cases cetaceans actively seek out fishing gear to feed on the entangled fish, while in others they are not attracted to the gear but encounter it incidentally while feeding or travelling. These different behavioural patterns have implications for the methods of mitigation that might be effective. Most of the behavioural descriptions of these interactions needed to devise optimal, low-energy, mitigation methods is lacking, and, in particular, little is known of the patterns of sonar use in different gear interactions.
3. The use of battery-powered 'pingers' to frighten away porpoises is at present the only method of by-catch reduction in gill-net fisheries that has been shown to work, at least in the short term, but it has serious drawbacks, and no method of reducing midwater trawl by-catch has been discovered.
4. A self-contained submersible device to log cetacean sonar clicks is described as a new research tool to reveal cetacean responses to net features or to potential aversive or alerting stimuli.
Keywords: by-catch, gill-net, pelagic trawl, sonar, cetacean, porpoise, dolphin

Introduction

Cetaceans are greatly valued, in aesthetic terms, by the public in many countries and by-catch is often seen as a problem even when it does not threaten to reduce population sizes significantly (Jones, this volume, Chapter 23). Table 17.1 shows some of the major by-catches known in the North Sea and north-east Atlantic and their possible significance to the cetacean populations. Assessing whether these by-catches threaten a population requires knowledge of the population size, stock identity, by-catch rate and the potential growth rate of the population.

Population sizes in the North and Celtic Seas were assessed by the multinational SCANS survey (Hammond *et al.*, 1995), which gave us a single population estimate from the summer of 1994 and is the basis for the figures in Table 17.1. The porpoise *Phocoena phocoena* population in the North Sea, including a large area north of

Table 17.1   Some known annual cetacean by-catches in Europe, excluding the Mediterranean

Species	Fishery	Location	Number	Significant?
Harbour porpoise	Bottom-set gill-nets (2)	North Sea	4500 by Danish gill-netters	2.6% local population; 1.7% whole North Sea pop.
		Celtic Sea	2300 by UK and Irish offshore netters	6.2% Celtic Sea pop.
		By-catch occurs everywhere that set gill-nets and porpoises coexist. It has been recorded or studied in Iceland, Faeroes, Norway, Sweden, Germany, France, Portugal, Spain		
	Drift nets (1&2)	Some recorded by-catch in Baltic, Danish Belt Seas, North Sea, West coast of Ireland, Norwegian waters		
Common dolphin	Pelagic trawls (3)	Especially in fisheries for mackerel, clupeoid fishes, tuna. Correlated with stranding events, some > 600 animals in 2 weeks, on west coasts of Europe. Total probably more than 1000		
	Tuna drift nets (1&2)	Biscay albacore fishery 400 +		
	Bottom-set nets (2)	Recorded from Portugal, Spain, Celtic Sea		
Pilot whale	Pelagic trawls (3)	Observed in North Sea. Possibly indicated by strandings record south-west UK		
Striped dolphin	Tuna drift nets (1&2)	Biscay albacore fishery 1100 +		
Whitesided dolphin	Pelagic trawls (3)	Especially in mackerel fisheries SW of Ireland		

Numbers in brackets refer to by-catch mechanisms listed in Table 17.2. Data compiled after Anon. (1995), Collet & Mison (1995), Couperus (1998), Fertl & Leatherwood (1998), Goujon *et al.* (1993), Hammond *et al.* (1995), Kirkwood *et al.* (1997), López-Fernández & Valeira-Mata (1997), Morizur *et al.* (1995), Tregenza *et al.* (1997), Vinther (1994).

Scotland, was 270 000 with 36 000 found in the Celtic Sea. None was seen in the English Channel and very few in the south-westerly parts of the North Sea.

Stock identities remain poorly understood for common dolphins *Delphinus delphis*, but some evidence from morphological, biochemical and genetic studies exists for porpoises, in which a range of studies has shown a difference between the eastern and western sides of the British Isles (Walton, 1997). There is evidence of a further distinction of these stocks from the larger animals of the Iberian peninsula.

By-catch rates

The occurrence of by-catches is now documented in more detail as more observer programmes are carried out. For porpoises caught in bottom-set gill-nets, an unresolved problem is that approximately half of the animals fall out of the net during hauling and some are only seen as 'floaters' alongside the boat. It must be presumed that some are missed at the surface and some never reach the surface at all. Nevertheless, the size of this fraction of the by-catch remains unknown (Tregenza, 1997).

By-catches in pelagic trawls present a problem of sample size, as the evidence from reports of strandings points towards occasional large by-catches, which have not been revealed by any of the small fishery observation programmes conducted so far (Tregenza & Collet, 1998). These observations (Morizur *et al.*, 1995) have shown by-catches of dolphins at about 1 per 100 h of trawl tow. Larger by-catches may nevertheless occur as the interaction between dolphins and pelagic trawl fisheries is one in which there is strong spatial and temporal clumping of three key elements – the prey, the fishing effort and the predator. It may be that high cetacean by-catches occur in years when prey numbers are low or their spatial distribution is exceptionally restricted. Such sporadic very high by-catch rates are known in gill-net fisheries in Norway, where catches of many thousands of seals have occurred in years when depletion of food resources has caused a major southward movement of the seals into the main fishery areas.

To date the potential growth rates of cetacean populations have been modelled on better-known terrestrial mammals including man, and upper limits to growth rates of 4–10% have been obtained (Barlow & Boveng, 1991; Woodley & Read, 1991). There are drawbacks to this approach; for example, it takes no account of density-dependent effects that result from the direct depletion of cetacean stocks by fishing. Furthermore, the damaging effects of pollution might actually cause a negative growth rate. Estimation of potential growth rates is important for management purposes, and efforts are under way to establish the necessary biological parameters directly from cetacean studies.

Criteria for conservation action

To define when costly or difficult conservation action might be needed, various biological criteria have been set by a number of management or advisory structures. These have defined stocks that are either deemed to have 'favourable conservation status' (ASCOBANS; The agreement on the conservation of small cetaceans in the Baltic and North Seas), or are at a level that is 'above half of their carrying capacity' (European Commission STCEF), while limits for incidental takes have been variously set at 1% (IWC Scientific Committee) or 2% (European Commission STCEF).

The most sophisticated criterion in use is the potential biological removal (PBR) defined for the implementation of the requirements of the Marine Mammal Protection Act in the USA. The PBR was formulated explicitly to consider uncertainty to base management on parameters that could be estimated and to provide incentives to gather better data. The PBR uses a minimum population estimate, which allows for uncertainty in such estimates (the lower 5% of the distribution of the abundance estimate), multiplied by half the estimated potential growth rate. The PBR also incorporates a recovery factor that can be adjusted between 0.1 and 1.0 to allow for chance occurrences that might affect a population's potential growth rate, e.g. the outbreak of a population-threatening virus. If the PBR is exceeded by reported by-catches, management can be taken against the responsible fisheries to prevent further by-catches. The PBR has become established as a management tool where the previous criterion of maintaining populations above their maximum net productivity level had failed.

In Europe, there is no legislation comparable to the Marine Mammal Protection Act, hence action has been consequent on a moving compromise between bodies of the European Commission. Definitive action has been enforced mainly with respect to drift net fisheries, which have been recently restricted. ASCOBANS currently addresses the need for a clearer framework for conservation actions with respect to small cetacean species.

Are cetacean numbers below carrying capacity?

Trends in cetacean numbers are of great interest, but are known only for a few stocks around the world where intensive studies have been undertaken in recent years. The population recoveries that occurred in response to the end of whaling have been conclusively shown for some great whales, notably the grey whale, which is thought to be now approaching its carrying capacity in the eastern Pacific. Previously, this species was greatly depleted in the Pacific and entirely eliminated from the Atlantic.

Small cetaceans in Europe have shown historical declines in many areas, with notable declines of porpoises in the Baltic, southern North Sea and English Channel. A retrospective survey of sightings by residents of Cornwall in the south-west of England showed a decline of over 95% in sighting rates of small cetaceans from the coast in the last 50 years (Tregenza, 1992). This appeared to involve both porpoise and bottlenose dolphin sightings, and precedes the use of monofilament gill-nets. Historical data confirm this trend, with, for example, records of river management boards in the south-west of England that paid for the destruction of porpoises as recently as the 1940s in rivers in which they cannot now be seen even on prolonged searching. The relationship of these inshore sighting rates to total numbers remains unknown.

Monitoring cetacean populations as a means of assessing their need for conservation action is not an effective approach for small populations, as the uncertainty of population estimates will rise even further as their population size falls.

By-catch mechanisms

Four patterns of incidental capture have emerged which relate to whether the cetaceans are attracted to the gear and the behaviour of the cetaceans. These are shown in Table 17.2 with notes on the use of sonar by cetaceans in each situation. These differences are significant as they determine both the identity of the species at risk and the type of behavioural modification that would be needed to prevent capture of cetaceans. A further category occurs outside Europe, where the net is actively moved towards the cetaceans, as in the encircling of spinner and other dolphins in the tuna purse-seine fishery in the eastern tropical Pacific.

A new tool for by-catch research

Cetaceans are difficult to observe in proximity to nets, and the deployment of on-line hydrophones to record their use of sonar is a complex task. Satellite tags can show cetacean surface positions and dive depths, but not their position relative to nets at any point in time. In order to obtain some data on cetacean behaviour around nets,

Table 17.2   A behavioural classification of cetacean by-catch in fisheries

Type	Example	Species caught	Cetacean sonar	Notes
No attraction to gear				
(1) Travelling	Surface driftnets for salmon or tuna	Every species in the locality of surface gill-nets	Often not in use, and Mysticete species do not have sonar	By-catch rates in small odontocetes increase with sea state
(2) Feeding	Bottom-set gill-nets	Harbour porpoise, less often common dolphin	Not known, presumed to be usually 'on'	Prey species may not include any target species of the fishery
With attraction to gear				
(3) Predation on fish in trawls	Pelagic trawls in all areas	Social odonto-cetes that approach boats, e.g. common dolphin	Not known – some sonar use has been recorded	Usually some cetacean prey species and target species are the same, but common dolphins in the tuna fishery are an exception
(4) Predation on fish on lines	Porpoise on long lines in Faeroes	Various odontocete species may learn this feeding method	Not known	Entanglement of cetaceans uncommon; but retaliation by fishers may be more significant where fish damage is common

the Cornwall Wildlife Trust has developed a fully submersible, self-contained sonar click logger with support from the European Commission and the Body Shop Foundation. This device – 'the POD' or porpoise detector (more details can be viewed on the Internet at http://www.chelonia.demon.co.uk) – can log clicks detected in three different frequency bands over a period of weeks at depths down to 200 m. The POD is potentially capable of logging clicks from bottlenose dolphins and common dolphins as well as from porpoises. Data from each period of immersion, including ambient noise levels, are downloaded to a PC after retrieval.

The POD has been designed to require no special expertise at the time of deployment and has already been sent to sea on commercial gill-netters working in the Celtic Sea. In the Celtic Sea, which has a porpoise density of around 0.18 porpoises km^{-2}, porpoise clicks are typically detected for periods of a few minutes to an hour on several occasions each day. Some trips record porpoise activity solely at the surface, but, more often, clicks are logged also at the seabed. In addition, the occurrence of clicks at the surface and at the seabed can coincide. This vertical segregation is not surprising, as cetacean sonar is highly directional, and it indicates that porpoises are usually, but not always, feeding close to the bottom where they are at risk of entanglement in bottom-fishing set gears. The frequency of encounters with porpoises at the seabed discounts the view that entanglement is a usual outcome of a porpoise encountering a net. A further discovery from the initial trial deployments of the POD has been that, at inshore locations, click rates have been much higher during darkness.

By-catch factors and mitigation

Little is known of how cetaceans use sonar in the circumstances of different by-catch mechanisms, or the extent to which they utilise ambient sound to perceive their environment and prey. Considerable variations relevant to by-catch mitigation seem likely.

Detectability of gill-nets

Observational estimates (Hatakeyama *et al.*, 1988) suggested that monofilament nylon gill-nets could be detected by small porpoises at ranges of only 2 m or less. For dolphin species that have more energetic sonar clicks, theoretical ranges have been calculated that are considerably greater (Au & Jones, 1991). The difficulty associated with gill-net detection is supported by the observations of tuna drift nets in the Bay of Biscay, which have shown a significant positive relationship between sea state and dolphin by-catch rates. This concurs with expectations because, in rougher seas, the acoustic detection of nets should be impaired by higher levels of background noise and by increased absorption of high-frequency sounds by bubbles in the water. Tuna catch, in contrast to dolphins, does not increase progressively with sea state, but peaks at moderate sea states.

Preliminary data from the POD fitted to nets in the Celtic Sea suggest that a circle of less than 0.5-km radius will be entered by porpoises about three times each day, but by-catch occurs only about once every 50 km day^{-1} of net immersion. This indicates that porpoises succeed in escaping or avoiding nets in more than 99% of encounters, but more work is needed to define the level of porpoise activity. By-catch of porpoises in gill-nets is not confined to monofilament nets, and was commented on in the early 1800s as a regular feature of the arrival of large 'shoals' of porpoises on the north coast of Cornwall in pursuit of shoals of pelagic fish.

Behavioural responses to gill-nets

Drift net by-catches give a clear indication that dolphins encountering the nets swim along them very close to the sea surface, as entanglements are strongly concentrated at the top of the net. There is also evidence that dolphin groups swimming along the net near the surface have an increased risk of entanglement in the vicinity of major irregularities of the netting sheet. Dolphins are rarely, if ever, reported to leap over drift nets, although they have been observed to swim under them. They have been seen feeding close to drift nets for prolonged periods without the occurrence of entanglement.

Attraction to trawl nets

Cetaceans caught in midwater trawls are certainly attracted to them, as the normal density of cetaceans is far too low for the net mouth to encounter one cetacean per 100 h of towing (Morizur *et al.*, 1995). In addition, there are reports of visual, video and acoustic observations of pilot whales and smaller dolphins within and around midwater nets. The volume of data is small and it may be that devices such as the POD will be able to yield a much larger volume of data from which a description of the frequency and determinants of this behaviour can be built up.

Attraction of bottlenose dolphins to bottom trawls has often been reported (Fertl & Leatherwood, 1998) but without by-catch, and it appears clear that the dolphins feed outside the trawl on disturbed or injured fish – sometimes they have been observed in the wake of trawlers throwing flatfish around like frisbees. Dolphins appear to be aware of the trawling process, as they arrive only when trawling begins.

Response to 'pingers'

Pingers are battery-powered devices that emit brief noises at intervals of a few seconds to scare away cetaceans. Pingers were devised to reduce the entanglements of humpback whales in cod traps on the west coast of Newfoundland (Lien *et al.*, 1990). Subsequent trials in gill-net fisheries have shown that an active pinger every *c.* 100 m along the headrope of a gill-net can greatly reduce by-catch of porpoises (Kraus *et al.*, 1995). Studies subsequently reported to the International Whaling Commission have included three controlled trials that showed great reductions in by-catch and some observation of 'commercial' deployment without controls in areas of known high by-catch rates, the results of which varied according to season. Studies of the response of captive porpoises to pingers (Kastelein *et al.*, 1995) show that, in general, porpoises flee from the source of the sound if is close to them. At greater distances, at which the pinger sound might be alerting rather than aversive, the porpoise would be unable to detect the net acoustically. Several trials of a commercially available pinger, the Dukane Netmark series, have shown reductions of 80% or more in by-catch rates, but one trial showed no effect. All trials have been of very short duration and observation of their use in a commercial setting has yet to show a great reduction in porpoise by-catch.

The success, at least in short-duration trials, of pingers has lead to the development of variations, such as the PICE pinger developed by Loughborough University. This uses a high-pitched 'ping' (20–140 kHz) at randomised intervals in the 4–30-s range. Significant gaps in the understanding of the potential benefits or pitfalls of using pingers at present include the following.

1. Will widespread deployment lead to habituation of whole populations of porpoises?
2. If some pingers fail, could the by-catch rate be increased by 'funnelling' porpoises into gaps in the noise barrier that actually contain net?
3. All pinger trials have been in situations in which cetaceans are not attracted to the net. Deployment has also been suggested for situations that involve attraction of cetaceans to the fishing gear. However, in these circumstances the pinger could act as a 'dinner bell', although in the case of midwater trawls the trawl itself is so noisy that it can probably be heard easily by cetaceans over 2 km away.

By-catch mitigation at sea

The best example of by-catch mitigation so far is that of the humpback whales, which used to become entangled in cod traps off Newfoundland. The extensive damage incurred means that the fishers profit from having fully functional pingers on their nets. By contrast in most situations by-catch is an infrequent event for the individual fisher and produces only a minor increase in net damage, while many pingers must be maintained to prevent it. In such cases, the motivation for costly and time-consuming long-term provision of fully active pingers is quickly going to wane, if it ever exists, and any effective monitoring and enforcement task will be huge and politically unpalatable.

High levels of damage to catches caused by cetaceans have been reported in some Indian Ocean longline fisheries. In such cases of catch stealing, the use of pingers might be effective because the level of damage is high but it would have to be very effectively maintained because of the 'dinner bell' effect working against the aversive effect. In this real-world context, fisheries divide into those in which:

1. The fishers are self-motivated, without outside pressure and commercial forces, and will develop better pingers and they will be used effectively.
2. Fishers don't have the motivation without outside pressure, and low-maintenance permanent gear modification solutions are needed. This rules out almost any device with batteries.
3. None of these solutions exist and fishery managers are faced with imposing regulatory measures.

In Europe, the first situation does not exist, the solutions required for the second situation have not been found, and fishery managers hope that the last situation does not become an option. In the USA, two fisheries have been required to use battery-powered pingers during a very brief seasonal swordfish drift net fishery in California and a set net fishery in the Gulf of Maine (situation 2). In the latter, pingers have been made a condition of access to specified areas of high porpoise density. Some fisheries have been closed as a result of marine mammal by-catch problems, such as set gill-nets in Monterey Bay, where sea otter by-catch occurred.

By-catch mitigation research

For gill-net by-catch, three themes exist in current research programmes:

1. Impact assessment – measuring by-catch, populations and their dynamics. The latter is the least understood at present, particularly with respect to the impact of pollution.
2. Aversive methods such as the use of pingers. Long-term studies of effectiveness of these devices in real applications are particularly lacking.
3. Enhancement of net detectability.

Two other areas have yet to be explored:

4. Alerting methods – active but non-aversive methods of stimulating cetaceans to start using their sonar when they are not doing so.
5. Finding passive net characteristics that make cetaceans choose to avoid them. This isn't the same as (3) above, as a cetacean may detect a net and then fatally ignore it.

Because aversive methods have such heavy drawbacks of enforcement, a method based on points (3), (4) or (5) would be a great leap forward. It is possible that the perception of a net nearby may act as a reward to a cetacean alerted to its presence by some low-intensity active signal, and this response would consequently be maintained. By contrast, a cetacean's response to an aversive signal seems unlikely to generate any reward and may therefore suffer from habituation. Alerting methods could require very much less energy than aversive pingers and it might be possible to derive this energy from relative movement of nets and water, or even from pressure changes.

Prospects

However, we need to ask ourselves whether there is much chance of passive or low-energy solutions working? A lot of unsuccessful efforts have been made in this area of research, but further efforts may be justified because the alternatives are so unattractive. Against the feasibility of passive or low-energy solutions are:

● The many unsuccessful attempts already made by increasing the detectability of nets using various methods including metal chains, air filled tubes, etc. (Jefferson & Curry, 1994).
● Evidence that wild porpoises will swim between vertical ropes only 1 m apart, that there is a large by-catch of dolphins in the heavy mesh shark nets used to protect bathers on beaches in South Africa, and that there was a by-catch in the natural fibre nets that preceded synthetics.

In favour of the feasibility of passive or low-energy solutions are:

● Evidence that, at least in the short term, there is an avoidance reaction to small floats that give a strong echo when placed on nets (Goodson *et al.*, 1994). Longer-term evidence that bottlenose dolphin by-catch in shark nets can be reduced by the addition of such floats.
● Evidence that porpoises are often not echo-locating (Kastelein *et al.*, 1995).
● Evidence that porpoises and dolphins are reluctant to cross linear features – ropes on the surface or even shadows cast by a rope above the surface (Kastelein *et al.*, 1995).
● The development for use in gill-net fisheries of highly buoyant ropes capable of maintaining their buoyancy at depth. These ropes are a workable linear acoustic feature that has been previously unavailable.
● Development of acoustically enhanced nylon nets – chemical treatment of nylon to generate stronger echoes.
● Availability of the POD to gather sufficient data to discover the responses of wild porpoises and dolphins to fishing gear of various types and to alerting or aversive sounds.

References

Anon. (1995) *Cetacean by-catch in the UK tuna driftnet fishery in 1995. Report to the Ministry of Agriculture, Fisheries and Food, UK, from the Sea Mammal Research Unit*, P639.2/11.95, 14 pp. House of Commons Library. London.

Au, W.W.L. (1994) Sonar detection of gillnets by dolphins: theoretical predictions. In: *Report of the International Whaling Commission* (Special Issue 15), pp. 565–71.

Au, W.W.L. & Jones, L. (1991) Acoustic reflectivity of nets: implications concerning incidental take of dolphins. *Marine Mammal Science*, **7**, 258–73.

Barlow, J. & Boveng, P. (1991) Modeling age-specific mortality for marine mammal populations. *Marine Mammal Science*, **7**, 50–65.

Caswell, H., Brault, S., Read, A., Smith, T. & Barlow, J. (1995) Uncertainty analysis of harbor porpoise population growth rate and by-catch mortality. Paper SC/47/SM28. Scientific Committee of the International Whaling Commission.

Collet, A. & Mison, V. (1995) *Analyse des échouage de cétacés sur le littoral française. Report* IFREMER-DRV-RH.94.2.511036. La Rochelle, France.

Couperus, A.S. (1998) Interactions between Dutch midwater trawl and Atlantic white-sided Dolphins (*Lagenorhynchus acutus*) southwest of Ireland. *Journal of Northwest Atlantic Fisheries Science*, **22**, 209–18.

Fertl, D. & Leatherwood, S. (1998) Cetacean interaction with trawls: a preliminary review. *Journal of Northwest Atlantic Fisheries Science*, **22**, 219–48.

Goodson, A.D., Mayo, R.H., Klinowska, M. & Bloom, P.R.S. (1994) Field testing passive acoustic devices designed to reduce entanglement of small cetaceans in fishing gear. In: *Report of the International Whaling Commission* (Special Issue 15), pp. 597–605.

Goujon, M., Antoine, L., Collet, A. & Fifas, S. (1993). *Approche de l'impact écologique de la pêcherie thonière au filet maillant dérivant en Atlantique nord-est. Rapport interne de la Direction des Resources Vivantes de l'IFREMER*, 47 pp. IFREMER, France.

Hammond, P.S., Benke, H., Berggren, P., Borchers, D.L., Buckland, S.T., Collet, A., Heide-Jørgensen, M.P., Heimlich-Boran, S., Hiby, A.R., Leopold, M.F. & Øien, N. (1995) *Distribution and abundance of the harbour porpoise and other small cetaceans in the North Sea and adjacent waters. Final Report to the European Commission* LIFE 92–2/UK/027, 240 pp.

Hatakeyama, Y., Ishii, K., Soeda, H., Shimamura, T. & Tobayame, T. (1988) *Observation of harbor porpoise's behaviour to salmon gillnet*, 18 pp. International North Pacific Fisheries Commission, Fisheries Agency of Japan, Tokyo. Japan.

Jefferson, T.A. & Curry, B.E. (1994) *Review and evaluation of potential acoustic methods of reducing or eliminating marine mammal-fishery interactions*, 59 pp. Report to the US Marine Mammal Commission.

Kastelein, R.A., Nieuwstraten, S.H. & Verboom, W.C. (1995a) Echolocation signals of Harbour Porpoises (*Phocoena phocoena*) in light and complete darkness. In: *Harbour Porpoises, Laboratory Studies to Reduce Bycatch* (eds P.E. Nachtigall, J. Lien, W.L. Au & A.J. Read), pp. 55–68. De Spil, Woerden, The Netherlands.

Kastelein, R.A. de Hann, D., Staal, C. (1995b) Behaviour of harbour porpoises (*Phocoena phocoena*) in response to ropes. In: *Harbour Porpoises, Laboratory Studies to Reduce Bycatch* (eds P.E. Nachtigall, J. Lien, W.L. Au & A.J. Read), pp. 69–90. De Spil, Woerden, The Netherlands.

Kastelein, R.A., Goodson, A.D., Lien, J. & de Haan, D. (1995) The effect of acoustic alarms on Harbour Porpoise (*Phocoena phocoena*) behaviour. In: *Harbour Porpoises, Laboratory Studies to Reduce Bycatch* (eds P.E. Nachtigall, J. Lien, W.L. Au & A.J. Read), pp. 157–67. De Spil, Woerden, The Netherlands.

Kirkwood, J.K., Bennett, P.M., Jepson, P.D., Kuiken, T., Simpson, V.R. & Baker, J.R. (1997) Entanglement in fishing gear and other causes of death in cetaceans stranded on the coast of England and Wales. *Veterinary Record*, **141**, 94–8.

Kraus, S., Read, A., Andersen, E., Baldwin, K., Solow, A., Spradlin T. & Williamson, J. (1995) A field test of the use of acoustic alarms to reduce the incidental mortality of harbor porpoise in gill-nets. Paper SC/47/SM17. International Whaling Commission Scientific Committee.

Kuiken, T., Simpson, V.R., Allchin, C.R., Bennett, P.M., Codd, G.A., Harris, E.A., Howes, G.J., Kennedy, S., Kirkwood, J.K., Law, R.J., Merrett, N.R. & Phillips, S. (1994) Mass mortality of common dolphins (*Delphinus delphis*) in south west England due to incidental capture in fishing gear. *Veterinary Record*, **134**, 81–9.

Lien, J., Todd, S. & Guigne, J. (1990) Inferences about perception in large cetaceans, especially humpback whales, from incidental catches in fixed fishing gear, enhancement of nets by 'alarm' devices and the acoustics of fishing gear. In: *Sensory Abilities of Cetaceans: Laboratory and Field Evidence* (eds J.A. Thomas & R. Kastelein), pp. 347–62. Plenum Press, New York.

López-Fernández, A. & Valeiras-Mata, X. (1997) Causes of mortality and suspected by-catches by gross post-mortem examination of cetacean strandings on the Galician coast (NW Spain). *European Research on Cetaceans*, **11**, 42–4.

Lowry, N. & Teilmann, J. (1994) Bycatch and bycatch reduction of the harbour porpoise (*Phocoena phocoena*) in Danish waters. *Report of the International Whaling Commission* (Special Issue **15**), pp. 203–9.

Morizur, Y., Tregenza, N., Heessen, H., Berrow, S. & Pouvreau, S. (1995) *By-catch and discarding in pelagic trawl fisheries. Final Report to European Commission* DGXIV, BIOECO/93/017, 182 pp.

Read, A.J., Kraus, S.D., Bisack, K.D. & Palka, D. (1993) Harbor porpoises and gillnets in the Gulf of Maine. *Conservation Biology*, **7**, 189–93.

Tregenza, N.J.C. (1992) Fifty years of cetacean sightings from the Cornish coast, SW England. *Biological Conservation*, **57**, 65–71.

Tregenza, N.J.C., Berrow, S., Leaper, R. & Hammond, P.S. (1997a) harbour porpoise, *Phocoena phocoena* L., by-catch in set gill-nets in the Celtic Sea. *ICES Journal of Marine Science*, **54**, 896–904.

Tregenza, N.J.C., Berrow, S. & Hammond, P.S. (1997b) Attraction of common dolphins to boats setting gill-nets. *European Research on Cetaceans*, **11**, 47–9.

Tregenza, N.J.C. & Collet, A. (1998) Common dolphin *Delphinus delphis* bycatch in the pelagic trawl and other fisheries in the northeast Atlantic. *Report of the International Whaling Commision*.

Vinther, M. (1994) Incidental catches of the harbour porpoise (*Phocoena phocoena*) in the Danish North Sea gill-net fisheries: preliminary results. Paper presented to the *Conference on the State of the North Sea*, Ebeltoft, 11 pp.

Walton, M. (1995) A report on the genetic structure of harbour porpoises *Phocoena phocoena* in the seas around the UK, Eire and the Netherlands. Paper SC/47/SM42. Scientific Committee of the International Whaling Commission, 1995.

Waring, G.T., Gerior, P. Payne, M.P. Parry, B.L. & Nicolas, J.R. (1990) Incidental take of marine mammals in foreign fishery activities off the northeast United States, 1977–88. *Fishery Bulletin*, **88**, 347–60.

Woodley, T.H. & Read, A.J. (1991) Potential rates of increase of a harbour porpoise (*Phocoena phocoena*) population subjected to incidental mortality in commercial fisheries. *Canadian Journal of Fisheries and Aquatic Sciences*, **48**, 2429–35.

Chapter 18

Effects of fishing on non-target species and habitats: identifying key nature conservation issues

M.L. TASKER[1], P.A. KNAPMAN[2] and D. LAFFOLEY[2]

[1]*Joint Nature Conservation Committee, Dunnet House, 7 Thistle Place, Aberdeen, AB10 1UZ, UK*
[2]*English Nature, Northminster House, Peterborough, PE1 1UA, UK*

Summary

1. This paper summarises the key nature conservation issues arising from the effects that fishing may have on the marine environment in north-western European seas.
2. Nature conservation issues arise as a result of the localised effects caused by fishing as well as cumulative impacts that result at the ecosystem level. Such concerns have both given rise to a growing body of research and also contributed towards international and national agreements, conventions and directives aimed at conserving biodiversity and putting uses of the environment on an ecologically sustainable basis. Such initiatives have increased the pressure for change and helped to focus when nature conservation issues arise.
3. The paper concludes by making a number of suggestions about how fisheries and nature conservation could be brought closer together for the benefit of fishermen, the industry as a whole and nature conservation interests. This paper is accordingly very much a discussion paper and should not be taken as a position paper of any nature conservation organisation.

Keywords: nature conservation, biodiversity, ecological integrity, Common Fisheries Policy, *Natura 2000*, Sea Fisheries Committees.

Introduction

At the broadest level, nature conservation is shaped and guided by the concept of ecologically sustainable development, the key principles of which are the maintenance of biological diversity (the variety of life) and ecological integrity (the ability of our environment to support and maintain the full range of habitats and species that occur in a natural state).

The UK and many other nations around the world have committed themselves to the maintenance of biological diversity and ecologically sustainable development, through agreements made, for example, at the Earth Summit held in Rio de Janeiro in 1992. Maintaining the biodiversity and ecological integrity of our seas is accordingly no longer an option but an objective, which now must be integrated into the full range of uses society makes of this resource.

Examination of how uses of the environment interact with biodiversity and its conservation and management have become a priority, reflected in many recent changes in legislation. The general aim of nature conservation is to reduce and preferably minimise the effects of human activities on nature. This chapter briefly highlights the nature conservation issues arising from fishing activities that are covered extensively elsewhere in this book. In addition, it is clear that nature conservation in the marine environment is not as advanced as that in the terrestrial environment and our understanding and implementation of the methods of reducing or minimising the impacts of fisheries are still in their infancy.

Identifying key nature conservation issues

Fishing activities can have a profound effect on marine ecosystems, mainly on those species targeted by fisheries, but also on non-target species and habitats. Nature conservation issues arise as a result of:

1. mortality of target fish species, non-target fish species, marine mammals, seabirds and benthic organisms;
2. disturbance to the seabed and consequent damage to benthic habitats and organisms as a result of impact by fishing gear;
3. input of fishery wastes (predominantly offal and discarded organisms) into the sea – these wastes may provide food for scavenging organisms such as seabirds;
4. fisheries-related litter that leads to 'ghost fishing' – the ability of lost fishing gear to continue to fish in an uncontrolled manner (Eno *et al.*, 1996; Kaiser *et al.*, 1996).

The direct impact of fisheries activities on benthic ecosystems will be dependent on the distribution, abundance, and sensitivity of the habitats and species that are impacted. Nearshore areas tend to have more data available on the distribution of habitats and species, resulting from studies such as the Marine Nature Conservation Review, conducted by the Nature Conservancy Council and, more recently, the Joint Nature Conservation Committee, between 1987 and 1998. Work on sensitivity of habitats and species is now a rapidly expanding field, complementing this distribution and map information. The initiatives of the OSPAR (Oslo Paris Convention) working group and the MarLIN (Marine Life Information Network) project, currently in progress under the auspices of the Marine Biological Association of the United Kingdom, will help focus some of the conservation and fisheries concerns more precisely than has been possible in the recent past.

Managing the effects of fisheries on marine biodiversity

Another way of viewing the nature conservation interactions with fisheries is to examine the important role fisheries have in the context of sustainable development and biodiversity of the marine environment. This has been recognised in a number of international agreements and conventions and, in themselves, they have given impetus to the need to develop new approaches to the way fisheries are managed:

1. The Convention on Biological Diversity (CBD).
2. The United Nations Convention on the Law of the Sea (UNCLOS).
3. The United Nations Convention on Straddling Fish Stocks and Highly Migratory Species.
4. FAO Code of Conduct of Responsible Fisheries.
5. Agreement on the Conservation of Cetaceans of the Baltic and North Seas (ASCOBANS).
6. Council Directives 92/43/EEC and 79/409/EEC – The Conservation of Natural Habitats and of Wild Fauna and Flora, and, the Conservation of Wild Birds (the Habitats and Species Directive and the Birds Directive, respectively).
7. The Esbjerg Declaration, from the Fifth International Conference on the Protection of the North Sea.

In addition to these, following its mid-term review in 1992, the Common Fisheries Policy (CFP) listed the protection and conservation of the marine environment as its key objective. In particular, Article 2 of the framework Regulation 3760/92 states that:

the general objectives of the Common Fisheries Policy shall be to protect and conserve available and accessible marine aquatic resources, and to provide for rational and responsible exploitation on a sustainable basis in appropriate economic and social conditions for the sector, taking account of its implications for the marine ecosystem, and in particular taking account of the needs of both producers and consumers.

Article 130r(2) of the Maastricht Treaty also establishes that environmental protection requirements must be integrated into the definition and implementation of other Community policies. Furthermore, Article 130s states that the precautionary principle must be taken into account when formulating Community environmental policies. It is plain that the first part of this aim is not currently being met in many areas. The second part of the aim has so far manifest itself in two regulations specifically for the environment, and both of these relate to reducing the impact of specific fisheries on dolphin populations.

The Convention on Biological Diversity defines sustainable use as:

the use of components of biological diversity in a way and at a rate that does not lead to the long-term decline of biological diversity, thereby maintaining its potential to meet needs and aspirations of present and future generations.

A consequence of this convention is that biological criteria must be taken into account and given greater weight in management decisions. In a fisheries context, not only should fish stocks be exploited in a sustainable manner, but there should be as few adverse effects on biological diversity as possible. The original Convention has been further refined since it came into force. Action has been recommended to:

1. integrate coastal and marine area management;
2. establish marine and coastal protected areas;
3. implement agreed guidelines on sustainable fisheries (essentially including a precautionary approach);
4. manage mariculture and its potential impacts;
5. manage the risks from the introduction of alien species.

The general aims of this convention are partially supported in both international and national law, but, overall, the legal framework of most states is not at present capable of meeting fully the needs of nature conservation. This is particularly the case in the marine environment.

Despite these problems, it is plain that there is both a legal and a general public wish to ensure the survival of healthy populations of marine animals, plants and their habitats. There is equally a need to ensure that marine ecosystems are utilised in a sustainable and, from the exploiter's point of view, profitable manner. The challenge is to bring these two together within a management framework. The evidence is that existing fisheries management frameworks in north-western European seas are not meeting this challenge. Many commercially exploited fish stocks generate significant conservation issues, as they are at either an overexploited or fully exploited level, and there is widespread evidence of non-sustainable ecosystem damage.

So what does the future hold for integrating fisheries and nature conservation interests? Any management scheme requires agreed objectives, stated as precisely as possible. From the foregoing discussion, one could derive two extreme objectives: (i) avoid all damage to the ecosystem, and (ii) do not hinder the profitability of the fishery in any way. These are both plainly unrealistic. However, there are a number of potential options for objectives somewhere between the two, their implementation having varying scales of difficulty and cost implications. For example, we could:

1. Aim to understand the wider environmental consequences involved with various fisheries management options. The ICES Advisory Committee on Fisheries Management (ACFM) already provide alternative management options for some fish stocks with their anticipated consequence for the stock. It might be possible to expand these options to include marine ecosystems.
2. Define the best current fishing practices that would harvest fish rationally with minimum disturbance to the ecosystem. This may require an element of risk management and potential trade-offs, i.e. if fisheries were managed within their limit reference points, could the potential continued risk to species such as the common skate be acceptable?
3. Estimate appropriate fishing mortality for target and non-target species and levels of degradation for certain habitats. Limit reference points for some fish stocks beyond which they should not be exploited have been introduced. Should we aim to provide similar limits, even if they are arbitrary, for other species and habitats?
4. Establish a series of no-take zones (NTZs) to provide for improved ecosystem integrity. While it is recognised that we do not know what a pristine

environment is, the establishment of NTZs would allow us to study how ecosystems alter subsequently, would protect specific habitat and species, and could reduce by-catch and guard against damaging fishing practices.

All these and other options are currently under consideration, but the agreement on these objectives is plainly a political process.

In March 1997, a meeting of North Sea Ministers of the Environment and of Fisheries agreed three main objectives for fisheries and the environment in the North Sea (IMM, 1997):

1. to ensure sustainable, sound and healthy ecosystems in the North Sea, thereby restoring and/or maintaining their characteristic structure and functioning, productivity and biological diversity;
2. to achieve sustainable exploitation of the living marine resources, thereby securing a high yield of quality food';
3. to ensure economically viable fisheries.

A series of strategies were also agreed to meet these objectives with a number of agreed actions. Key to these actions was agreement that overall fishing effort needed to be reduced dramatically: this would have the common benefit of both easing pressures on exploited fish stocks and reducing impact on the marine environment. Ministers considered that a more precautionary approach needed to be taken to fish stock management in future, which would give fish stocks a greater buffer against overexploitation. It was further considered that greater integration of fisheries and ecosystem management objectives (the ecosystem approach) would place fisheries management on a more ecologically sensible basis.

Achievement of an ecosystem approach, however, has yet to be realised in practice. Indeed, the cost of achieving it is potentially prohibitive. As we move toward an ecosystem approach, the cost of providing the appropriate information becomes untenable, and the scale of realism becomes idealistic. While this model provides a simple and powerful indication of where this form of management sits in comparison with existing options, it does not include a precautionary element. Use of a precautionary approach will result in 'ecosystem management' moving down the scale and into a more realistic area of funding. When information is deficient, management measures should reflect this lack of information, and, as more information becomes available, management measures can be altered to reflect the greater understanding of the effect of the fishery.

While these approaches may provide a better management framework, it must be recognised that some of the underlying problems within fisheries management at present derive from non-compliance. If any management is to succeed, compliance with, and preferable co-ownership of, any regulations must be regarded as vital. It is important that fisheries scientists, fishermen and those concerned with ecosystem management join with fisheries managers in a more integrated framework in future (Lindeboom, this volume, Chapter 19). This is probably the area that requires greatest thought and attention by all parties if future management of fisheries is to move onto a more sustainable basis (both for stocks and the environment) in the

Table 18.1   Some gear types used in the North Sea Fisheries in relation to target species and by-catches of target and non-target species (adapted from Svelle *et al.*, 1997)

Fishing gear	Fisheries	By-catch
Demersal active gear		
Otter trawl (human consumption fisheries)	*Nephrops*, roundfish and some pelagic species	Unwanted sizes of target and non-target species of fish and other vertebrates
Otter trawl (industrial fisheries)	Small fish species (sandeel, Norway pout, sprat)	Human consumption fish species
Demersal seines: single and pair	Human consumption fish (roundfish and flatfish)	Unwanted species and sizes of fish
Beam trawl: light nets equipped with bobbins	Brown shrimp	Significant by-catch of flatfish and benthic organisms
Beam trawl: heavy gear equipped with chains	Flatfish (mostly sole and plaice)	Juvenile target species, non-target fish and benthic organisms
Dredges	Molluscs	Flatfish; damage to target and non-target benthic species
Pelagic active gear Purse seines	Shoaling pelagic species	Low by-catch of non-target species
Pelagic trawl, single and pair	Shoaling pelagic species	Dolphins in some areas
Passive gear nets: gill-nets, demersal set nets, drift nets	Human consumption fish species (cod, turbot and other species)	Seabirds, harbour porpoise
traps: portable baited traps and coastal trap nets	Crustacean shellfish and salmonids	Undersized and non-target shellfish, seabirds in some areas
longlines	Deep-water demersal fish species	Scavenging seabirds

longer term. It is the implementation of legal nature conservation obligations that may come close to providing the first broad-scale working example of this desired model.

The Habitats and Species Directive and the Birds Directive require the establishment of a network of wildlife conservation areas to be known as *Natura 2000*. In the marine situation in northern European waters, these are required for

Table 18.2   Northern European marine habitats and species whose conservation requires the designation of Special Areas of Conservation under the Habitats Directive

Habitats	Species
Sandbanks (shallow)	Otter *Lutra lutra*
Estuaries	Grey seal *Halichoerus grypus*
Intertidal mud and sandflats	Common seal *Phoca vitulina*
Lagoons	Bottlenose dolphin *Tursiops truncatus*
Large shallow inlets and bays	Harbour porpoise *Phocoena phocoena*
Reefs	Lampern *Lampetra fluviatilis*
Marine columns in shallow water made	Sea lamprey *Petromyzon marinus*
by leaking gases	Sturgeon *Acipenser sturio*
Submerged or partly submerged sea caves	Houting *Coregonus oxyrinchus*
	Shads *Alosa* spp.

most regularly occurring migratory seabirds and coastal birds, and for those species and habitats listed in Table 18.2. These protected areas are in the process of being established by the Member States of the European Union. Once established, each will have a management plan to ensure that the special interest for which the area is designated is maintained at a favourable conservation status. In the UK, these management plans will be developed and implemented by the statutory nature conservation agencies and those authorities that have statutory responsibilities within the marine environment. While fishermen will not be directly represented on the management groups for the UK sites, they will have an input to the management process via an advisory group that informs the management group. Also, in England and Wales, local inshore fisheries management is undertaken by Sea Fisheries Committees, which have within their constituents representatives from the fishing industry. Sea Fisheries Committees will play a key role within the management group by providing local knowledge of the fishing industry and the capability of monitoring the condition and the compliance of any environmentally related fisheries management measure. It is recognised that the difficulty of undertaking something similar outside territorial limits with other Member States would be significantly increased. However, this is a genuine example of how the integration of nature conservation and fisheries management is bringing fishermen closer to the decision-making process.

Conclusions

Mechanisms should be established to provide for a more local management of fisheries, including those outside territorial waters. It may be that the principle of equal access would need to be redefined, although not on a nationalistic basis. The best available biological as well as fisheries information should be available to implement such mechanisms, and decision-taking should be made in the light of potential effects on the ecosystem (in much the same way as decision-taking in relation to oil industry developments require an Environmental Impact Assessment). Fishing activity should probably be zoned further in both a time and spatial sense; these zones would include some no-take zones, where nature could develop with minimal interference from humans. All fisheries, not just a selected few, should be fully managed including, for instance, deep-water and industrial fisheries. Management should be on a precautionary basis, fisheries and environmental conservation measures must be designed on the best scientific evidence available. Thus, where evidence is poor, conservation measures should reflect the lack of information, but as more evidence becomes available, conservation measures can be altered to reflect the greater understanding of the effect of a fishery. There are encouraging signs already that measures to implement this in relation to fish stocks are being developed.

Perhaps though, the greatest challenge is in obtaining the agreement of some sectors of the fishing industry to these changes. Present attitudes and approaches are essentially very short term. Much of this appears to be due to the economics of the industry, whereby large bank loans require immediate servicing. A move away from this situation by removing public subsidies and over-capitalisation in the medium term is required. It should also be recognised that the essential aims of a more environmentally friendly management of fisheries is to ensure a long-term future of the fishing industry.

References

Anon. (1995) *Report on the study group on ecosystem effects of fishing activities. ICES Co-operative Research Report*, 200.

ASCOBANS (1997) *Cetacean By-catch Issues in the ASCOBANS Area.* ASCOBANS, Cambridge.

Camphuysen, C.J. & Garthe, S. (1997) An evaluation of the distribution and scavenging habits of northern fulmars (*Fulmarus glacialis*) in the North Sea. *ICES Journal of Marine Science*, **54**, 654–83.

Eno, N.C., MacDonald, D. & Amos, S.C. (1996) *A study on the effects of fish (Crustacea/Mollusc) traps on benthic habitats and species. Final Report to European Commission* DG XIV, 94/076, 43 pp.

Garthe, S., Camphuysen, C.J. & Furness, R.W. (1996) Amounts of discards in commercial fisheries and their significance as food for seabirds in the North Sea. *Marine Ecology Progress Series*, **136**, 1–11.

IMM (1997) *Statement of Conclusions: Intermediate Ministerial Meeting on the Integration of Fisheries and Environmental Issues*, 13–14 March 1997. Bergen, Norway. Ministry of the Environment, Oslo, Norway.

Kaiser, M.J., Bullimore, B. Newman, P. & Gilbert, S. (1996) Catches in 'ghost fishing' set nets. *Marine Ecology Progress Series*, **136**, 1–11.

Larkin, P.A. (1997) The costs of fisheries management information and fisheries research. In: *Proceedings of the Second World Fisheries Congress* (eds D.A. Hancock, D.C. Smith, A. Grant & J.P. Beumer), pp. 713–18. Brisbane, Australia.

Maguire, J.-J. & Sinclair, A.F. (1997) The precautionary approach, with special emphasis on the report of the Lysekil (Sweden) technical consultation (6–13 June 1995). In seminar report: *The precautionary approach to North Sea fisheries management. Fisken og havet* No 1. Institute of Marine Research, Bergen.

Svelle, M., Aarefjord, H., Heir, H.T. & Verland, S. (eds) (1997) *Assessment report on fisheries and fisheries related species and habitat issues.* Fifth North Sea Conference Secretariat, Ministry of the Environment, Oslo, Norway.

Chapter 19
The need for closed areas as conservation tools

H.J. LINDEBOOM
Netherlands Institute for Sea Research, NIOZ, PO Box 59, 1790 AB Den Burg, Texel, The Netherlands

Summary

1. A large body of evidence indicates that the long-term changes in benthic communities observed in the North Sea have been caused to a large extent by the direct and indirect effects of fishing activities and not solely by eutrophication, climatic fluctuations and/ or pollution.
2. In order to minimise the effects of fisheries, and to move towards the sustainable use and protection of the marine ecosystem, it is necessary to reduce fishing effort, modify gear design and create areas closed to fisheries.
3. The rationale for the creation of closed areas includes: protection of specific species, habitats or juvenile fish, creation of a more natural population age-structure, and the prevention of continuous heavy impacts of certain fishing techniques slowly changing the entire ecosystem. An example for the North Sea is worked out in the text.
4. Closed areas are also needed for scientific and monitoring purposes. Without them it will be very difficult to study the natural trends in the marine ecosystem or to ascertain which human activity has influenced the ecosystem the most. Furthermore, there may be no value in data that have been collected from areas with an unknown level of fishing disturbance.
5. The size of protected areas should be determined by the objectives of the closure and by the behaviour of species that are characteristic to that area. In such areas, where fisheries and inputs of pollutants will be prohibited or restricted, scientific research into the species composition, abundance and age distribution of different populations should be carried out and trends established.
6. The successful implementation of protected or closed areas requires the definition of clear objectives for the closure. In addition, stakeholders should be included from the beginning of the planning process to design proper, manageable and legally controllable boundaries. Regular monitoring and evaluation programmes should be executed to see if the objectives are met, and to redesign the areas if necessary.

Keywords: closed areas, conservation, management objectives.

Introduction

From the previous chapters, it has become clear that there are many signals that fishing activities affect the marine ecosystem on local and sometimes regional scales. Stocks of economically important species and populations of non-target species have declined. In the Dutch sector of the North Sea, at least 25 species have decreased considerably in numbers or have disappeared completely (Bergman *et al.*, 1991;

Philippart, 1998). In contrast, populations of opportunistic species have increased in numbers.

However, it is not clear that all these changes relate to fisheries (Frid & Clark, this volume, Chapter 13). On the Dutch continental shelf, fishing activities are now so intensive that every square metre on average is trawled at least once to twice a year (Rijnsdorp *et al.*, 1998). Results of a recently completed international research programme (IMPACT II) led to the conclusion that changes observed in the North Sea ecosystem over the past 100 years can, to a large extent, be attributed to the activities of fisheries (Lindeboom & De Groot, 1998). If the present-day fishing activities continue as they have done over recent decades, it is likely that certain species will disappear completely from the seabed in the Dutch part of the North Sea, as has already occurred for the common whelk in the Wadden Sea.

Sustainable use and the precautionary principle are frequently invoked with reference to management of the marine ecosystem. However, it is necessary to be clear about the objectives of environmental management. For example, should all species be protected in all areas, do we wish to restore the system to its former condition, or do we accept that man has caused permanent change? The biggest challenge of all may be to find a workable compromise between the aims of sustainable fisheries and protection of the marine environment. In this chapter, I discuss the long-term impacts of fisheries in more detail, and then address the issue of why and how closed areas might be established, and their role in future management of marine systems.

Long-term effects of fisheries

Long-term shifts in the infaunal benthic invertebrate communities found in the North Sea have been suggested by several studies reviewed by Frid & Clark in Chapter 13 of this book. Several studies indicated a dominance of opportunistic short-lived species and a decrease of long-lived senile organisms such as large bivalve species. Perhaps one of the most revealing studies to date was an investigation of long-term records of deliveries of by-catch species that were supplied by fishermen in exchange for payment (Philippart, 1998). These specimens were delivered by fishermen to the Zoological Station in Den Helder between 1947 and 1981. A fish catchability model was applied to the occurrence of several demersal fish and invertebrate by-catch species in the south-eastern North Sea and revealed declining trends in the occurrence of certain species, which could be attributed to the introduction and use of different fishing gears (Philippart, 1998). The catchability estimates for otter trawls were higher for fish than for invertebrate species. According to the model results, otter trawling resulted in a *c.* 95% decline of roker (*Raja clavata*) and greater weever (*Trachinus draco*) in the sampling area between 1947 and 1960. Smooth hound (*Mustelus mustelus*), common skate (*Raja batis*) and angler (*Lophius piscatoris*) decreased by more than 75%, whilst the lesser spotted dogfish (*Scyliorhinus canicula*), stingray (*Dasyatis pastinaca*), European lobster

(*Homarus gammarus*) and edible crab (*Cancer pagurus*) decreased by more than 50% during this 14-year period. The slender spindle shell (*Colus gracilis*), velvet swimming crab (*Necora puber*) and the dahlia anemone (*Urticina [Tealia] felina*) were hardly affected by otter trawling, but declined rapidly from 1960 onwards, and reached less than 20% of their original population size by the end of the study period. These declines coincided with an increase in beam trawling effort and resulted in a further reduction of smooth hound, roker, stingray, angler, red whelk (*Neptunea antiqua*) and lesser octopus (*Eledone cirrosa*) to less than 5% of their original abundance as recorded in 1947.

The observed variation in annual numbers of fish and invertebrates delivered to the Zoological Station were found to be related to the changes in gear and fishing effort of bottom trawlers. Otter trawlers caught relatively more fish than invertebrates, whilst beam trawlers caught invertebrate species (i.e. velvet swimming crabs and slender spindle shell) that were hardly ever caught by otter trawlers. The results of the catchability model implied that bottom fisheries had a considerable impact on the marine ecosystem by reducing several demersal fish and benthic invertebrate species to very low levels of abundance within 35 years. Beam trawls had a far more adverse effect on populations of non-target sessile species than otter trawls.

Analyses of trends in the log-transformed relative abundance of demersal fish as derived from different surveys between 1969 and 1993 in the south-eastern North Sea indicated that, on average, the relative species composition appeared to have changed during the last decades. A decrease in the abundance of several flatfish species such as plaice (*Pleuronectes platessa*) and sole (*Solea solea*) and benthic invertebrates such as the sea potato (*Echinocardium cordatum*) and common whelk was observed, whilst other species such as grey gurnard (*Eutrigla gurnardus*), dab (*Limanda limanda*), starfish (*Asterias rubens*) and, in particular, dragonet (*Callionymus lyra*) increased in numbers. The observed changes concur with the hypothesis that demersal fisheries affect mainly commercial flatfish species and vulnerable benthic invertebrate species.

It is clear that changes in population sizes and distribution have occurred for mammal, avian, fish and invertebrate species in the North Sea (Daan, 1989; Daan *et al.*, 1990; Dunnet *et al.*, 1990; Reijnders & Lankester, 1990; de Vooys *et al.*, 1991; Camphuysen *et al.*, 1995; Walker, 1998). Although other factors such as a rise in sea temperature, eutrophication, wind force and direction, and intra- and interspecific interactions may play a role, the observed changes seem to be explained to a great extent by increasing fisheries mortality for some species and improved circumstances for growth and survival for others.

Fishing intensity

The actual long-term effects of fisheries depend on the fishing intensity and the techniques used in the area of concern. Monitoring data indicate that, since the early

1980s, on average every square metre of the Dutch sector of the North Sea was trawled at least once or twice a year. However, not every area is trawled with the same intensity. Rijnsdorp *et al.* (1998) described the spatial distribution of fishing efforts on a micro-scale and concluded that vessels do not trawl at random, but concentrate their efforts on restricted fishing grounds. They estimated that during their 4-year study period conducted in eight of the most heavily fished ICES rectangles of the North Sea, 5% of the surface area was trawled less than once in 5 years and 29% less than once per year. Thirty per cent of the surface area of the seabed was trawled between once and twice year, while 9% was trawled five times per year. The surface area trawled more than five times in a year was estimated at 9%. It is clear that the distribution of beam trawl effort is patchy; however, the conclusion that significant areas of the North Sea remain untrawled, giving refuge to the benthic species vulnerable to trawling, remains premature.

For example, we were unable to find unfished reference areas in the Dutch sector. This was illustrated by our experiences in the Borkum Riff (BEON, 1992). The Borkum Riff is an area regarded by Dutch fisherman as one of the few places in the southern North Sea where beam trawling is seriously hampered by the presence of rocks on the seabed. However, when a side-scan sonar recording of the area was made, 70% of a 3-km transect was covered with trawl tracks rendering the area unsuitable for the planned study on the long-term effects of beam trawling. Attempts to discover other 'untrawled' areas in the Dutch sector were also unsuccessful. With present-day fishing techniques, it is very likely that all areas in the Dutch part of the North Sea where exploitable amounts of fish are found will be trawled regularly by the Dutch fishing fleet. Bergman *et al.* (1991) concluded that there is virtually nowhere in the Dutch sector of the North Sea where benthic communities can develop undisturbed.

Marine protected areas for conservation purposes

The majority of recent marine management documents focus on the concept of a sustainable use of marine resources. As stated by Agardy (1997):

It [the sustainable use of marine resources] is now touted the world over as the solution to real and prospective global, regional and local environmental problems. However, sustainable use as a concept is rarely defined. Prolonged economical gain, ecologically sound development, low-level use of renewable resources, or parity among all resource users, are terms often expressed. The most common meaning of ecological sustainability has to do with the ecosystem function. For an activity to be sustainable within the functional limits of an ecosystem, that activity must not interfere with the workings of that system and its ability to keep critical parameters within homeostatic limits. That is, the activity must not cause environmental degradation in the systems sense. Removing organisms from an ecosystem or interfering with its critical processes can only be sustained over time if the system's functioning is not adversely impacted.

However, a major problem is to define the way in which these systems function. Both a pristine environment with many trophic levels that might include marine mammals, and an anaerobic mudflat with phototrophic bacteria can be perfectly functional and natural ecosystems. In turn, human activities can result in alternative marine systems that could fall anywhere between these two extremes. This implies that there is no general recipe of sustainability that applies to every marine area. Sustainability needs to be defined for specific areas, depending on their original and current status, the use of the area and the desired ecosystem functioning of that area, including the occurrence of individual species or groups of species. This includes clear definitions of sustainable protection of non-target species and the definition of thresholds beyond which the risk of changes in the coastal environment are considered unacceptable. One of the great challenges is to set these definitions for the marine environment on local, regional and global scales. Fishing mortality of both target and non-target species must be limited to levels that do not cause a decline in and eventual collapse of the defined ecosystem properties.

Planning and management for sustainable use requires basic knowledge about the functioning of the system, as well as about the actual and potential uses of its components and the effects of exploitation. This is especially true because ecosystems are not static, unchanging entities, but rather a complex and dynamic web of interactions, which are affected by cumulative impacts (Agardy, 1997). In order to tackle effectively the substantial marine conservation problems we face today, we need to ask clear questions about the mechanism by which we undermine ecosystem function and biodiversity, how we can continue to use living resources sustainably and how we can modify our behaviour to achieve that goal.

This goal may be reached in part by creating specific areas where the constant pressure of human activities is minimised by creating no-take zones or closed areas. The so-called 'precautionary principle' may be important in this context. Although an often misused term, it implies that actions that produce irreversible change to ecosystems (e.g. extinction and permanent restructuring of food webs) must be avoided, and risks and uncertainties must be taken into account. As long as we are not certain about the long-term effects of fisheries, the creation and maintenance of relatively undisturbed areas may be an important part of a precautionary approach. Following the approach on land, the time has come seriously to consider the creation of real nature conservation areas in the open sea, where the marine ecosystem may develop without continuous human harvesting pressures.

Creation of closed areas

There may be different reasons to create protected or closed areas (see also Horwood, this volume, Chapter 20).

To protect specific species or groups of species

Species for which it may be important to establish protected or closed areas include: species in imminent danger of extinction; species that play a central role in ecological communities, often called 'keystone species'; species that may serve as indicators of the ecological condition of a system; and species that may help to raise public awareness (Agardy, 1997). In the Dutch sector of the North Sea, rays have disappeared from the coastal zone, most likely as a result of fishing (Walker, 1998). Even if fisheries did not cause their complete disappearance, present-day fishing practices make it very unlikely that rays will ever be able to re-establish their populations. Using tagging experiments, Walker (1998) showed that rays do not range throughout the North Sea but remain mostly within 20 km of their point of release. She recommended closed areas about the size of ICES rectangles in which local ray populations might re-establish themselves. Such closed areas might also have additional benefits for species such as oysters and lobsters.

Protection of juvenile fish

The 'plaice box' is a good example of a closed area designed to protect juvenile fish (Piet & Rijnsdorp, 1998). This area along The Netherlands, German and Danish coast, established in 1989, was closed from 1 April until 30 September to beam and otter trawlers with engines exceeding 300 hp (221 kW). The 'box' was intended to cover the major distribution area of the main commercial demersal fish species such as plaice, sole and, to a lesser extent, cod. A reduction of fishing mortality for the juveniles of these species was expected. In 1997, the box was closed to trawlers > 300 hp for the whole year. Comparing the 'box' with a reference area, Piet & Rijnsdorp (1998) showed that the overall size structure of the commercially exploited fish species was affected by the change in trawling effort unlike that of the non-target species. The marketable size range of commercial fish increased. The species composition was not significantly affected. Other trends that were observed both within and outside the 'box' included a general increase in species richness due to the influx of southerly species, and a decrease in the relative abundance of plaice. The latter led to the fishermen's opinion that the 'plaice box' did not function effectively as a tool for protecting fish stocks. However, it is possible that other causes such as natural variation led to a decrease in the plaice population during the 10-year life span of the 'box'. Lindeboom *et al.* (1996) indicated large changes in the Wadden Sea and North Sea ecosystem in the late 1980s, leading to smaller biomasses of shellfish in the Wadden Sea and possibly plaice in the North Sea. Lessons that can be learned from the plaice box are that the removal of fishing pressure leads to measurable changes in the marine ecosystem. However, initial 'positive' effects may be completely overshadowed by other trends that take place in the natural system. To overcome this problem, long periods of closure and continuous monitoring are needed.

Restore natural age structure of fish populations

One of the features of overexploited fish populations is a shift in the age distribution towards younger specimens. In the past, fish such as cod could grow to an age of 40 years or more; presently specimens older than 6 or 7 years are rare. Similar age shifts have been recorded in non-target species (van der Veer *et al.*, 1990). These age shifts may influence the capability of populations to sustain sudden collapses caused by, for example, cold winters or diseases. Within closed areas, fishes that stay in the area can grow until their natural death, thus increasing the mean age of the population, which may render the population less vulnerable to natural variations.

Habitat protection

The best examples of reserves designed to preserve marine habitats occur in the tropics, e.g. Great Barrier Reef in Australia, the Galapagos Islands and the Saba Natural Reserve in the Dutch Antilles. Often these parks are multi-user protected areas where certain functions, such as fisheries, anchoring, diving etc., are either permitted with strict regulation or prohibited in certain areas. Craik *et al.* (1990) states that 'the selection of sites usually owes more to the fact that they are not in demand for more obvious economic priorities than the intrinsic nature of the ecosystem'. In the North Sea, potential sea-grass fields and stony areas may require protection from bottom-fishing activities.

Prevention of the effects of chronic disturbance

In many areas, we should give up our traditional preoccupation with conserving structures or specific species, and instead direct ourselves towards safeguarding the critical ecological processes and properties that are responsible for maintaining the desired ecosystem. In this approach, we take the direct impact of the fisheries as the starting point. Depending on the fisheries intensity and the direct effects on target and non-target species, managers may decide that this is not tolerable. As part of a 'precautionary approach', the creation of areas where the impact of fisheries is negligible may then be a good conservation option.

Scientific research and monitoring purposes

There are various reasons for establishing protected areas for marine research (Lindeboom, 1995). Ten years ago, a biological monitoring programme was started in the Dutch sector of the North Sea (Duineveld, 1992). Benthic samples are taken annually at 100 sites and analysed for infauna and small epifauna. The aim is to establish possible trends in the development of this fauna during a period of

5–10 years. But what do these data mean if beam trawls ploughed the sampling area an unknown amount of times prior to the sampling? For example, what would happen when in one year the area remained untouched by fisheries, whereas in another year fishermen made it their favourite fishing area? As fisheries cause detectable short-term changes in benthic communities, they may also influence the data collected in monitoring programmes, rendering these data useless for establishing possible trends caused by, for example, eutrophication or pollution. If trends caused by actions other than fisheries are to be monitored in the Dutch sector of the North Sea, the sampling sites should be off-limits to fisheries. The results of several other scientific programmes knowingly, or more often unknowingly, may have been influenced by fisheries. Studies of the settlement and survival of benthic organisms, studies of sediment–water exchange or the transport of suspended matter, and even the benthic mapping carried out by ICES members in 1986 are possible examples (Künitzer *et al*, 1992).

The comparison of the effects of fisheries with the effects of other anthropogenic influences will be a major task of applied scientific research. This will be especially true when an economic recession forces governments to direct the available money to measures that will be most effective for the sustainable development of the marine ecosystem, and questions of what measures are most effective are being raised. Studies to answer these questions are becoming more and more important. However, as discussed before, it is almost impossible to estimate quantitatively the individual effects of fisheries, eutrophication and pollution in a certain marine area. The establishment of a protected region in such an area may provide the practical means of studying the effects of different anthropogenic activities.

Dutch North Sea: an example

In the 1990s, we conducted a study into the necessity and feasibility of the designation of protected areas in the Dutch sector of the North Sea as a contribution to the conservation and, where possible, rehabilitation of a natural diversity of ecologically valuable areas (Bergman *et al*., 1991; Lindeboom, 1995). The objectives of such a designation would be (1) to preserve, rehabilitate and develop natural values by limiting the effects of human activities that cause detectable changes, and (2) to protect animals that are an integral part of the Dutch sector of the North Sea.

First, we developed four criteria that may be used for the designation and selection of areas that qualify for protected status. The first criterion addresses the extent to which specific activities have developed into a threat to the existence or normal functioning of groups of animals or species. The second criterion addresses the question of whether a prohibition or restriction of certain human activities would reduce this threat. The third criterion is the use of ecological criteria, such as diversity, integrity and vulnerability, to identify the areas most suitable for protected status. Finally, the fourth criterion addresses the question of whether there are adequate legal instruments to ensure effective protection of the selected areas.

Taking into account the effects of different human activities described earlier in this chapter and on the basis of the above criteria, it was concluded that an area directly north-west of the Frisian Islands qualifies for protected status. This area contains coastal waters, sandy bottoms, the Frisian Front area, muddy areas and limited stony areas, hence it will be possible to protect different types of benthic communities, including both invertebrates and fish. As a result, the following protective measures have been proposed for the area: (1) close the area to all types of fisheries throughout the year; (2) prevent or minimise oil-containing discharges from offshore mining installations; (3) take area-specific measures with respect to offshore mining, shipping, military activities, sand extraction, dumping and the laying of cables and pipelines whenever the situation in the area calls for such measures; and (4) consider additional measures if the area is to be used as a reference area for scientific research.

Following the publication of Bergman *et al.* (1991), the Dutch government debated the initiatives to establish a protected area in the Dutch sector of the North Sea in order to study the actual beneficial effects of such an area. However, owing to strong opposition from the fisheries sector, lack of political motivation and the lack of support at the European level, the idea was temporarily abandoned. We now realise that it was a mistake not to involve all stakeholders in the discussion from the start. The media presentation of the ideas behind the protected area provoked a very hostile response from fishermen who, in turn, influenced the politicians. Although one wonders whether the politicians would have reacted at all if we had used another approach, involving the fishing community from the onset could have avoided many of the antagonistic reactions. However, recently these discussions have resumed on a new footing.

Successful protected areas

Closed areas and multiple-use marine protected areas are two possible tools that move marine management away from largely ineffective sectoral control towards true conservation that benefits both humans and the natural environment. The principles for the success of marine protected areas are listed as follows (after Agardy, 1997).

1. Clearly define specific objectives for marine protected or closed areas at the onset.
2. Get as much input from stakeholders as possible. The involvement of the stakeholders, in this case the fishermen, is crucial for different reasons. Stakeholders have traditional knowledge about resource dynamics and ecosystems that will be important in determining levels of sustainable use. Stakeholders can increase the public awareness and promote good marine stewardship, including use, responsibility and protection.
3. Make the planning process truly participatory, as opposed to allowing user groups to comment on a plan developed by a single stakeholder (usually a government agency).

4. Design zoning to maximise protection for ecologically critical areas, while allowing sustainable use in less sensitive, vulnerable or important areas. If non-destructive fishing techniques are a feasible alternative, they could be allowed in (part of) the area. It may even be possible that more environmentally friendly fishing techniques may become economically profitable if destructive techniques are banned in larger areas.

5. Design marine protected area boundaries so that they reflect ecological reality as much as possible (avoid squares and other 'unnatural' shapes, encompass estuaries and landward sides of coastal zones, etc.). To enforce the protection of the area, the positions of fishing vessels could be monitored by the use of 'black box' position recorders.

6. It should be possible to alter the design or the management regime in light of new ecological and sociological information.

7. Design the marine protected area and develop its management plan with feasibility in mind, and look for ways of self-financing the management operation from the onset.

8. Obtain international recognition of the protected area, and assure a world-wide adopted legal status. Important instruments in this context include: United Nations Convention on Law of the Sea (UNCLOS); United Nations Environment Programme (UNEP); UNESCO's Biosphere Reserve Programme; Agreements from Agenda 21 of the Rio Meeting; and the RAMSAR Convention on Wetlands of International Importance.

9. Develop monitoring and evaluation methodologies that are appropriate to the specific objectives and include these in design criteria. Hereby, both the monitoring of biological, economical and social parameters and the prioritisation of research needs should be closely linked to the management objectives (FAO, 1998).

10. Form an independent, non-partisan or multi-user group to manage the marine protected area and to monitor its effectiveness using established benchmarks.

11. Undertake valuation exercises periodically under a broader public to ensure that the full value of the marine protected area is being realised.

12. Use the marine protected area as a way of raising awareness and stimulating education.

Conclusions

There are undoubtedly many potential benefits that might be derived from the creation of protected areas in the marine environment. Nature conservation calls for them, scientific research desperately needs them and even fisheries might benefit from them. However, the establishment of such areas in the open seas of Europe will demand the approval of the European Community. In addition, local economies may be adversely affected by the creation of fishing-free areas; hence, a long and difficult political process lies ahead, during which sociocultural issues will have to be taken into account (Fiske, 1992). Only an approach that integrates the needs and priorities of all managers, users and the scientists involved will facilitate the successful creation of protected and closed areas.

References

Agardy, T.S. (1997) *Marine Protected Areas and Ocean Conservation.* R.G. Landes/Academic Press, Austin, TX.

BEON (1992) *Effects of beamtrawl fishery on the bottom fauna in the North Sea. III – The 1991 studies. BEON Report* 16, 27 pp. Netherlands Institute of Sea Research, Den Burg, Texel, The Netherlands.

Bergman, M.J.N. & Lindeboom, H.J. (1999) Natural variability and the effects of fisheries in the North Sea: towards an integrated fisheries and ecosystem management? In: *Biogeochemical Cycling and Sediment Ecology* (eds J.S. Gray *et al.*). Kluwer, Dordrecht, pp. 173–84.

Bergman, M.J.N., Lindeboom, H.J., Peet, G., Nelissen, P.H.M., Nijkamp, H. & Leopold, M.F. (1991) *Beschermde Gebieden Noordzee noodzaak en mogelijkheden. NIOZ Report* 1991-3, 195 pp. Netherlands Institute of Sea Research, Den Burg, Texel, The Netherlands.

Bergman, M.J.N., Daan, N., Lanters, R.L.P., Salz, P., Smit, H., de Vries, I. & Wolff, W.J. (1997) Kansen voor natuur en visserij in de Noordzee. *Werkdocument IKC Natuurbeheer* nr. W-41, 33 pp.

Craik, W., Kenchington, R. & Kelleker, G. (1990) Coral-reef management. In: *Ecosystems of the World 25: Coral Reefs* (ed. Z. Dubinsky). Elsevier Science, Amsterdam.

Daan, N. (1989) The ecological setting of North Sea fisheries, 1989. *Dana,* **8**, 17–31.

Daan, N. (1996) *Desk study on medium term research requirements in relation to the development of integrated fisheries management objectives for the North Sea. RIVO-DLO Report* C054/96, 83 pp. Daan, N., Bromley, P.J., Hislop, J.R.G. & Nielsen, N.A. (1990) Ecology of North Sea fish. *Netherlands Journal of Sea Research,* **26**, 343–86.

Duineveld, G.C.A. (1992) *The macrobenthic fauna in the Dutch sector of the North Sea in 1991. NIOZ Report* 1992-6: 19 pp + appendices.

Dunnet, G.M., Furness, R.W., Tasker, M.L. & Becker, P.H. (1990) Seabird ecology in the *Netherlands Journal of Sea Research,* **26**, 387–425.

FAO (1998) *Integrated Coastal Area Management and Agriculture, Forestry and Fisheries.* Food and Agriculture Organization, Rome.

Fiske, F.J. (1992) Sociocultural aspects of establishing marine protected areas. *Ocean & Coastal Management,* **18**, 25–46.

Künitzer, A., Basford, D., Craeymeersch, J.A., Dewarumez, J.-M., Dörjes, J., Duineveld, G.C.A., Eleftheriou, A., Heip, C., Herman, P., Kingston, P., Niermann, U., Rachor, E., Rumohr, H. & de Wilde, P.A.W.J. (1992) The benthic infauna of the North Sea: species distribution and assemblages. *ICES Journal of Marine Science,* **49**, 127–43.

Lindeboom, H.J. (1995) Protected areas in the North Sea: an absolute need for future marine research. *Helgoländer Meeresunters,* **49**, 591–602.

Lindeboom, H.J. & de Groot, S.J. (eds) (1998) *The effects of different types of fisheries on the North Sea and Irish Sea benthic ecosystems. NIOZ Report* 1998-1/*RIVO-DLO Report* C003/98, 404 pp Netherlands Institute of Sea Research, Den Burg, Texel, The Netherlands.

Lindeboom, H.J., van Raaphorst, W., Beukema, J.J., Cadée, G.C. & Swennen, C. (1996) (Sudden) Changes in the North Sea and Wadden Sea: Oceanic influences underestimated? In: *Scientific Symposium on the North Sea Quality Status Report*, Ebeltoft, Denmark, pp. 79–85.

Philippart, C.J.M. (1998) Long-term impact of bottom fisheries on several by-catch species of demersal fish and benthic invertebrates in the south-eastern North Sea. *ICES Journal of Marine Science*, **55**, 342–52.

Piet G.J. & Rijnsdorp A.D. (1998) Changes in the demersal fish assemblage in the south-eastern North Sea following the establishment of a protected area ('plaice box'). *ICES Journal of Marine Science*, **55**, 420–429.

Rauck, G. (1985) Wie Schädlich ist die Seezungenbaumkure für Bodentiere? *Information Fischereiwissenschaft*, **32**, 165–8.

Reijnders, P.J.H. & Lankester, K. (1990) Status of marine mammals in the North Sea. *Netherlands Journal of Sea Research*, **26**, 427–35.

Rijnsdorp, A.D., Buys, A.M., Storbeck, F. & Visser, E.G. (1998) Micro-scale distribution of beam trawl effort in the southern North Sea between 1993 and 1996 in relation to the trawling frequency of the sea bed and the impact on benthic organisms *ICES Journal of Marine Science*, **55**, 403–19.

Rumohr, H., Ehrich, S., Knust, R., Kujawski, T., Philippart, C.J.M. & Schroeder, A. (1998) Long term trends in demersal fish and benthic invertebrates. In: *The effects of different types of fisheries on the North Sea and Irish Sea benthic ecosystems* (eds H.J. Lindeboom & S.J. de Groot), pp. 280–353. *NIOZ Report* 1998-1/*RIVO-DLO Report* C003/98. Netherlands Institute of Sea Research, Den Burg, Texel, The Netherlands.

Van der Veer, H.W., Creutzberg, F., Dapper, R., Duineveld, G.C.A., Fonds, M., Kuipers, B.R., van Noort, G.J. & Witte, J.IJ. (1990) On the ecology of the dragonet *Callionymus lyra* L. in the southern North Sea. *Netherlands Journal of Sea Research*, **26**, 139–50.

De Vooys, C.G.N., Witte, J.IJ., Dapper, R., van der Meer, J.M. & van der Veer, H.W. (1991) *Lange termijn veranderingen in zeldzame vissoorten op het Nederlands continentaal plat van de Noordzee*. *NIOZ Report* 1991-6, 81 pp. Netherlands Institute of Sea Research, Den Burg, Texel, The Netherlands.

Walker, P.A. (1998) Fleeting images: Dynamics of North Sea ray populations. Ph.D. thesis, University of Amsterdam.

Chapter 20

No-take zones: a management context

J.W. HORWOOD

Centre for Environment, Fisheries and Aquaculture Science, Lowestoft Laboratory, Pakefield Road, Lowestoft, NR33 0HT, UK

Summary

1. Examples of the theoretical and empirical evaluation of closed areas for fisheries management are described. The potential benefits of the closure of the North Sea Plaice Box, and the actual benefits of the closure of the Mackerel Box off the south-west of England are demonstrated.
2. Closed areas to protect juveniles fish, especially those areas with high discards, will be of benefit to the stocks and fisheries.
3. Other closed areas require a case-specific evaluation, and the results will be sensitive to the biology of the fish, the behaviour of the fishermen and the other fishery regulations in operation. Closed areas, which divert fishing elsewhere, taking the same weight of fish, are unlikely to have any significant benefits to the fish or fishery.
4. It is shown that closed areas may require monitoring over a considerable time to demonstrate empirically any benefits on naturally highly variable populations of fish.
5. No-take zones are recognised as a special case of closed areas, and are amenable to *a priori* evaluation provided the objectives for management are specified.
6. Examples where no-take zones may have a utility are described, but in many cases areas with fishery restrictions may give similar results with less local disruption.
7. The single most important measure for the management of our commercial fisheries is to restore the balance between the size of the resource and the size of the fishing fleets.

Keywords: no-take zones, fisheries management, closed areas.

Introduction

'No-take zones' (NTZs) have been variously defined. Most generally they are recognised as areas in which extractive use is prohibited. Such prohibited use would cover, for example, all fishing and gravel extraction. It would not seemingly prohibit dumping or impact by, for example, sewage discharge. For the current purpose, it is assumed that NTZs are areas of nil or negligible anthropogenic impact. It is doubtful if any such areas of significant size exist in UK waters.

In a fisheries context, the NTZs are a tool for management: they are not an end in themselves. Immediate questions are posed. What is being managed? What is the objective of the management? Will NTZs deliver any benefit? How much is the benefit? How much is the cost? Are there cheaper or better alternatives? Fisheries

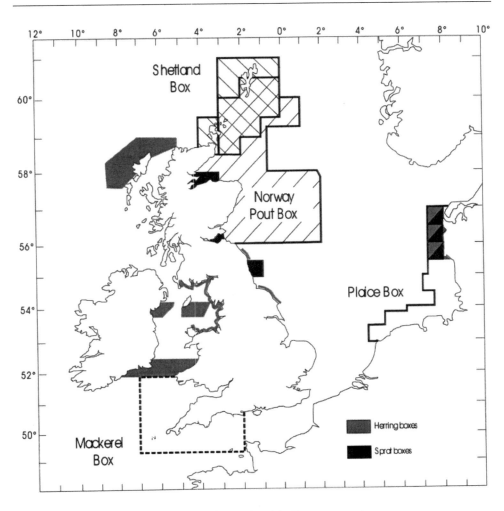

Fig. 20.1 Major areas where fishing is restricted in European waters

science has developed methods to assist with these questions, and fisheries scientists have an extensive experience both in the use of closed areas as a management tool and in the evaluation of the utility of closed areas. Figure 20.1 shows some of the areas in which fishing is restricted under European Union (EU) regulations.

The use and evaluation of closed areas in fisheries on the European Shelf are described here. Such evaluations are undertaken by specialist groups of scientists. The precise methodologies used for the evaluations are often not explicitly described, and some models for evaluating the utility of closed areas are presented. The discussion then highlights the importance of the prior evaluations, but also notes the limited nature of the class of problems addressed, and the sensitivity of the results to the other management measures that may be in place.

Examples of evaluations

The Plaice Box

The Plaice Box is an area, of about 38 000 km^2, off the Danish, German and Dutch coasts where trawling is restricted to protect juvenile plaice (*Pleuronectes platessa*) and sole (*Solea solea*) (Fig. 20.1). The Plaice Box was introduced in 1989, and its existence confirmed in the latest European Commission Regulation EC 850/98: for the conservation of fishery resources through technical measures for the protection of juveniles of marine organisms. As is frequently the case in other areas, not all fishing is prohibited in the Plaice Box, and there are derogations for small local vessels.

The Advisory Committee on Fishery Management (ACFM) undertook an evaluation of the utility of the Box in 1994 (Anon., 1994). It compared the long-term landings and spawning stock biomass (SSB) of plaice and sole, with different management options, relative to the then *status quo* of restrictions on the larger vessels in quarters 2 and 3 of the year. The results for plaice are summarised below. It can be seen that if the Box were removed, long-term landings and SSB would decline by 8–9%, whereas, if the prohibition was extended to all the year and to all vessels, landings and SSB would increase by 24–29%. The main reason for the impressive utility of the Plaice Box is that previous discarding rates within the Box averaged 83%. Improvements in relative selectivity of the population would also have reduced growth-overfishing. The new Regulation (EC 850/98) indicates that the Plaice Box was extended to all year but with significant derogations, which permit selected vessels to continue fishing, and particularly those fishing for shrimps.

The North Sea cod

The North Sea cod has been at, or near, historically low levels for about a decade, and has only recently shown signs of recovery (Fig. 20.2). Fishing rates were such that the equilibrium SSB per recruit was reduced to less than 5% that at zero fishing. In 1993, the EU, through the International Council for Exploration of the Sea, and

Table 20.1 Percentage increase (%) in long-term landings spawning stock biomass (SSB) relative to 1994 *status quo* Q2 and Q3 closures

Management option	Landings (%)	SSB (%)
Remove box	−8	−9
Extend to Q4	+11	+14
Extend to all year	+14	+17
All year + no discarding fleets	+24	+29

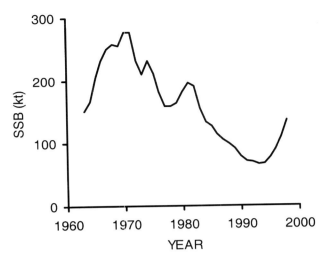

Fig. 20.2   The trends in spawning stock biomass (SSB) of North Sea cod (kt) by year.

its own Scientific, Technical and Economic Committee on Fisheries, investigated the potential of closing large areas of the North Sea in order to protect the cod. Given the current understanding of fish movements and the behaviour of fishing fleets, it was concluded that closing areas, even as large as the one-quarter of the North Sea, would do little or nothing to protect the widely dispersed and mobile cod (Anon., 1993).

The Trevose spawning grounds

UK fishermen are concerned about the state of fish stocks. In 1993, the UK's National Federation of Fishermen's Organisations (NFFO) published '*Conservation, an Alternative Approach*' (NFFO, 1993), in which they advocated restricting fishing in certain areas. One such area was the Trevose spawning ground, off the north coast of Cornwall, England. This is a spawning ground for most of the important commercial fish in the Bristol Channel and northern Celtic Sea (Horwood *et al.,* 1999). An evaluation of the proposal was undertaken by UK fisheries scientists at CEFAS. To the disappointment of the fishermen, and similar to the results for the cod, it was found that prohibiting fishing on the spawning grounds over spring provided no benefits to cod, sole or plaice populations in terms of long-term catch or SSB.

The reason for the lack of benefits in the latter two examples is that the fisheries are managed by quotas called total allowable catches (TAC). If the fish are not caught in one locality, such as that proposed for a closure, then the same fish will be caught elsewhere, as the areas used for management by TACs will generally correspond to the biological ranges of the populations.

The evaluation models

Approaches to modelling the effects of restricted areas have been presented in various places, but most of the work within ICES has been given in Working Group reports, and these often lack the mathematical detail necessary fully to comprehend the analyses (e.g. Anon., 1994). Horwood *et al.* (1999) presented models for the evaluation of closed areas, conditional on the type and amount of biological and fishery data typical of the European waters. The following briefly introduces some of the models.

A rule of thumb

We can consider the relative sensitivity of an equilibrium biomass (B) of the mature component of the population to proportional changes in adult (Z_a) and juvenile (Z_j) mortality. It can be shown that the relative sensitivity of mature biomass to unit changes in adult relative to juvenile mortality, $(dB/dZ_a)/(dB/dZ_j)$, is $(t_d Z_a)^{-1}$, where t_d is the time from recruitment to age at maturity.

When the sensitivity is unity, the spawning biomass benefits equally by an incremental decrease in either Z_a or Z_j. For higher values, the greater proportional benefits accrue from a decrease in the adult mortality and vice versa. If we take $t_d = 2$ years, for adult mortality rates Z_a below 0.5 year^{-1}, a greater benefit to the spawning biomass will be achieved by a reduction in adult mortality rather than in juvenile mortality. For many stocks $t_d Z_a > 1$, indicating that the mature stocks would benefit more from a proportional protection of juveniles.

A model with some spatial catch data

Key information for any evaluation of a spatial restriction on fishing is the knowledge of the biological character (e.g. age composition) of the fish caught in an area, and of the fish caught in any other area into which the fishing effort is displaced. Even if we could accurately predict the behaviour of fishermen, seldom, even in our relatively data-rich region, do we have biological data on small enough space scales. Thus, we have to make assumptions in order to fill these gaps in knowledge.

Horwood *et al.* (1999) assumed that typical fishery data would include the total and fishing mortality rates at age in the population (Z_a, F_a), the total international catch weight (C_Σ), and that of a potentially restricted fleet inside (CIT_Σ) and outside (COT_Σ) the Box. It was assumed that the total catches, by age, of the target (CT_a), and total (C_a) fleets are known, but cannot be allocated by area. Under these conditions, further assumptions are needed to evaluate the effects of a closure; in particular, the 'relative availability' of the fish by age, to the target fleet, in the Box (A_a). For example, if the Box is a spawning ground, immature fish will be scarce and an estimate of the 'relative availability' could be the proportion mature at age. The assumption allows an estimate of the catch at age from the target fleet in the box. A second assumption is that effort can be related to catch and catch rate, if known, and reallocated accordingly. We need to estimate the partial fishing mortality of the target fleet operating in the Box (FIT_a), and then calculate the effects on the equilibrium yield (Y_e) and spawning biomass (B_e) per recruit.

Defining $Y_e(F) = \sum_a F_a \varphi_a$, it can be shown that $FIT_a = ((CIT_\Sigma/C_\Sigma)Y_e A_a/\sum_a A_a \varphi_a)$, which implies knowledge of FOT_a. We can now calculate a revised $Y_e(F)$ and $B_e(F)$ with $F_a = F_a - FIT_a$ to give yield and biomass recruit if the fishing of the target fleet in the Box were stopped and not redeployed. In practice, fishing effort is likely to be redeployed. If effort of the target fleet in and out of the Box is unknown, we can assume that the relative effort is related to the catch and catch rates (CR), such that the new total fishing mortality at age becomes

$$\hat{F}_a = F_a - FIT_a + FOT_a(CIT_\Sigma/COT_\Sigma)(CR_o/CR_i)$$

and we can calculate $Y_e(\hat{F}_a)$ and $B_e(\hat{F}_a)$.

Such a model was applied to evaluate the benefits of closing the Trevose spawning ground.

Empirical evaluations

The models above permit an *a priori* evaluation of the effects of implementing restricted fishery closures. However, few examples exist of empirical *a posteriori* evaluations. One such case is that of the Mackerel Box (Fig. 20.1), which was established in 1981 off southern England and Ireland to protect the relatively high concentrations of juvenile mackerel found in the area. Table 20.2 shows the relative selectivity (as $F(age)/F(4–8)$) for the period 1972–1980 compared with the post-Box period of 1982–1996. After the introduction of the Box, the relative mortality on 0-, 1- and 2-year-old mackerel reduced by 83%, 60% and 20%, respectively.

More generally, the Organisation for Economic Co-operation and Development (Anon., 1997) recently reviewed the benefits to fisheries of 52 restricted areas. It found that, in 32 cases, stocks declined or had major oscillations. In only 16 cases had stocks either increased or remained the same.

Table 20.2 Fishing mortality at age, relative to that for $F(4–8)$, for Western mackerel for the period 1972–1980 compared with 1982–1996

	1972–1980	1982–1996	% Reduction
Age 0	0.023	0.004	83
Age 1	0.277	0.111	60
Age 2	0.503	0.404	20

The application of closed areas on Georges Bank has been described recently by Murawski *et al.* (1999). Whilst increases in whitefish stocks occurred at the same time as closed areas were implemented, the cause of the increases remains unclear, as other management measures changed over the period of the evaluation. Notably, however, the scallop stocks increased considerably.

In order to appreciate the difficulty of undertaking such empirical studies, consider the example of the North Sea Plaice Box. Potential benefits are relatively large; of the order of 20%. It is not possible to monitor SSB directly, and increased landings are not a good indicator as they can occur at the expense of the stock. Benefits will most directly be judged through apparent increases in recruitment. Recruitments can be measured with low precision from annual surveys, but they will be more accurately estimated as part of the annual assessment process. The logged-recruitment is approximately normally distributed, and independent from year to year; the variance of plaice log-recruitment is about 0.15 (SD 0.39). Under these conditions, it would take over 30 years before we had a 90% probability of recognising a 20% improvement in mean recruitment with 5% statistical significance. A similar statistic for detecting a 15% improvement would be about 60 years! It can be recognised that the benefit from any one closed areas is unlikely to be clearly demonstrated over, say, a decade, unless the changes are very large.

Discussion

We return here to relate the information above to the debate over NTZs. The use of closed areas, or NTZs, has been the subject of much international debate (e.g. Anon., 1990; Ballantine, 1991; Roberts & Polunin, 1993; Rowley, 1994; Gubbay, 1996; Pullen, 1996; Russ & Alcala, 1996; McGlade *et al.*, 1997; Roberts, 1997; Schmidt, 1997; Walters *et al.*, 1997; Watson, 1997; Clarke, 1998; Crosby, 1998). Clarke (1998) specifically makes the point that NTZs can be viewed as a fishery management tool, and not just as a tool for enhancing the marine environment as for example would a Marine Protected Area.

There have been significant claims for the benefits of NTZs, including: spawning biomasses can double or quadruple over 2–10 years; increased catches; increased diversity; enhanced habitat; benefits demonstrated by field studies; and all delivered with closures of the order of 10–20%. It is therefore no surprise that fishing interests and conservation organisations are looking towards NTZs as an answer to the pressing problems of growth and recruitment overfishing. However, the examples of quantitative analyses of closed areas given in this paper demonstrate that these benefits will not be realised for the wide-ranging European stocks that are essentially regulated by TACs.

From the examples outlined, we can see that, in general, restricting fishing on nursery grounds will give benefits to the stock, and especially so if there is significant discarding of small fish. Protecting spawning grounds, where the stock is managed by a TAC, is likely to divert effort and negate any benefits. In fact, more fish may well be killed because the fishing may be displaced onto smaller fish. Why then the positive claims for NTZs?

Most of the positive benefits attributed to NTZs have been associated with the management of reef fisheries. These are more characterised by an effort management. Adult fish, which inhabit the reefs, are relatively immobile, but their spawning products can be widely dispersed. The NTZs act as a source of recruits for other adjacent areas. This is not the biology for the fish used in the quantitative examples and implicitly modelled in those examples. The difference in results between the analyses described here and those more positive results for other regions can thus be put down to differences in biology and in management measures. Clearly, one must conclude that the benefits from reef fisheries do not directly translate to marine temperate fisheries managed by TACs.

Nevertheless, there are examples where NTZs should attract serious consideration, in European waters. There are vulnerable species, caught as by-catch, which are not, and cannot be, fully protected by TACs. Rays are such an example. They are wide fish, caught from their first year of life in trawls. They have low fecundity, and mature at relatively late ages (10 years for *R. clavata;* Walker & Hislop, 1998), making them potentially vulnerable to overfishing as a by-catch species. NTZs may be an important and practical tool in giving protection to such species. Such protection has to be at a population level. A NTZ would have to cover most or all of the distribution of the population to afford that population full protection. In many cases, this would be at significant cost to local or international fisheries. North Sea populations of *R. clavata* are ill defined, but appear to be localised. A local NTZ would impact considerably on local small fishing vessels not capable of fishing elsewhere. Complete protection is also unnecessary. What is necessary is to reduce the fishing by-catch mortality, on the population, to an acceptable level that ensures its sustainability. This might be achieved: (1) by a reduced area for the NTZ; (2) by reducing the total fishing effort in the area; (3) by reducing effort at particular times; (4) by restricting the use of some types of fishing gear; or (5) by some combination of these measures. The NTZ, then, is seen as an extreme, and perhaps not essential, form of fishery closed area. The effect of the different management options should be open to quantitative evaluation and the benefits weighted against the costs of alternative management regimes.

Some areas of the seabed are clearly vulnerable to fishing, such as those inhabited by delicate slow-growing organisms such as corals. If we want these areas and organisms to be protected, it is difficult to imagine that anything else other than a NTZ will work. Indeed, we may need something stronger than a NTZ – perhaps a no-impact zone.

There are also species that behave like reef-species, in that they are essentially sedentary, disperse their progeny over wide areas and are vulnerable to over-exploitation; scallops might be such an example. It is possible to imagine a management system wherein key areas of scallops are protected in order to seed adjacent areas – just like in the reef fisheries. Nevertheless, the challenge remains to identify such species and areas.

There are areas with concentrations of fish that, if afforded protection, would benefit a wide range of species. Horwood *et al.* (1999) demonstrated that if an area from St Ives to Lundy Island out to about 30 miles was closed during March and April, some protection against capture, damage and disturbance at spawning would be afforded to: plaice, sole, turbot, thickback sole, lemon sole, solenette, long-rough dab, topknot, Norway topknot, cod, ling, whiting, bass, sprat, rocklings, argentine, dragonets and the gurnards. Most of these species are not regulated under TACs, but are caught as by-catch in other commercial fisheries.

Finally, if we want to study the dynamic behaviour of fish benthic habitats, we will need to monitor the natural changes that occur in the seabed over decades. NTZs or no-impact zones would be essential for this purpose.

For all the potential uses considered above, it is possible to quantify the effects on stocks and fisheries of proposed closed areas. Such results will be sensitive to the behaviour of fish and fishermen, and to the other management systems in place. But in all cases, it is essential to define the purpose of the management so that the best tool for management can be identified.

It has to be recorded that there are potentially significant downsides to the advocacy of NTZs as a key tool for the conservation of commercial fish stocks. Many of our important commercial stocks are at historically low levels, caused in the main by historically high levels of fishing mortality. Further technical measures such as increased mesh sizes and closed areas, unless taken to draconian levels, will have only a modest effect on stock sizes. The single and most important management measure is to restore an appropriate balance between the size of the resources and the size of the fleets. Permanent reductions in fishing mortality, and associated fishing effort, will increase stock size, increase catch rates, increase profitability, give greater stability to the industry and be less intrusive on the environment. NTZs must not be seen as an easy and false option to restoring this balance.

Acknowledgements

This work was funded by MAFF Fisheries Division III, and a reduced version of this paper was presented at the ICES Theme Session 'Evaluation of marine protected areas as management tools' in 1998. Thanks are extended to Mr Mike Nicholson for advice on the statistics.

References

Anon. (1990) *Potential of marine fishery reserves for reef fish management in the US southern Atlantic. Report, National Oceanic and Atmospheric Administration, Technical Memorandum* 261, pp. 51. National Marine Fisheries Service, Southeast Fisheries Science Center.

Anon. (1993) Report of the North Sea Cod Task Force. Unpublished report for the EU Commission from meetings held in October and November 1993, 29 pp. + tables & figures.

Anon. (1994) Report of the Study Group on the North Sea Plaice Box. *International Council for the Exploration of the Sea*, C.M.1994/Assess: 14, 52 pp.

Anon. (1997) *Towards Sustainable Fisheries. Economic Aspects of the Management of Living Marine Resources.* Organisation for Economic Co-operation and Development, Paris.

Ballantine, W.J. (1991) Marine reserves for New Zealand. *University of Auckland, Leigh Laboratory Bulletin*, 25, 1–196.

Clarke, B.M. (1998) *No Take Zones (NTZs) a Realistic Tool for Fisheries Management?* 51 pp. Marine Conservation Society, Ross-on-Wye, UK.

Crosby, M.P. (1998) Moving towards a new paradigm for interactions among scientists, managers and the public in marine and coastal protected areas. In: *Proceeding of the Second International Symposium and Workshop on Marine and Coastal Protected Areas*, July, 1995 (eds M.P. Crosby, D. Laffoley, G. O'Sullivan & K. Geenen), 247 pp. Office of Ocean and Coastal Resource Management, National Oceanic and Atmospheric Administration, Silver Spring, MD, USA.

Gubbay, S. (1996) *Marine refuges. A report.* Worldwide Fund for Nature UK, Godalming, Surrey, UK.

Horwood, J.W., Nichols, J.H. & Milligan, S. (1999) Evaluation of closed areas for fish stock conservation and the Trevose spawning ground. *Journal of Applied Ecology.*

McGlade, J., Price, A., Klaus R. & Metuzals, K. (1997) *Recovery plans for the North Sea ecosystem, with special reference to cod, haddock and plaice. A report.* Worldwide Fund for Nature UK, Panda House, Godalming, Surrey, UK.

Murawski, S.A., Brown, R., Lai, H.L., Rago, P.J. & Hendrickson, L. (1999) *Bulletin of Marine Science.*

NFFO (1993) *Conservation, an Alternative Approach.* National Federation of Fishermen's Organisations, Fish Docks, Grimsby, UK.

Pullen, S. (1996) The role of marine protected areas and fisheries refuges. *Marine Update*, Vol. 28, 4 pp. World Wildlife Fund.

Roberts, C.M. (1997) Ecological advice for the global fisheries crisis. *Trends in Ecology and Evolution*, 12, 35–8.

Roberts, C.M. & Polunin, V.C. (1993) Marine reserves: simple solutions to managing complex fisheries? *Ambio*, 22, 363–8.Rowley, R.J. (1994) Marine reserves in fisheries management. *Aquatic Conservation: Marine and Freshwater Ecosystems*, 4, 233–54.

Russ, G.R. & Alcala, A.C. (1996) Do marine reserves export adult fish biomass? Evidence from Apo Island, central Philippines. *Marine Ecology Progress Series*, 132, 1–9.

Schmidt, K.F. (1997) 'No-Take' zones spark fisheries debate. *Science*, 277, 489–91.

Walker, P.A. & Hislop, J.R.G. (1998) Sensitive skates or resilient rays? Spatial and temporal shifts in ray species composition in the central and north-western North Sea between 1930 and the present day. *ICES Journal of Marine Science*, 55, 392–402.

Walters, C., Christensen, V. & Pauly, D. (1997) Structuring dynamic models of exploited ecosystems from trophic mass-balance assessments. *Reviews in Fish Biology and Fisheries*, 7, 139–72.

Watson, M. (1997) Where fish may safely graze. *New Scientist*, 153 (2069), 46.

Part 6
Socio-economic implications and mechanisms for reducing the impacts of fisheries

Dolphin by-catch in a tuna fishery. (Reproduced with the permission of Nick Tregenza.)

Chapter 21
Economic incentives to discard by-catch in unregulated and individual transferable quotas fisheries[1]

S. PASCOE

Centre for the Economics and Management of Aquatic Resources, University of Portsmouth, Locksway Road, Portsmouth, PO4 8JF, UK

Summary

1. The decision to discard is primarily a function of the economic incentives facing the fisher. Even in the absence of regulation, fishers will only land catch if the price received exceeds the costs of landing it. Hence, discarding may occur even in the absence of fisheries management.
2. Management can alter the incentives offered to fishers, and thereby affect the level of discarding. Individual transferable quotas (ITQs), in particular, have been criticised for leading to increased levels of discarding.
3. In this chapter, the economic incentives to discard are examined in both an open access fishery and a fishery managed by ITQs.
4. It is concluded that, while ITQs provide additional incentives to discard, fishers are faced by a number of incentives to discard even in unregulated fisheries.

Keywords: discarding, economic incentives, ITQs.

Introduction

The incidental capture and subsequent discarding of unwanted marine life is a common feature of many fisheries around the world, and is one of the major biological and political issues facing modern fishing (Gillis *et al.*, 1995a). In total, approximately 27 Mt of fish are thought to be discarded annually, representing roughly 25% of the total marine harvest (Alverson *et al.*, 1994).[2] As a wide variety of fish species occupy the same habitat, fishers are generally unable to catch individual species without some unintended catch of other species. Where the by-catch has little or no commercial value, the cost involved in landing the fish (such as storage, icing and freight costs) may exceed the price received. In such cases, the fisher is financially better off disposing of the fish rather than landing it.

The decision to discard unintended catch (or by-catch) is a function of the relative costs and benefits of retaining and landing the fish or discarding it. For non-commercial species, this decision is fairly straightforward, as there are no benefits associated with landing the species. For commercial species, the decision to discard

---

See end of chapter for notes indicated by superscript numbering.

will depend on a number of factors and will vary from fishery to fishery. These include the expected price received, the cost of landing and the opportunity cost of storing the fish on-board. In addition, the decision to discard commercial species will depend on the management regulations imposed on the fishers. Fisheries management can take a number of forms. These range from no regulation at all (free and open access) to varying combinations of input and/or output controls. These management options influence the level of by-catch or discards by changing the incentives faced by fishers (and hence affecting their behaviour; Gillis *et al.*, 1995b) or by changing the type of fishing technology used (Dewees & Ueber, 1990).

In this chapter, the economic incentives to discard by-catch in an unmanaged fishery and under individual transferable quotas (ITQs) will be examined. It is intended to demonstrate that discarding is not strictly a result of fisheries management, but can be affected by fisheries management. Details of the effects of other fisheries management options on the incentives to discard (such as mesh size restrictions, limited entry and days at sea restrictions) are reported elsewhere (Pascoe, 1997).

Discarding of commercial species in an unregulated multispecies fishery

It is generally assumed that, while all non-commercial species would be discarded under open access, there is no incentive to discard commercial by-catch species. This, however, is not necessarily the case. References to discarding date back to at least biblical times (Alverson *et al.*, 1994; Corey & Williams, 1995).[3] Hence, historically the taking and discarding of unwanted fish was part of the process of getting marketable fish to shore (Corey & Williams, 1995) and is not just a modern phenomenon relating to fishing technology and management regimes.

Discards of commercial species in an unregulated fishery can occur for two reasons. First, the price received for the fish does not compensate the fisher for the costs involved in its handling and dispatch to market; second, the boat faces a storage capacity constraint and the skipper is better off utilising the available storage capacity for higher-valued species. These forms of discarding are referred to as high-grading (Copes, 1986). That is, the value of the catch is maximised by landing only the higher-valued components.

Price-related high-grading

The first form of high-grading occurs largely in a fishery with different grades of fish receiving different prices. For discarding to occur, the different grades need to be detectable by fishers and so need to be defined in terms of some physical characteristic of the fish (Arnason, 1994). For example, each grade may be defined in terms of sex, size, skin damage, colour or some other obvious feature of the fish.

The incentives to land or discard the fish of each grade will depend on the price received, the costs of landing the fish and the costs associated with discarding the fish. Arnason (1994, 1995) developed a discard rule describing the conditions under which discarding is an economically rational activity. Following Arnason (1994, 1995), consider a fishery consisting of $i = 1, ..., I$ grades of fish. These each attract a separate price on the market, p_i. The total catch (C) is the sum of the catch of the individual grades, such that

$$C = \sum_i C_i(e, x, i) \qquad (1)$$

where $C_i(e, x, i)$ is the catch of grade i, a function of the level of effort e, the stock biomass x and the grade composition of the catch i. The catch of each grade may be either landed or discarded, where the quantity landed is given by l_i and the quantity discarded is given by d_i. The relationship between landings and discarding can be expressed by

$$l_i = C_i(e, x, i) - d_i. \qquad (2)$$

Both landing and discarding impose a cost on the fishery. These can be defined by non-decreasing, convex cost functions $CL_i(l_i)$ and $CD_i(d_i)$, where $CL_i(l_i)$ is the cost of landing grade i, a function of the quantity of landing of grade i; and $CD_i(d_i)$ is the cost of discarding grade i, a function of the quantity of discards of grade i. The costs of landing the fish include activities such as preliminary fish processing (e.g. the labour involved in gutting and gilling), storing (e.g. ice costs) and handling (e.g. crate costs) as well as actual costs involved in landing the fish (e.g. landing levies and transport costs to the market). The costs of discarding are expected to be relatively low for low quantities of discards as discarding is relatively easy, but would increase with the quantity of discards as a greater proportion of the crew time would be involved in the discarding process.[4]

The profit of the fishing operation is given by

$$\pi(e, d, x, p) = \sum_i p_i l_i - CE(e) - \sum_i CL_i(l_i) - \sum_i CD_i(d_i) \qquad (3)$$

where $CE(e)$ is the cost of effort, taken as a function of the level of effort e. From this, total profit is a function of the level of effort e, the set of discards d, the stock size x, and the set of prices p. Maximising profits with respect to the level of discards, d_i, results in the necessary condition relating to the optimal level of discards. From this, discarding fish of grade i is economically rational if:

$$p_i + CD_i(0) < CL_i(C_i(e, x, i) - 0) \quad \forall i. \qquad (4)$$

The left-hand side of the inequality represents the marginal cost of discarding. This comprises the forgone price received on the market (the opportunity cost of the discarded fish) and the cost of discarding the fish itself, evaluated at the zero discard level. The right-hand side of the inequality is the marginal benefit of discarding. This is the landing costs not incurred if the fish are discarded. These are also evaluated at the zero discarding level.

From this rule, Arnason (1994) develops a discarding function, given by

$$\Gamma_i = CL_i(C_i(e, x, i) - 0) - p_i - CD_i(0).\tag{5}$$

If the value of the discarding function for a particular grade is positive, the catch of that grade will be discarded. Conversely, if the value of the discarding function is negative, the catch will be retained. The value of Γ_i is hence a measure of the propensity to discard or retain the catch.

The discarding function may take a number of shapes. For example, in Fig. 21.1a, grades to the left of i^* would be discarded, while fish in grades above i^* would be retained. Hence, i^* can be considered to be the minimum economic size of the fish. In contrast, in Fig. 21.1b, catch of size grades below and above grades i_l and i_u, respectively, would be discarded, while the middle grades would be retained. This latter situation might occur for fish that are primarily consumed in the restaurant trade, where 'plate-sized' fish attract a premium price. Fish too small or too large for this market would receive a substantially lower price.

In most cases, different size classes are the main cause of price-related high-grading, although damaged fish would also be discarded for the same reasons. Given that fishers are able to modify their gear to target particular size classes, the incentives to high-grade need to be compared with the incentive to modify the gear in order to avoid the capture of the smaller fish.

Arnason (1995) extended the above analysis to incorporate varying size selective fishing gear. In this case, the harvesting function can be represented by

$$y_i = C_i(e, x, i)(1 - a_i)\tag{6}$$

where y_i is the harvesting function of grade i, $C_i(e,x,i)$ is the 'unselective' harvesting function, and a_i is the selectivity factor ($0 < a_i < 1$). When a_i is equal to 1, none of the size class i is caught. Conversely, when $a_i = 0$, all of the size class i are harvested by the gear.

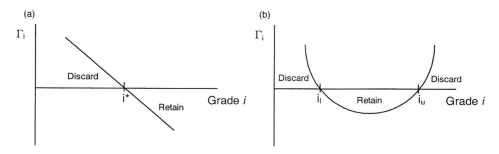

Fig. 21.1 Examples of discarding functions. (a) Illustrates the case where price increases with fish size. All fish smaller than size i^* are discarded. (b) Illustrates the case where price increases across some size classes, but decreases for larger fish. Fish that are smaller than i_l or larger than i_u are discarded. Taken from Arnason (1994, 1995).

   Adopting this selective gear is not costless. Arnason (1995) assumed that the cost of adopting the selective gear, $CS_i(a_i)$, is a function of the selectivity parameter a_i. The profit function can be redefined as

$$\pi = \sum_i p_i[C_i(e, x, i)(1 - a_i) - d_i] - CE(e)$$
$$- \sum_i CL_i(l_i) - \sum_i CD_i(d_i) - \sum_i CS_i(a_i). \tag{7}$$

Differentiating this profit function with respect to effort, discards and selectivity results in the necessary first-order conditions for profit maximisation by the fisher:

$$CE'(e) = \sum_i p_i\big[C_i'(e, x, i).(1 - a_i)\big] - \sum_i CL_i'(l_i(e)) \tag{8}$$

$$-[p_i - CL_i'(l_i)] \leq CD_i'(d_i) \tag{9}$$

$$-[p_i C_i(e, x, i)] - CL_i'(l_i) \leq CS_i'(a_i). \tag{10}$$

From the first condition, a profit-maximising producer will continue to apply effort to the fishery until the marginal benefit of fishing (the revenue from catching the marginal fish less the cost of landing the marginal fish) was equal to the marginal cost of the additional unit of effort evaluated at the total effort level e (i.e. $CE'(e)$). The second condition is a restatement of the previous condition in equation (4), where the net price received from landing the fish must be both negative and less than the cost of discarding before discarding is an optimal option.[5] The final condition indicates that in order to employ more selective gear rather than discard, the net benefits received from landing the size class i (this is the price times the quantity caught, assuming no discards) must be both negative and less than the marginal cost associated with achieving that level of selectivity.

   From these conditions, Arnason (1995) demonstrated that if the marginal cost of discarding (evaluated at the level of discarding) is less than the marginal cost of selectivity (evaluated at zero selectivity) (i.e. $CD_i'(d_i) < CS_i'(0)$), then the profit-maximising fisher will only discard and not employ selective gear. Conversely, if the marginal cost of discarding evaluated at zero discards is higher than the marginal cost of selectivity (i.e. $CS_i'(a_i) < CD_i(0)$), then the profit-maximising fisher will only employ selective gear and avoid unwanted size classes of fish, hence removing the need to discard. Finally, both selectivity and discarding may occur simultaneously. The optimal combination of use of selective gear and discarding is given by the condition

$$CS_i'(a_i) = CD_i'(d_i). \tag{11}$$

Following Arnason (1995), this relationship can be demonstrated graphically. In Fig. 21.2, the marginal cost of discarding and selectivity intersect the benefits of not

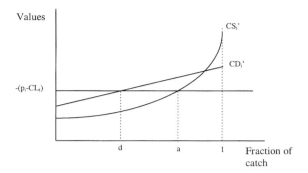

Fig. 21.2 Optimal discarding and selectivity. CS_i' is the marginal cost of selectively removing size class i, while CD_i' is the marginal cost of discarding size class i. The intercept term, $-(p_i - CL_i)$, is the marginal benefit of landing the fish in each size class i. The proportion of the potential catch of size class i 'removed' though improved selectivity is a, while the proportion of the catch that is discarded is d. Taken from Arnason (1995).

landing (the negative of the net price) at d and a, respectively. From this, the optimal selectivity will be a, and the proportion discarded (after selectivity) will be d. Hence, the quantity of discards of grade i would be $C_i(e, x, i)(1 - a)d$ (Arnason, 1995).

In reality, these conditions would vary trip by trip and the size (or other grade) composition will vary haul by haul. Catch compositions can vary according to season, location, type of gear and the way the gear is employed (Anderson, 1994). Even if fishers were able to estimate accurately the catch composition, it is most likely that the cost of changing the gear for each trip would be prohibitive (regardless of the expense of having sufficient types of gear to meet the optimal requirements for each trip). Hence, it is unlikely that fishers could ever operate at the optimal mix of selectivity and discarding. However, it is likely that fishers would have some predefined expectations about the average catch composition and may incorporate selective gear on that basis.

The above analysis does demonstrate, however, that it is perfectly rational behaviour for fishers to discard some grades of fish in the absence of regulation. While it is possible to employ selective gear in some cases so that the unwanted catch is avoided, this may not necessarily be the most rational behaviour from the perspective of the individual fisher.

Capacity-related high-grading

High-grading can also occur when there is a limit on the amount of fish that can be stored on-board or limitations on the amount of ice that can be produced to keep fish fresh (Cunningham, 1993). Gillis *et al.* (1995a) suggest that such discarding only takes place on the last haul. In such a case, the size-class distribution of the discards is the same as that which occurs in the catch of the last haul. However, it may be optimal to discard certain grades of fish that have been caught in earlier hauls in order to maximise the value of the trip.

For each boat with a constraint on the quantity of fish it can return to shore, the critical decisions faced by the skipper are how long to fish and what types of fish to keep (Anderson, 1994). Anderson (1994) examined the problem using a constrained optimisation approach to estimate the optimal decision rule for discarding. Following Anderson (1994) (but maintaining the notation used above), the trip profit function may be given by

$$\pi = \sum_i p_i[C_i(e, x, i) - d_i] - CE(e) - \sum_i CD_i(d_i). \tag{12}$$

In this case, grade i can represent a particular size grade, or may represent a by-catch species that has a different value to the target species. The profit function in this case is constrained by two factors. First, the amount of landings must be less than or equal to the volume that can be stored in the hold; and second, the amount of discards must be less than the total quantity caught. These can be expressed respectively as

$$\sum_i (C_i(e, x, i) - d_i) \leq B \tag{13}$$

$$d_i \leq C_i(e, x, i) \tag{14}$$

where B is the hold volume expressed in terms of the amount of fish that can be stored.

These constraints can be incorporated into the maximisation process through the Lagrangian, given by

$$L = \sum_i p_i[C_i(e, x, i) - d_i] - CE(e) - \sum_i CD_i(d_i)$$
$$+ \lambda \left[B - \left(\sum_i C_i(e, x, i) - d_i \right) \right] + \sum_i \gamma_i(C_i(e, x, i) - d_i) \tag{15}$$

where λ and γ_i are Lagrangian multipliers associated with the constraints. If the constraint is non-binding, these take the value of zero. However, if the constraints are binding, they take on a non-zero value representing the shadow value of the constraint. This is the value of an additional unit of the constraint. For example, λ in this case is the value of one additional unit of hold capacity, while γ_i is the value of discarding an additional unit of fish.

The necessary conditions for profit maximisation are derived by differentiating the Lagrangian with respect to effort and discards of each grade i. These are given by[6]

$$\frac{\partial L}{\partial e} = \sum_i p_i C_i'(e, x, i) - CE'(e) - \lambda \sum_i C_i'(e, x, i) + \sum_i \gamma_i C_i'(e, x, i) \leq 0 \qquad (16)$$

$$\frac{\partial L}{\partial d_i} = -p_i - CD'(d_i) + \lambda - \gamma_i \leq 0 \quad \forall i. \qquad (17)$$

With a large number of species or grades, solving this set of conditions is difficult. For the purposes of demonstration, assume that there are only two species that are caught in constant proportion. Also, if we assume that the catch per unit of effort is constant and the cost per unit of discarding is constant and the same for both species, the two necessary conditions can be expressed as

$$\frac{\partial L}{\partial e} = p_1 \alpha y + p_2(1 - \alpha)y - CE'(e) - \lambda y + \gamma_1 ay + \gamma_2(1 - a)y \leq 0 \qquad (18)$$

$$\frac{\partial L}{\partial d_i} = -p_i - D + \lambda - \gamma_i \leq 0 \quad \forall i \qquad (19)$$

where y is the average catch per unit of effort, α is the proportion of catch that is comprised of species (or grade) i and D is the marginal cost of discarding a unit of either species.

In practice, it will be optimal to high-grade only one species (or size class). If the price of one species is less than that of the other, there is no benefit in discarding the more valuable species in order to land the less valuable species. Hence, the shadow price associated with discarding for one species will be zero. For illustrative purposes, assume that the second species (or size grade) has a higher price than the first and is not discarded, such that $\gamma_2 = 0$.

The amount of catch required to fill the hold with the higher-valued species is equivalent to $B/(1-\alpha)$.[7] That is, the capacity of the hold divided by the proportion of the catch that consists of the higher-valued species. If total catch is less than B, the shadow value associated with the hold is zero (i.e. $\lambda = 0$). From equation (19), the shadow value of discarding under such conditions is given by

$$\gamma_i = -p_i - D. \qquad (20)$$

That is, the shadow price of discarding is the forgone price that could be received and the cost of discarding itself. As this is negative, discarding an additional unit of fish will decrease profits. Hence, discarding under such conditions would not be rational behaviour. This condition is fairly intuitive and did not need such complex calculus to determine. However, it does indicate that the model provides intuitive answers under such conditions, validating the technique.

If the capacity constraint were binding, however, the shadow price associated with the hold constraint would be positive. Equation (18) can be simplified by dividing through by the catch rate y, and taking into account the zero shadow price of the second species, giving

$$p_1\alpha + p_2(1 - \alpha) - \frac{CE'(e)}{y} - \lambda + \gamma_1\alpha \leq 0. \tag{21}$$

These equations can be solved simultaneously to derive the shadow prices associated with a unit of storage space in the hold and discards of each species. Solving equation (21) for λ gives

$$\lambda = p_1\alpha + p_2(1 - \alpha) - \frac{CE'(e)}{y} + \gamma_1\alpha. \tag{22}$$

Substituting this back into equation (19) and solving for the lower-valued species (the species that may be discarded, assumed to be species 1) gives

$$\gamma_1 = p_2 - p_1 - \frac{1}{(1 - \alpha)}\left[\frac{CE'(e)}{y} + D\right]. \tag{23}$$

The shadow value of discarding the first species is consequently related to the net benefit of landing only the higher-valued species. This is largely a function of the price difference that can be received by landing only the higher-valued species (i.e. $p_2 - p_1$). However, this benefit is reduced by the additional costs incurred in replacing the discarded species 1 by fishing longer to catch more of species 2 and the costs associated with discarding themselves. The costs of capturing the additional fish and the costs associated with discarding vary inversely with the proportion of the catch that is comprised of the target species, $(1-\alpha)$.

From this, it can be seen that discarding lower-valued species may be rational under unregulated fisheries where the boats are subject to hold-capacity constraints. These constraints will vary from trip to trip, depending on the various catch rates and catch compositions encountered. For some trips, it may not be rational to discard at all if the hold capacity is not reached. However, some fishers may discard part of their catch early during the trip in anticipation of filling the hold (Gillis *et al.*, 1995a). In other cases, fishers may choose to store lower-valued fish and discard these only when the hold capacity is met. For example, Japanese longliners operating in the Atlantic have been observed to discard albacore and other lower-valued species after processing and storage in the freezer if the hold space is subsequently required by more valuable species such as yellowfin or blue fin tuna.

Individual transferable quotas

ITQs involve landing limits on individual fishers rather than on the fishery as a whole. While the potential economic benefits of such a system have been widely recognised (see for example, Squires *et al.*, 1995; Hannesson, 1996), the system has also been criticised for the incentives it creates to discard fish (Copes, 1986).

With individual quotas, fishers are permitted to land only a certain quantity of each quota species given by their quota holdings. Like the aggregate quota, a total allowable catch is determined for each species to be managed. This total quota, however, is subdivided into disaggregated units, which are allocated to individual fishers to remove the need to race for fish. These individual quotas can then be traded between fishers, so that each fisher can adjust his or her quota holdings to best suit their fishing operation.

Discards of over-quota catch

Catches of quota species that are not covered by the fisher's quota holdings must be discarded. These species may not be covered by the fisher's quota holdings if the fisher has already filled his or her quota for that species or has never held quota for that species. In theory, ITQs can address this problem through quota trading. That is, either by purchase or lease of additional quota (Dewees & Ueber, 1990). In many cases, however, the fisher is unable (or unwilling) to buy or lease more quota. This might be because the total allowable catch has been taken or the trading price for quota is greater than the fisher is willing to pay.

When the quota of one of the species is finished, the fisher will fish in the location where profits are maximised excluding the revenue from the over-quota species. This may result in a reduction in the catch of the over-quota species as in the examples above. However, if the over-quota catch is of a species that is relatively minor in the fishers' normal fishing operations, running out of quota may not necessarily induce them to change fishing location.

Under an ITQ system, the fisher is able to plan the use of their quota over the year. The fisher will choose the fishing strategy that maximises their profits over the year, subject to the management, environmental and social constraints they face. This may also involve an overall reduction in discards compared with an aggregate quota system. Willen (1990) suggests that, at least in a number of fisheries, by-catch problems are reduced (if not eliminated) under ITQs. Discards of all species by Alaskan sablefish vessels were found to decrease from 24.5% of the catch to 11.7% of the catch following the introduction of ITQs, suggesting that fishers are better able to avoid unwanted species in the ITQ fishery (Warren *et al.*, 1997).

High-grading

Fishers have an increased incentive to high-grade under ITQs (Anderson, 1994). Both Anderson (1994) and Vestergaard (1996) estimated the set of conditions under which high-grading would increase in an ITQ fishery. Anderson (1994) assumed for simplicity that hold constraints are not binding, while Vestergaard (1996) incorporated hold constraints into the analysis.

Following Vestergaard (1996) and assuming (for simplicity) a single-species fishery comprising several size grades, the total catch that can be taken by the fisher is limited by the quota constraint Q, such that

$$N\sum_{i}(a_i yE - D_i) \le Q \tag{24}$$

where again N is the number of trips, a_i is the proportion of the catch of size class i, y is the catch per unit of effort, E is the level of effort and D_i is the level of discards of size class i. As quota can be traded, the catch is constrained by the final level of quota held, which may differ from the initial allocation $\bar{Q}$. Quota can be either purchased or leased in on an annual basis. For simplicity, it is assumed that the cost of these are the same on an annual basis.[8]

The profits that can be obtained from fishing over the year as a whole can be given by

$$\pi = N\left\{\sum_{i}[np_i(a_i yE - D_i) - c_d D_i]\right\} - N(CE(E)) - s(Q - \bar{Q}) - F \tag{25}$$

where np_i is the net price received for size class i (that is, price less the cost of landing the size class i), c_d is the cost of discarding one unit of catch, $CE(E)$ is the total cost of effort each trip (a function of the level of effort), F is the fixed costs associated with the fishing vessel and s is the average annualised price of quota. Hence, $s(Q - \bar{Q})$ represents the costs (or benefits) of buying (or selling) quota.

As well as the quota constraint, the fisher may also be subject to a hold constraint, given by

$$\sum_{i}a_i yE - D_i \le B \tag{26}$$

where B is the hold constraint. As before, a further constraint is added to the model for completeness to ensure that the level of discards cannot exceed the quantity caught, given by:

$$a_i yE - D_i \ge 0. \tag{27}$$

As in the previous analysis, the Lagrangian function can be represented by

$$L = N\left\{\sum_{i}[np_i(a_i yE - D_i) - c_d D_i]\right\} - N(CE(E)) - s(Q - \bar{Q}) - F$$
$$+ \lambda_1\left(Q - N\sum_{i}(a_i yE - D_i)\right) + \sum_{i}\lambda_{2,i}(a_i yE - D_i) + \lambda_3\left[B - \sum_{i}(a_i yE - D_i)\right] \tag{28}$$

where λ_1 is the shadow price of an additional unit of quota, $\lambda_{2,i}$ is the shadow price associated with the discarding constraint for each size class i and λ_3 is the shadow price of an additional unit of hold capacity. Differentiating this function with respect to the number of trips, the length of each trip, the level of discarding and the amount of quota held results in the necessary conditions for profit maximisation:

$$\frac{\partial L}{\partial N} = \sum_i \left[np_i(a_i yE - D_i) - c_d D_i \right] - CE(E) - \lambda_1 \left(\sum_i (a_i yE - D_i) \right) \leq 0 \qquad (29)$$

$$\frac{\partial L}{\partial E} = N \sum_i \left[np_i a_i y - CE'(E) \right] - \lambda_1 N \sum_i a_i y + \sum_i \lambda_{2,i} a_i y - \lambda_3 \sum_i a_i y \leq 0 \qquad (30)$$

$$\frac{\partial L}{\partial D_i} = -N(np_i + c_d) + \lambda_1 N - \lambda_{2,i} + \lambda_3 \leq 0 \qquad \forall i \qquad (31)$$

$$\frac{\partial L}{\partial Q} = -s + \lambda_1 \leq 0. \qquad (32)$$

From the Kuhn–Tucker conditions, if N, E, D and Q are greater than zero (as would be expected), the four inequalities in equations (29)–(32) can be replaced with equal signs. From equation (32), the fishing operation is at an optimum when the shadow value of a unit of quota is the average annualised price of quota. From equation (29), the shadow value of a unit of quota is also equivalent to the average short run profit[9] per unit of landings. This is obtained by rearranging equation (29), giving

$$\lambda_1 = \frac{\sum_i \left[np_i(a_i yE - D_i) - c_d D_i \right] - CE(E)}{\left(\sum_i (a_i yE - D_i) \right)}. \qquad (33)$$

From this, if the price of quota is less than the average profit per unit of landings, it is worthwhile buying more quota.

If the fisher is not constrained by the hold capacity, such that $\lambda_3 = 0$, then from equation (31), the shadow price associated with discarding size class i ($\lambda_{2,i}$) is given by

$$\lambda_{2,i} = -N(np_i + c_d) + sN. \qquad (34)$$

Discarding is worthwhile only if $\lambda_{2,i}$ is positive. In such a case, equation (34) can be expressed as

$$s \geq (np_i + c_d). \qquad (35)$$

From this, if the price of quota is greater than the net price received plus the cost of discarding (effectively representing the opportunity cost of landing the size class i and the cost of discarding it), then the fisher is better off discarding the fish than landing it and using the quota to land more of the higher grade (Anderson, 1994; Vestergaard, 1996). If the hold constraint is binding, then the condition for discarding becomes

$$s \geq (np_i + c_d) - \lambda_3/N. \tag{36}$$

As λ_3 will be positive if the constraint is binding, it is likely that this condition will hold for more grades than if the hold constraint is not binding. Hence, a higher level of discards could be expected when both quota and hold constraints are binding.

Arnason (1995) noted, however, that fishers are able to change their gear and hence alter the size composition of their catch. The incentives to do this were examined under conditions of free and open access earlier. These incentives are also affected by the use of ITQs.

Following Arnason (1995), the profit function in equation (7) used earlier can be redefined to include the effects of ITQs.

$$\pi = \sum_i (p_i - CL_i - \Omega)[C_i(e, x, i).(1 - a_i) - d_i] - CE(e) - \sum_i CD_i(d_i) - \sum_i CS_i(a_i) \tag{37}$$

where p_i is the price received for size class i, CL_i is the unit cost of landing[10] size class i, Ω is the opportunity cost of quota (i.e. the value of the quota consumed by landing the size class i), $C_i(e,x,i)$ is the total catch of grade i, $CE(e)$ is the cost of effort, taken as a function of the level of effort, e, and $CD_i(d_i)$ is the discard cost function. Associated with the gear is a selectivity cost function, $CS_i(a_i)$, which is a function of the selectivity parameter a_i.

Differentiating this profit function with respect to effort, discards and selectivity results in the necessary first-order conditions for profit maximisation by the fisher:

$$CE'(e) = \sum_i (p_i - CL_i - \Omega)\left[C_i'(e, x, i)(1 - a_i)\right] \tag{38}$$

$$-[p_i - CL_i - \Omega] \leq CD_i'(d_i) \tag{39}$$

$$-[p_i - CL_i - \Omega]C_i(e, x, i) \leq CS_i'(a_i). \tag{40}$$

These conditions are similar to those in equations (8)–(10) in the earlier analysis, with the exception that the opportunity cost of quota has become an important component of the conditions. From the first condition, a profit-maximising producer will continue to apply effort to the fishery until the marginal benefit of fishing (the revenue from catching the marginal fish less the cost of landing the marginal fish) is equal to the marginal cost of the additional unit of effort evaluated at the total effort level e, (i.e. $CE'(e)$). The benefit from applying effort is now less than before, as the fisher has the option of selling the quota rather than catching the fish.

The second condition is a restatement of the previous condition in equation (4). A net cost must be incurred by landing the fish (i.e. the landing costs exceed the price received) and this must be greater than the cost of discarding before discarding is an optimal option. The final condition indicates that in order to employ more selective gear rather than discard, the net benefits received from landing the size class i (this is the net price including the opportunity cost of the fish landed times the quantity caught, assuming no discards) must be both negative and less than the marginal cost associated with achieving that level of selectivity.

The optimal combination of selectivity and discarding will change with the introduction of ITQs owing to the effects of the opportunity cost of using the quota to land small fish. In Fig. 21.3, the marginal cost of discarding and selectivity intersect the benefits of not landing (the negative of the net price) at d and a, respectively, for the unregulated fishery. From this, the optimal selectivity will be a, and the proportion discarded (after selectivity) will be d. Hence, the quantity of discards of grade i would be $C_i(e,x,i)(1-a)d$ (Arnason, 1995).

The introduction of ITQs changes the incentives to adapt selective gear and discard as discussed above. The additional opportunity cost of quota results in the optimal level of discards rising to d_1, but also results in an increase in the use of selective gear to a_1. Hence, the new quantity of discards would be $C_i(e,x,i)(1-a_1)d_1$. As a_1 is greater than a, less smaller fish would be caught under ITQs. However, as d_1 is greater than d, a higher proportion of the smaller fish that were caught would be discarded under ITQs than under open access.

The net effect of the introduction of ITQs could therefore be to either increase or decrease discarding, taking into account the incentive to change gear selectivity. For example (following Arnason, 1995), if the initial discard and selectivity parameters were 0.4 and 0.5, respectively, and the new selectivity parameters were 0.6 and 0.7, respectively, the level of discards changes from $D = 0.2C$ to $D = 0.18C$. Hence, in this example, the introduction of ITQs would lead to a reduction in high-grading.

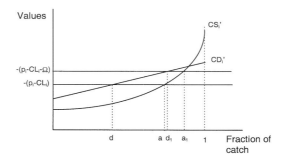

Fig. 21.3 Optimal discarding and selectivity under ITQs. CS_i' is the marginal cost of selectively removing size class i, while CD_i' is the marginal cost of discarding size class i. The two intercept terms $-(p_i - CL_i)$ and $-(p_i - CL_i - \Omega)$ are the marginal benefit of landing the fish in each size class i, respectively, under open access and ITQ management. The proportion of the potential catch of size class i 'removed' though improved selectivity under ITQs is a_1, while the proportion of the catch that is discarded is d_1. Taken from Arnason (1995).

Conclusions

By-catch occurs because different species are caught together as a result of the non-selectivity of the gear. These species may be both commercial species and non-commercial species. In an unregulated fishery (i.e. free and open access), fishers have an incentive to discard smaller-sized fish if the expected net price (that is, the price less landing costs) is negative and the resultant costs incurred in landing the fish are greater than the costs incurred in discarding the fish. This is termed high-grading. Small fish often do not receive a high price on the market relative to larger fish of the same species.

The incentives to discard in an unregulated fishery are increased if the boat has a limited hold capacity. In such cases, it is rational to discard low-valued sizes and also low-valued species in order to utilise the hold for the more valuable-sized fish or the more-valuable species.

Individual transferable quotas can result in additional incentives to discard, but can also result in lower levels of discarding. Under an ITQ system, a fisher is better able to plan his or her harvesting strategy. This could result in a fishing pattern that lowers discards or increases them, depending on the spatial distribution of the stock. The opportunity cost associated with the quota can also change the incentives to high-grade catch. While the incentive to discard is generally increased, the incentives to adopt more selective gear are also increased. Hence, while a greater proportion of the catch of small fish may be discarded, fewer small fish may be caught and, hence, overall discards may be lower. The actual direction of change will vary from fishery to fishery.

Increasing concerns over the level of discarding, particularly discarding that results from fisheries management, has led to the development of a number of policies and instruments to reduce discarding. In most fisheries, however, the total elimination of discards will be neither feasible nor desirable. There is an optimal level of discarding that takes into account the benefits produced by discarding (i.e. the catch of the commercial species and the savings in storage, preservation and transport costs) and the costs imposed by discarding. This will only be zero under extreme circumstances. Reducing discards below this optimal level may result in society's not achieving the greatest benefits possible from the use of the fisheries resources.

Notes

1. This paper is based on part of a review of the economics of discarding commissioned by the FAO (Pascoe, 1997).
2. This does not include the benthic invertebrates or marine mammals, sea birds and reptiles taken as incidental catches (Alverson *et al.*, 1994).

3. '... like a net that was thrown into the sea and caught fish of every kind. When it was full, they drew it ashore and put the good into baskets but threw out the bad' (Matthew 13:47–8).
4. Arnason (1994) also points out that if discarding is illegal or socially frowned upon, the cost of discarding may be substantially higher.
5. In absolute value terms, the net cost of landing the fish must be greater than the cost of discarding before discarding is economically rational.
6. Two further necessary conditions not presented are that $e(\partial L/\partial e) = 0$ and $d(\partial L/\partial d) = 0$. Hence, either the variable or the derivative with respect to that variable must take on the value of zero. As the value of effort and discarding can be assumed to be non-zero, the derivatives must equal zero. This fact will be utilised in the following analysis.
7. This is also the profit maximising level of catch given the hold constraint B (Anderson, 1994).
8. In theory, the purchase price of quota should reflect the net present value of the long-run average resource rent associated with a unit of quota. This would include the cost of capturing the fish. The lease price should reflect the annualised cost of this value, and hence, theoretically, should be equivalent on an annual basis. In practice, however, the lease price is likely to reflect the short-run marginal value of an additional unit of quota. The marginal value of a unit of quota for fish that has been caught will be the price of the fish less the cost of landing it. Hence, the lease cost could be substantially higher than the annualised purchase price, particularly at the end of the season when the demand for leased quota may exceed the supply.
9. This is the short-run profit as it excludes fixed costs associated with the boat.
10. In the earlier analysis this was assumed to vary with the level of landings. For simplicity, it will be assumed to be a constant unit cost.

References

Alverson, D.L., Freeberg, M.H., Murawski, S.A. & Pope, J.G. (1994) *A global assessment of fisheries bycatch and discards. FAO Fisheries Technical Paper* 339. Fisheries and Agriculture Organisation, Rome.

Anderson, L.G. (1994) An economic analysis of highgrading in ITQ fisheries regulation programs. *Marine Resource Economics*, **9**, 209–26.

Arnason, R. (1994) On catch discarding in fisheries. *Marine Resource Economics*, **9**, 189–207.

Arnason, R. (1995) On selectivity and discarding in an ITQ fishery. Paper presented at the *5th European Association of Fisheries Economists' Bioeconomic Modelling Workshop*, Edinburgh, 24–27 October 1995.

Copes, P. (1986) A critical review of the individual transferable quota as a device in fisheries management. *Land Economics*, **62**, 278–91.

Corey, T. & Williams, E. (1995) Bycatch: whose issue is it anyway? *Nor'easter*, **7**, 1–7.

Cunningham, S. (1993) Outcome of the workshop on individual quota management. In: *The Use of Individual Quotas in Fisheries Management* (ed. S. Cunningham). OECD, Paris.

Dewees, C.M. & Ueber, E. (1990) Introduction. In: *Effects of different fishery management schemes on bycatch, joint catch and discards: summary of a national workshop* (eds C.M. Dewees & E. Ueber), pp. 7–8. *California Sea Grant College Report* No. T-CSGCP-019. California Sea Grant College, University of California, La Jolla.

Gillis, D.M., Peterman, R.M & Pikitch, E.K. (1995a) Implications of trip regulations for high-grading: a model of the behavior of fishermen. *Canadian Journal of Fisheries and Aquatic Science*, **52**, 402–15.

Gillis, D.M., Peterman, R.M & Pikitch, E.K. (1995b) Dynamic discarding decisions: foraging theory for high-grading in a trawl fishery. *Behavioural Ecology*, **6**, 146–54.

Hannesson, R. (1996) *Fisheries Mismanagement: the Case of the North Atlantic Cod.* Blackwell Science, Oxford.

Pascoe, S. (1997) *Bycatch management and the economics of discarding. FAO Fisheries Technical Report* 370, Fisheries and Agriculture Organisation, Rome.

Squires, D., Kirkley J. & Tisdell C. (1995) Individual transferable quotas as a fisheries management tool. *Reviews in Fisheries Science*, **3**, 141–69.

Vestergaard, N. (1996) Discard behaviour, highgrading and regulation: the case of the Greenland shrimp fishery. *Marine Resource Economics*, **11**, 247–66.

Warren, B., Ess, C. & Swenson, E. (1997) Managing bycatch and discards: a review of progress and challenges in the United States. In: *Papers Presented at the Technical Consultation of Reduction of Wastage in Fisheries* (eds. I. Clucas & D. James). *FAO Fisheries Report* No. 547, pp. 69–88. Supplement, Fisheries and Agriculture Organisation, Rome.

Willen, J.E. (1990) ITQs and the bycatch problem. In: *Effects of different fishery management schemes on bycatch, joint catch and discards: summary of a national workshop* (eds C.M. Dewees & E. Ueber), pp. 50–1. *California Sea Grant College Report* No. T-CSGCP-019. California Sea Grant College, University of California, La Jolla.

Chapter 22
Options for the reduction of by-catches of harbour porpoises (*Phocoena phocoena*) in the North Sea

J.M. McGLADE and K.I. METUZALS
Centre for Coastal and Marine Science, The Hoe, Plymouth, PL1 3DH, UK

Summary

1. A governance framework was developed to address the problem of reducing by-catches of harbour porpoise (*Phocoena phocoena*) in gill-net fisheries in the North Sea. To support the framework, independent cost-effective methods were developed to derive estimates of by-catch using data from national fisheries statistics, observer programmes and in-depth interviews; the methods were tested for fleets in Grimsby (UK) and Denmark.

2. The UK test study was based on 27 gill-netters from Grimsby, fishing for cod and other species and using bottom-set and wreck-netting methods. Detailed analyses, including examination of effort, fishing tracks, landings and by-catch statistics, were undertaken on 12 vessels. The Danish fleet analysis was based on interviews with skippers from the western coastal ports of Esbjerg, Hvide Sande and Thorsminde, where the target species included plaice, sole, turbot and cod. Data from 30 licensed gill-netters (approximately 10% of the total fleet) were used.

3. Independent estimates of porpoise by-catch from fisheries data, in-depth interviews and observers showed a high degree of consistency: the by-catch for the UK gill-net fleet in Grimsby using observer data in conjunction with a detailed spatio-temporal analysis of the fishery was estimated to range from 81 to 193 over the period 1990–1997, and from interviews the estimate ranged from 95 to 202 for 1997–1998. Estimates for the Danish fleet from interviews for 1998 ranged from 3500 to 4500. The estimate of 4629 reported by Vinther for 1993 matched the upper level of the range. From the test studies, it was concluded that interviews represented an extremely cost-effective method of assessing levels of by-catch.

4. The development of more effective management structures also required the identification of areas, times and fishing operations associated with a high risk of by-catch. These were determined from spatio-temporal analyses of fishing effort by fleet and for individual vessels, in relation to oceanographic features, information from interviews and previously published studies. The results showed a high coincidence of by-catch of porpoises with (i) seasonal patterns of fishing for cod associated with fronts in the southern and central North Sea and with tidal-mixing in the summer along the inner waters of the Danish coast, and (ii) specific fishing operations, such as the height of nets and long soak times in deeper waters.

5. An analysis of the governing needs required to support the adoption of the Code of Conduct for Responsible Fisheries was undertaken for the gill-net fisheries studied. It was concluded that fishing practices, price competition and a lack of participation in decision-making often led to situations where by-catches occurred. Changes to the administrative and market structures were considered necessary for these fisheries to remain sustainable. In particular, participatory or co-management processes were

recommended in order to support stronger controls on specific fishing operations where there was a high risk of by-catch occurring.

Keywords: governance, by-catch, porpoises.

Introduction

The abundance of harbour porpoise, *Phocoena phocoena*, in the North Sea and adjacent waters has been estimated to be approximately 340 000 animals (Hammond *et al.*, 1995; SCANS, 1995). Evidence suggests that in the eastern North Atlantic including the Baltic, five populations exist: the Irish Sea/Wales, UK North Sea, Netherlands North Sea, Danish North Sea and inshore Danish Waters (Hammond *et al.*, 1995a, b; Vinther, 1995; International Whaling Commission (IWC), 1996; Andersen *et al.*, 1997).

The impact of incidental by-catches on these different populations is largely unknown, but information presented to the signatories to the Agreement on the Conservation of Small Cetaceans in the Baltic and North Seas (ASCOBANS) at their first meeting in 1994 indicated that levels of by-catch in the Danish gill-net fisheries alone were already sufficiently high (1.7% of the entire North Sea) to be of great concern (ICES, 1996). Given that harbour porpoises are also known to be taken in large numbers in other areas, such as in the gill-net fisheries operating on the Celtic Shelf (Berrow *et al.*, 1994), and that the harbour porpoise is listed as an Annexe II species (i.e. one whose conservation requires the designation of special areas of conservation), the issue of by-catch is especially critical.

In order to develop an effective management approach directed towards reducing by-catches without damaging the viability of an individual fishery, the levels of by-catch and their impacts on the various populations first need to be estimated. Then the effects of constraining or altering different aspects of the fisheries concerned (e.g. fishing operations, gears and local management practices) need to be examined. Finally, a system of effective and sustainable governance that will ensure and encourage adherence to regulation needs to be developed. The aim of this study was to develop and test various methodologies for improving the determination of levels of by-catch in fisheries, evaluate those factors affecting by-catch, and develop a model framework of effective governance and management.

Despite the fact that all member states of the European Union (EU) are required to monitor the incidental capture of all small cetaceans, data are often sparse and few monitoring or observer programmes exist. As costs are often the issue when discussing broad-scale monitoring programmes, it was thought appropriate to look at surrogate measures or other less costly ways of obtaining information to estimate by-catch. In this study, we chose to focus on two specific fisheries known to have by-catches of harbour porpoise, and used data from official national fisheries records, interviews *in situ*, observer programmes and published studies. Retrospective analyses were undertaken on the spatial dynamics of fishing fleets and individual vessels in relation to target species, fishing experience, other socio-economic drivers, key metocean phenomena and information about the migratory patterns of harbour porpoises. Given the sensitive nature of marine mammal by-catches amongst fishermen, the results of the analyses were cross-checked in order to obtain data that were as unambiguous as possible. The interviews were also used to determine the likely acceptance of various governance and management models.

The study fleets were based in Grimsby, UK and along the North Sea coast of Denmark. The bottom-set gill-net fleets from both areas are known to experience high by-catches of porpoises when fishing in the southern North Sea and off the coast of Denmark. The area off Denmark is particularly important, as high densities of porpoises occur in this area and large by-catches have been reported in the summer months (Hammond *et al.*, 1995a).

Methods

UK study fleet analysis

The UK study fleet was located in Grimsby, a town on the east coast of the UK. Fisheries data were obtained from the CEFAS (Centre of Environment and Fisheries and Science) Lowestoft Laboratory. Interviews were held *in situ* with most of the skippers in the fleet, with only one refusal.

Currently, the fleet comprises 17 active vessels, 10 of which fish full-time. From the 50 vessels listed in the CEFAS database, 27 were selected for detailed fleet analysis. Those vessels that only participated in gill-netting for 1 or 2 years were excluded from the analysis to ensure that the analysis was based on active fishing vessels only. All vessels were given a number, and the only identifying characteristic provided was the total length class.

On a small vessel (length 15–20m) there was usually a skipper and three crew members; on larger vessels there were usually four to five crew members when gill-netting and seven when long-lining. On average, there were about four fishermen associated with a single vessel; however, some crew members were part-time and the number could change between trips. Gill-net skippers were all responsible for at least five people. With 17 vessels classified as gill-netters in 1997, approximately 70 fishermen were fishing out of Grimsby in the study fleet.

The most common fishing practice in the fleet was bottom-set gill-netting (i.e. the net is set at the bottom of the seabed and then the upper part floats with the tide). Gill-netting is a very selective way of fishing in that only large and older fish are caught. Since the early 1960s, Danish gill-netting, wrecking or wreck netting, has evolved. This method uses known shipwrecks and relies on the fact that they act as a habitat for cod and other species.

According to the data collected in the interviews, fishing trips lasted for up to a week, but could range from a few days to 2 weeks. Usually, the length depended on the current market prices of cod. The agent arranged for the best price while the skipper was still at sea. If there were better prices to be obtained in The Netherlands, the vessel would land there and extend the trip for a day or so. Most often, Grimsby gill-netters would land at the Grimsby market. However, the Anglo-Dutch fleet has its administrative offices in Grimsby, although it operates mainly out of Holland.

Analysis of the total fishery

Total landings per individual trip per vessel were obtained for 1983–1997 for gill-net vessels fishing out of Grimsby. Total landings per vessel varied from a low of 1.4 t to a maximum of 277 t in 1985. Average landings of all species per vessel were calculated in order to estimate the potential earning power per vessel. Higher average landings were obtained by vessels with skippers who had more fishing experience (Fig. 22.1) and the highest total landings were obtained by vessels whose skippers had from 7 to 12 years' fishing experience in gill-netting.

An analysis of total landings by month for the period 1985–1997 (Fig. 22.2) shows that monthly landings from September to December have remained relatively stable, but in recent years a winter fishery in January–February has emerged. Although not shown here, cod comprise 84–97% of the catch per vessel, as fishermen actively seek this species in preference to others. Given this, the fleet analysis in relation to by-catch concentrated on cod as the main target species.

Analysis of the directed cod fishery

Cod-directed landings are generally highest in the third quarter of year (Q3) for the period 1985–1995 (Table 22.1). In 1987, 1995 and 1997, highest cod landings occurred in the first quarter, and in 1988, in the last quarter. Cod spawn in January and February, and a lucrative roe fishery has developed for this period of the year. However, the mean landings over 12 years show that highest mean monthly landings occur in the summer from June to September, with July and August having almost identical mean monthly landings for cod. Despite gaps in the data for the earlier part of the time series, there is a still a noticeable trend in landings. In the early 1990s,

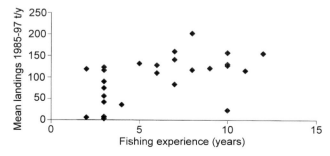

Fig. 22.1 Average landings of all species by individual vessels from 1985 to 1997 against years of fishing experience of the vessel skipper.

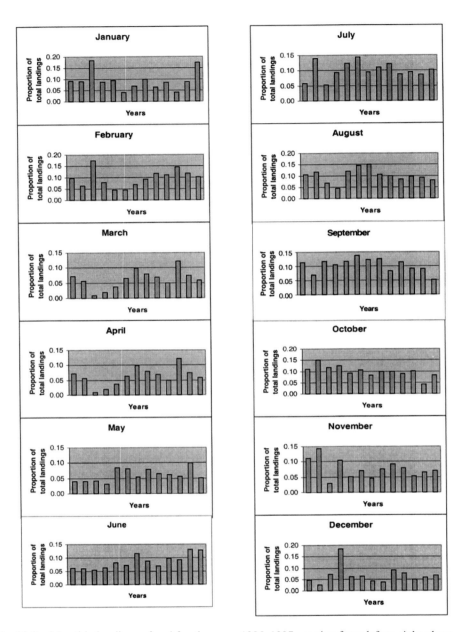

Fig. 22.2   Monthly landings of cod for the years 1985–1997 running from left to right along the *x* axis.

there was a significant increase in landings, which peaked at 2241 t in 1996. Data from the individual vessel logs and cod landings data show that one or two areas, designated by ICES rectangles, are always heavily fished regardless of season. The

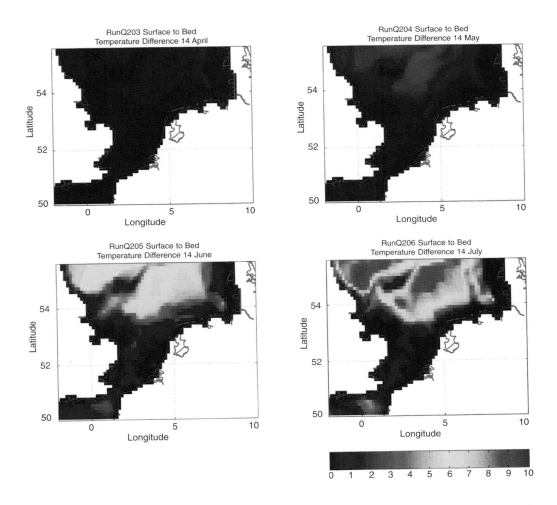

Fig. 22.5 Results from the 3D baroclinic model runs showing the monthly position of fronts and surface-to-seabed temperature difference (°C), in the North Sea for April–July.

Table 22.1 Total landings of cod (expressed in tonnes and % of total year's catch) by month and quarter for the Grimsby fleet from 1985 to 1997

	Jan	Feb	Mar	Apr	May	Jun	Jul	Aug	Sep	Oct	Nov	Dec	Total
1985	94.55	98.14	73.82	40.05	61.85	58.22	106.32	114.33	111.45	112.56	47.38	89.99	1008.73
	9.37	9.73	7.32	3.97	6.13	5.77	10.54	11.33	11.05	11.16	4.70	8.92	%
		Q1 = 26.42			Q2 = 15.88			Q3 = 32.92			Q4 = 24.78		%
1986	59.58	42.02	37.11	26.29	39.23	90.66	77.16	45.13	98.13	93.26	17.65	27.60	653.86
	9.11	6.43	5.68	4.02	6.00	13.87	11.80	6.90	15.01	14.26	2.70	4.22	%
		Q1 = 21.22			Q2 = 23.89			Q3 = 33.71			Q4 = 21.19		%
1987	143.60	135.43	6.44	32.78	42.97	41.43	53.85	91.12	92.17	23.00	56.94	60.30	780.08
	18.41	17.36	0.83	4.20	5.51	5.31	6.90	11.68	11.82	2.95	7.30	7.73	%
		Q1 = 36.60			Q2 = 15.02			Q3 = 30.40			Q4 = 17.98		%
1988	82.90	74.11	17.66	30.41	59.21	87.36	41.62	98.93	117.05	98.64	172.95	60.42	941.29
	8.81	7.87	1.88	3.23	6.29	9.28	4.42	10.51	12.44	10.48	18.37	6.42	%
		Q1 = 18.56			Q2 = 18.80			Q3 = 27.37			Q4 = 35.27		%
1989	113.98	52.93	43.50	101.83	120.14	147.72	141.97	138.93	110.29	60.43	74.34	109.03	1215.09
	9.38	4.36	3.58	8.38	9.89	12.16	11.68	11.43	9.08	4.97	6.12	8.97	%
		Q1 = 17.32			Q2 = 30.42			Q3 = 32.19			Q4 = 20.06		%
1990	31.25	35.61	50.21	63.50	56.26	111.07	113.17	107.16	82.46	54.50	49.19	25.10	779.53
	4.01	4.57	6.44	8.15	7.22	14.25	14.52	13.75	10.58	6.99	6.31	3.22	%
		Q1 = 15.02			Q2 = 29.61			Q3 = 38.84			Q4 = 16.52		%
1991	134.76	132.06	195.1	108.29	227.08	186.21	294.91	240.52	161.02	89.83	83.60	106.39	1959.74
	6.88	6.74	9.96	5.53	11.58	9.50	15.05	12.27	8.22	4.58	4.27	5.43	%
		Q1 = 23.57			Q2 = 26.61			Q3 = 35.54			Q4 = 14.28		%
1992	99.30	91.25	79.08	79.06	86.92	110.47	104.64	124.96	97.39	75.88	36.33	7.62	992.96
	0.88	10.12	8.77	8.77	9.64	12.25	11.61	13.86	10.80	8.42	4.03	0.85	%
		Q1 = 19.78			Q2 = 30.66			Q3 = 36.27			Q4 = 13.29		%
1993	102.71	188.89	109.86	106.20	111.93	194.73	159.58	133.55	157.38	146.90	148.42	48.621	1608.89
	6.39	11.74	6.83	6.60	6.96	12.10	9.92	8.30	9.78	9.13	9.23	3.02	%
		Q1 = 24.96			Q2 = 25.66			Q3 = 28.00			Q4 = 21.38		%
1994	156.37	200.61	88.23	111.36	175.52	158.92	154.27	205.33	162.16	142.63	139.62	105.08	1800.15
	8.69	11.14	4.90	6.19	9.75	8.83	8.57	11.41	9.01	7.92	7.76	5.84	%
		Q1 = 24.73			Q2 = 24.77			Q3 = 28.98			Q4 = 21.52		%
1995	68.79	252.76	213.58	99.97	161.34	167.53	166.91	159.37	176.92	92.62	86.57	98.86	1745.26
	3.94	14.48	12.24	5.73	9.24	9.60	9.56	9.13	10.14	5.31	4.96	5.66	%
		Q1 = 30.66			Q2 = 24.57			Q3 = 28.83			Q4 = 15.93		%
1996	197.17	264.57	168.07	221.01	291.51	193.99	206.47	201.56	95.87	145.38	130.03	125.62	2241.29
	8.80	11.80	7.50	9.86	13.01	8.66	9.21	8.99	4.28	6.49	5.80	5.61	%
		Q1 = 28.10			Q2 = 31.52			Q3 = 22.48			Q4 = 17.89		%
1997	193.76	112.51	64.27	57.97	142.12	113.11	88.608	57.962	92.018	78.93	77.16	36.11	1114.55
	17.38	10.09	5.77	5.20	12.75	10.15	7.95	5.20	8.26	7.08	6.92	3.24	%
		Q1 = 33.25			Q2 = 28.10			Q3 = 21.41			Q4 = 17.25		%

highest cod landings were recorded in June, July and August from 1990 to 1994. There has also been a tendency to fish closer to Denmark and further south.

Effort distribution

Effort data in days are recorded by MAFF/CEFAS from fishermen's logbooks. Missing effort data are then apportioned to estimate number of days at sea per species. Effort data were plotted by ICES rectangles to help identify which fishing areas were most densely occupied and when. The results were compared with areas of highest fishing success, observations of porpoise migration, patterns of by-catch derived from interviews and other environmental phenomena, in order to identify areas and times of increased risk of harbour porpoise by-catch. The total 'number of days at sea' and 'number of trips' were used. Although the two values are likely to be autocorrelated, both were used to ensure that biases caused through misreporting or incomplete logbook entries could be identified; porpoise by-catch data were also available on a trip-by-trip basis. Rather than estimating by-catch from thousands of hours fished, or nets set, which can vary much more than the simple measure of days or trips, the average number of trips per vessel was calculated.

The average number of days per trip per ICES statistical rectangle (Fig. 22.3) per quarter per year from 1985 to 1997 for the fleet was calculated; from these statistics an overall average for each vessel for all species was derived (Table 22.2). Days per trip varied from 1 to 15 days, although in their interviews, most fishermen said that a normal fishing trip was about 5–7 days. Individually, most vessels were remarkably consistent in the number of days spent on each trip. Over time, effort expressed as the length of fishing trips, increased from 4.69 days in 1985 to 6.53 in 1997.

Taking all the effort data (50 vessels), it was evident that some vessels make only one trip, to try gill-netting and then drop out of the fishery. The number of trips per year from 1985 to 1997, whether regular gill-netters or not, ranged from one trip a year to a maximum of 38.5 trips per year with a mean of 15 trips per vessel.

Effort indices

Two types of effort statistics were generated: days at sea divided by the number of active vessels, and the average effective area fished (measured in ICES rectangles) divided by the number of vessels (Table 22.2). In 1985, a total of 1278 days were spent at sea, and 19 vessels fished over an area covering 30 ICES rectangles. If we assume that searching time is an important indicator of CPUE, then it is evident that the total area searched has more than doubled from 26 rectangles in 1983 to 99 rectangles in 1997 (Fig. 22.4). A similar trend is seen for the number of days spent at sea.

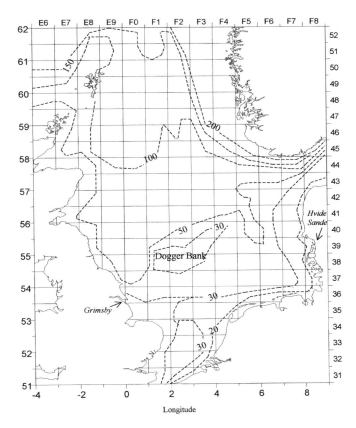

Fig. 22.3 Location of the study area showing the major depth contours and ICES statistical rectangles.

Spatial dynamics

By looking at the individual movements of vessels (data available from authors) as well as the effort in total days by the fleet, it was possible to show that the epicentre of fishing activities for this fishery remained constant over two periods, from 1983 to 1989 and then 1990 to the present. In 1983 during the first quarter of the year (Q1), it was 36F0, in Q2 37F2 and 37F1 with a return to 36F0 (Fig. 22.3) for the latter part of the year, Q3 and Q4. But in 1990 the fleet established a new type of fishing behaviour, in which more rectangles were searched and with a gradual movement southwards, to areas 36F2, 35F3 and 33F3.

Estimates of porpoise by-catch

An observer programme was undertaken in 1996–1998 (S. Northridge, Sea Mammal Research Unit, personal communication). These observations showed that the

Table 22.2 Summary table of the results of the effort analysis from the Grimsby fleet for 1983–1997 indicating the ICES rectangle where maximum effort occurred in each quarter, total number of days at sea and number of ICES rectangles with fish landings

Year	Q1	Q2	Q3	Q4	Total days	Total No. of ICES rectangles with fish landings	Vessels
1997	35F3	36F2	36F3	37F2	1133	99	17
1996	33F3	36F3	37F2	37F2	2010	99	20
1995	36F3	36F2	36F2	37F2	1901	98	27
1994	33F3	37F4	37F3	36F2	2399	97	24
1993	34F3	34F3	36F2	36F2	2470	93	25
1992	33F3	36F2	37F1	37F1	2702	85	19
1991	35F3	35F3	37F4	37F2	2416	76	27
1990	36F2	34F2	37F1	37F0	1636	58	13
1989	33F2	36F2	33F1	33F2	1308	50	24
1988	37F1	36F2	37F2	37F1	1070	45	19
1987	37F1	36F2	37F1	37F1	892	39	15
1986	37F1	36F2	36F1	37F1	976	39	17
1985	36F0	36F1	37F1	37F0	1278	35	19
1984	36F0	37F1	36F0	36F0	1201	30	?
1983	36F0	37F2	38F0	36F0	1196	26	?

Grimsby gill-net fleet took porpoise by-catches off the Shetlands/Orkneys in April in ICES rectangle 48E7 (one animal), and in the southern/central North Sea on the Dogger Bank in March in 38F3 (two animals) and 37F3 (one animal), July in 34 F2 (one animal), August in 39F3 (one animal), September in 40F1 and 40F2 (two animals), and October in 39F2 (one animal).

These data were then used to extrapolate the total by-catch in the fishery. In 1997, the Grimsby fleet was composed of 17 vessels; the total number of trips (i.e. total

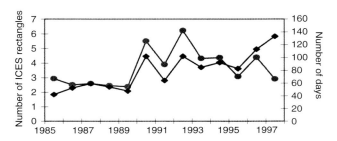

Fig. 22.4 Effort indices for the fleet in average number of days (●) and average number of ICES rectangles (◆) fished per vessel for 1985–1997.

number of days at sea/average number of days per trip) for the year was thus approximately 1133/7 or 162. If six trips out of the 18 observed took by-catch, then assuming a constant probability of encounter, harbour porpoises would have been caught on 54 trips in 1997. If the average number of porpoises caught on any one trip was 1.5, the estimate for 1997 for the Grimsby fleet would be 81 animals. However, the catch per unit effort in 1997 was very low compared with the previous 5 years. The skippers interviewed indicated that lower catch rates earlier in the year, forced some vessels to leave the fishery; those that remained individually fished over a wider number of the main areas. However, vessels with observers on board had by-catches within the epicentre of fishing activity for the period.

Given the above, estimates of harbour porpoises caught as by-catch by the Grimsby gill-netting fleet for the period 1990–1997 are 117, 173, 193, 176, 171, 136, 144, 81 (mean 149 per year). If, however, the number of porpoises caught on one trip was three (a maximum estimate provided by one skipper) per trip, then total estimates for the same period are 234, 345, 384, 351, 342, 270, 288, 162 (mean 297 per year).

Estimates of by-catch from interviews

To estimate the total porpoise by-catch in the gill-net fishery the individual values obtained from the eight gill-netters interviewed were used; for this subsample of the fleet, the estimate lay between 86 and 107 (Table 22.3). The fleet size was 17 in 1997 and 10 in 1998. Thus, in 1997 the estimate of porpoises taken as by-catch in the gill-net fishery per year would have been between 162 and 202, and for 1998 between 95 and 118. This number is reasonably close to the average estimate of 142 obtained from the analysis of observer and fisheries data, but is higher than the number for 1997, when the fisheries effort dropped off dramatically.

Table 22.3 Summary of annual by-catch estimates for porpoises in the UK study fleet ascertained from interviews

Fisherman and gear	Maximum porpoises per year
Crew gill-netter	12
Skipper gill-netter	0
Skipper gill-netter	48
Skipper gill-netter	2
Skipper gill-netter	1
Skipper gill-netter	2
Skipper gill-netter	8
Skipper gill-netter	0

Danish fleet analysis

There were a total of 114 licensed fishing vessels in Esbjerg, a small percentage (less than 5%) were gill-netters. There were 339 licensed vessels in Hvide Sande, most of which were gill-netters, and there were 126 registered vessels in Thorsminde (Year Book of Danish Fishery Statistics, 1996). Esbjerg, also known as the 'Sand-eel town of Denmark' had the larger vessels, 20–70 m in length. Hvide Sande, also known locally as the 'Cod town of Denmark', had vessels mainly between 20 and 25 m. It was in Thorsminde that the vessels were relatively small (<20 m) and had a wooden hull. The smaller boats were usually one-man boats, and the larger ones (>17 m) and had three or four crew on board (Flintegard, 1986).

Data from 30 gill-netters were obtained via a series of extended interviews during August 1998; this represented 10% of the 300 licensed gill-netters in the Esjberg, Hvide Sande and Thorsminde area. Most skippers who were approached for an interview agreed. Only five refused.

The gill-net fishery was concentrated along the western coast of Denmark. Throughout the year the fishermen targeted plaice, sole, turbot and cod. In the spring and summer the fishermen targeted sole and turbot. The preferred fish was cod, since it brought the highest price. Plaice were the second most preferred species, and the last choice was sole. Most gill-netters targeted cod, and the peak season was from October to December.

The fish is landed at Hvide Sande or Thorsminde after 3–4 days at sea for the larger vessels and daily for the smaller ones. The fish is sold fresh at the auction for the domestic market as well as other European countries.

Estimates of porpoise by-catch

From observer data it has been estimated that 5000–10 000 harbour porpoises are taken annually in the Danish gill-net fishery (Teilmann *et al.*, 1998). However, the only scientific estimate of by-catch available for the Danish fisheries was made in 1993 by Vinther (1994, 1995). At that time, by-catches were counted by observers on-board Danish gill-net vessels whilst fishing in the North Sea: the vessels came from the same ports as those employed in this study. A total of 117 porpoises were recorded from 51 trips made by 20 vessels. However, the programme had some problems, in that the study underrepresented the whole fleet. By simple extrapolation, an estimate of 4629 porpoises occurred as by-catch in the sole, turbot and cod fishery in 1993. Furthermore, Vinther then derived an estimate of 7000 porpoises for the total Danish North Sea gill-net fleet. Lowry & Teilmann (1994) consider that these data must be treated with caution until more extensive surveys with science-based sampling design and strategy are carried out.

Estimates of by-catch from interviews

Interviewees were asked to estimate the total number of porpoises taken as by-catch per year and, on how many trips per year these porpoises were present, as well as the maximum porpoises that they had ever caught (Table 22.4). The interviews provided an estimate of the extent of the by-catch problem, since many fishermen exploit different areas, at different depths and over different times. There were many fishermen who were aware of the issues and declared that porpoise by-catch was a real problem; however, they did stress that many times they had no by-catch whatsoever. Total by-catch estimates range from 346 to 442 porpoises per year: given that 10% of the fleet were interviewed the overall estimate from the gill-net fleet was of between 3500 and 4500. These figures almost exactly match those derived from the observer programme undertaken by Vinther (1994, 1995) on vessels from the same ports. The highest by-catches were associated with the turbot fishery, where the combination of large mesh sizes and long soak times may play a significant role in catching porpoises.

Identification of by-catch hotspots

One of the major elements in developing an effective form of governance is the identification of those fishing practices, areas and environmental conditions that could lead to an increase in the level of risk of by-catch.

The two major sightings databases, the Seabirds at Sea Team (SAST) (Northridge *et al.*, 1995) and the Small Cetacean Abundance in the North Sea (and adjacent waters) (SCANS) (Hammond *et al.*, 1995a, b) provided information on the seasonal and spatial distribution of harbour porpoises in the North Sea. In the first quarter, January–March, harbour porpoises were found in two major groupings; one in the deeper water of the north-western North Sea and one in the eastern area off Denmark. By the second quarter, these groups had dispersed and porpoises were found in the shallow waters in the eastern North Sea and along the western edge. During May and June, most porpoises appear to be along the western coast, from Yorkshire to the Shetlands (Northridge *et al.*, 1995); this period included the calving season (Fisher & Harrison, 1970). During the third quarter they were found throughout the North Sea, and by the fourth quarter the two major groups are apparent once more. Camphuysen & Leopold (1993) observed that in Dutch coastal waters, porpoises were most common between December and April. There were also dense aggregations of porpoises in the northern North Sea (i.e. north of 56°N) especially in July (Bjørge & Øien, 1990; Bjørge *et al.*, 1991). There was also evidence of migrations from the Baltic into the North Sea, presumably to escape the ice-conditions of the winter (Andersen, 1975). The SCANS survey of 1994 showed a similar overall distribution.

The distribution of UK-based gill-net fisheries in the North Sea was determined from official statistics (CEFAS Lowestoft). Throughout the years, large concentra-

Table 22.4 Summary of the by-catch estimates from interviews with the Danish gill-net fishermen

Vessel (m)	Target species				Porpoise by-catches			
	Cod	Plaice	Sole	Turbot	Total per year	Max./trip	Trips with by-catch	Max.
20–24	×	×		×	3–4			
	×	×			1–2		1–2	
	×		×					
	×		×		10–15		10	
	×	×		×	30–40		5–6	40
	×	×		×	1–0			
17–19.5	×	×		×	8–10		6–8	
	×	×	×		3–4			
	×	×		×	100	30	6–7	
	×	×	×		20			20
	×			×	50–100			
	×				1		1–2	
	×		×	×	10	100		100
	×		×		25		15	25
	×	×	×	×	5		10	
	×				2	0		
14–16.7	×	×			4–5			
	×	×	×	×	3		3–4	
	×		×		0	0		
	×	×		×	40–60			15
	×	×			24		4–5	24
	×	×	×				1–2	
10–12.4	×	×	×	×	1	0		
	×				0			
	×				0			
	×	×			1–2 or seldom			
	×	×		×	2		2	
	×	×			1	0		

tion of effort in all quarters was evident along the east coast of England, in the southern and central North Sea centred on the Dogger Bank and in the English Channel.

For the North Sea, by-catches of porpoises were reported off Shetland by the Sea Mammal Research Unit (UK) in 1990, and purse seiners and pair trawlers, targeting pelagic, shoaling fish in the deeper northern areas of the North Sea were also thought to take porpoises. But as estimates of by-catch in the UK gill-net fishery in the North Sea were thought to be in the low hundreds, no by-catch rates were calculated. Northridge (1988) and Berrow *et al.* 1994 reported that 2000 harbour porpoises were caught annually in the Irish and UK bottom-set gill-net fisheries for hake on the Celtic Shelf to the south-west of Britain and Ireland. Their results reinforce the fact that gear type and prey density play a key role in determining whether an area is likely to be a by-catch hotspot. The data from the UK observer programme in 1996–1998 showed that the Grimsby gill-net fleet had by-catches associated with wreck-netting for cod. The low number of animals taken reinforces the low estimates overall. However, an assessment of the fishery statistics on the Dogger Bank since 1950 has shown that the relative importance of this area for demersal species has increased compared with the whole of the North Sea (Purdom & Garrod 1990), thereby reinforcing the Dogger Bank as a by-catch hotspot.

Fronts as indicators of by-catch hotspots

Tidal and frontal distribution may influence porpoise abundance and behaviour. Frontal zones typically maintain a strong convergence of surface currents: in addition, fronts are associated with concentrations of fish as they are regions of high productivity (Bowman & Esaias, 1978). Fronts occur at the border of summer stratified and mixed water at many places on the north-west European continental shelf (Simpson & Bowers, 1981). The position of these fronts has been successfully predicted on the basis of water depth and tidal current speed (Pingree *et al.*, 1978). The Flamborough Head front is a well-known phenomenon (Pingree *et al.*, 1978). It runs roughly parallel to the north Yorkshire coastline (UK) for a considerable distance before branching eastward at Flamborough Head (Matthews *et al.*, 1993). Another frontal system is the Dogger Bank (van Aken *et al.*, 1987). It extends from the southern part of the bank right across the North Sea. Off Denmark, there is also a well-studied front called the Jutland coastal front (Pedersen, 1994).

Hindcasts from the Proudman Oceanographic Laboratory 3D baroclinic model for the North Sea indicated where regions of well-mixed waters, stratification and fronts normally occur. By using the difference between sea surface temperatures (SST) and near seabed temperatures, regions of high stratification could be detected across the southern and central North Sea. In general terms, the hindcasts show a front emerging in the spring months (Fig. 22.5, see colour plate). This is indicated by the narrow region (the lightest blue) separating the well-mixed waters (dark blue) from stratified waters (orange–yellow–red colours). Stratification was evident across the Dogger Bank from June to August, when it became most pronounced and then slowly broke down by the end of October. The model can predict the strength of stratification in the summers and the depth of the thermocline (Holt & James, 1999).

From the test run for 1989 (Fig. 22.5), the growth and decline of a tidal mixing front occurs, separating the well-mixed waters in the south and east from stratified waters in the deeper northern region. The fishing fleet expended most effort in 36F2 (2nd quarter 1988) and 37F2, 36F2 and 35F2 (2nd quarter 1989). Stratified regions extended to ICES rectangle 37F2, where most of the fleet concentrated its effort. In July, although some effort was concentrated further south, most of the fleet fished in 37F0, 37F1 and 37F2, just on the border of the stratified waters. In August and the beginning of September, the fleet's efforts were again concentrated in 37F0 and 37F2, where stratified water still occurred. These results show that that there may be a correlation in the North Sea between the formation of the front with fishing effort, especially in the gill-net fleet, and at a time of the year when by-catches were highest. Fronts may indicated by-catch hotspots.

Palka (1994) concluded that internal changes in surface temperature and fish density could have been the reasons for changes in the distribution and abundance of harbour porpoises in the North Atlantic. Given historical records of porpoise sightings and by-catches, in conjunction with the output of a range of tidal-mixing and 3D baroclinic models (Charnock *et al.*, 1994), it may be possible to predict when high aggregations of porpoises are likely to occur inshore.

Other areas noted for by-catches include known breeding or nursery areas. Sonntag *et al.* (1998) examined stranded animals from the island Sylt, along the North Sea coast of Schleswig Holstein, and found that 72% of all animals were less than 1 year old.

Management of the resource

Governance generally refers to the patterns of interaction between society and government. The term has a number of meanings: it can be the activity or process of governing, a condition of ordered rule, those people charged with the duty of governing or the manner/method/system by which a particular society is governed. The current use of governance does not treat it as a synonym of government, but rather signifies a change in the meaning of government. In reviewing the use of governance, at least six separate uses can be found for the term (Rhodes, 1996): the minimal state; corporate governance; the new public management; good governance; socio-cybernetic system; self-organising networks. All exist with respect to fisheries management within the EU.

More recently the governance term has been extended to include natural resources and endowments such as the oceans. A recent example is the Lisbon Principles of Sustainable Governance of Oceans (Costanza *et al.*, 1998) which lists the following principles: responsibility, scale-matching, precaution, adaptive management, full cost allocation and participation. All these principles are of relevance to the issue of by-catch of porpoises and are thus taken up in the next section.

One of the key principles in dealing with the by-catch of porpoises relates to precaution. The 1995 Food and Agricultural Organisation (FAO, 1995) Code of Conduct, incorporating the precautionary approach, requires that those critical factors that constitute the social and economic dimensions of the management system be understood, especially in terms of how revenues vary with the level of exploitation and relate to dynamic market forces, and how individuals or groups of fishermen behave in relation to the resource upon which they depend. Social and economic factors are of course intertwined, and changes in the distribution of income, the type and amount of employment and the degree to which interest groups can influence decision-making will all have an effect on the management regime. The case of by-catches of small cetaceans in fisheries is an example of this situation. Poor economic performance, as well as management failures, are frequently linked to such complex situations (i.e. by-catch issues, incomplete and multiple jurisdictions, irreconcilable objectives). It is therefore critical to put the appropriate governance framework in place, so that flexibility can be introduced into the management regimes available.

Sustainable institutional design

The problem of by-catches of small cetaceans is more than just an issue of better fisheries management, although the mechanisms and the causality lie within fisheries. Rather it is connected to the issue of human perceptions of the ecological value of a group of animals and the institutions in place to regulate the system. Mechanisms for reducing by-catch are dependent upon both economic and public perception criteria. Economic approaches generally frame the problem in terms of economic efficiency or lost revenues should the market fail to behave appropriately. In this category, we need to consider adequate pricing of fish taken at times or in ways that serve to reduce the risks of by-catch: the role of eco-labelling is important here. The public-oriented approaches are complementary and involve public participation in determining the importance of marine mammals to society, or ensuring that political institutions responsible for enforcing adequate regulations carry out their duties. Both approaches have, however, failed to recognise the significant changes that are required to ensure that fisheries are sustainable.

Generally speaking, avoiding overexploitation defeats both market-based and participatory-based approaches. The reason for this is quite simple; ensuring both sustainability of the resource and organisational reliability are tasks that cannot be tackled by applying classical regulatory approaches. Certainly, regulations and technical standards contribute to the likelihood of the resource's not being overexploited. Such regulations do not recognise that there are crucial differences between designing and implementing a technical measure and designing an organisation. While performance of a vessel or a specific gear can be measured, it cannot be conceived of in the same way once an organisation is included. Developing and enforcing procedural rules is a practical necessity in industries that operate in a hazardous and largely open environment. But this alone will not give any guarantee that the rules are applied in a responsible way.

One further consideration is the geographical scale over which jurisdiction lies. That fisheries problems often require different jurisdictional units is a well-known fact. However, an immediate consequence of this fact is almost never mentioned: it is not so much the decentralisation of authority *per se*; rather, flexible administrative units are needed. De Rougemont pointed this out in the context of regional administrative units (de Rougemont, 1983), and there is a trend towards this in Europe today. However, in the case of fisheries, it would probably be illusory to believe that, just by addressing the institutional problems involved in implementing flexible jurisdictional units, many of the existing problems would be effectively addressed. Instead, we have to go further, and recognise that many of the conflicts arise not at a local level, but within national and international markets and conventions. While conflicts and problems may be localised, addressing the issues will require jurisdictional authorities well beyond the scope of national government. The fact that many of the relevant collective actors operate in a national or international market suggests that, for fisheries, the most adequate unit of analysis and action is the industry itself.

Given this, it is important to examine the institutional framework within which the industry is situated. First, markets are not settings of anonymous, purely fictional social interaction. Real people interact with each other; buyers and sellers continuously adjust their behaviour patterns according to responses to the rules that they themselves have helped to evolve. Thus, competition is not simply a case of market performance, it is about how the players respond to the rules. For example, when fishing for cod on wrecks, or wreck fishing, most fishermen replied that they would investigate before going to the grounds who was fishing there before. The unwritten rule was 'first come first served'. All fishermen followed this behaviour. When the Danish fishermen were asked whether they had any problems with Grimsby fishermen, they unanimously replied that they had almost none.

The way in which such rules evolve is generally determined by the institutional structure of a given industry. Distinguishing between different forms of competition allows us explicitly to address the issue of sustainability and the economic valuation of resources. Generally speaking, price competition drastically reduces organisational slack, limiting the opportunities of economic actors to develop less-damaging production processes. With regard to fisheries, price competition tends to give rise to and reinforce risk-taking behaviour. Price competition reduces the planning horizon and provides incentives to overexploit the fishery. By contrast, product competition explicitly rewards innovation and longer planning horizons. As prices play a secondary role, economic actors can pay more attention to maintenance and prevention of overexploitation of the resource.

If we assume that competition enables markets to evolve, and we consider that in fisheries, market participants are often organisations, such as the Producer Organisations or supermarkets, rather than individuals, then participants will end up competing with each other in different forms. A market is not just a set of fish-buyers and sellers; there are many ancillary organisations that co-determine the rules of the game. Taken together, these organisations form the institutional framework of a certain market. Traditionally, these organisations have lobbied political bodies, but, more recently, many of them have taken on a different role with respect to their regulators: that of negotiators for industry-specific rules that support the principles of sustainability and help to spread the risk of economic and resource failure. They rely on an institutional framework and the trust that has been engendered within it by the different constituents.

The fishermen in Grimsby were asked how they would manage their own fishery, what their opinions were as to the Common Fisheries Policy and options if their fishery became non-viable. All fishermen expressed their dissatisfaction with the current management of the fish stocks. They all wanted to see simpler regulations applied across the board and more say in the decision-making concerning local issues. Regarding the by-catch problem, none of the fishermen wanted to catch porpoises in their nets, but fishing was a 'hunting activity' and there was an inevitable risk of by-catch and 'waste'. There were no real alternatives to fishing for most of the local Grimsby people; unemployment levels were already at more than 13.5%, so most would seek to leave the area if necessary. Given the numbers of people involved in the industry (approximately 200 directly and indirectly) and the current level of landings, the loss of revenues would exceed several million pounds sterling per year.

In Denmark, the industry developed along different lines from the UK in that the vast majority of fishing was undertaken not by larger company boats, but in smaller individually owned vessels. The advantage was that they were more flexible and less capital intensive, while at the same time using efficient modern technology. The larger vessels required large concentrations of fish to make fishing profitable, while, in general, the smaller boats could exploit less dense concentrations of fish. The North Sea and the coastal waters became the focus for these fishermen allowing them to switch rapidly from one species to another, each fisherman receiving his salary as a share of the catch (Hansen, 1997). Unlike the system in the UK, the fishermen were given monthly quotas and, on the whole, most fishermen were happy with the system. In this way, the burden of risks was spread across many individuals through time, rather than being concentrated at any one time in a few large organisations.

Devolved management systems are already under discussion in Denmark (Vedsmand & Nielsen, 1995; Vedsmand *et al.*, 1995). Four alternative models have been examined; these look at different ways in which there can be enhanced involvement of user-groups in the decision-making process, given different considerations of resource and market. Currently, the Danish decision-making process is centralised, with industry participation restricted to consultation through advisory boards. Devolution of authority and responsibility for monitoring, control and enforcement is a sensitive issue; Hallenstvedt (1993) has argued that fishing is now a privilege for a few rather than a right for many. As with all such public privileges, there is an obligation to protect this resource from overexploitation. The absence of fishermen from the formulation and implementation process of regulations appears to be one of the main reasons of non-compliance. Even though many barriers exist to the devolution of management, it is clear from the interviews in this study and the suggestions given below that examples of enhanced co-operation among user-groups are already emerging.

Industry suggestions to reduce by-catches

During interviews with fishermen, a number of suggestions were made on how to reduce by-catches. These included:

a. *A ban on certain types of gill-net fishing*; some compensation could be paid to the fishermen involved.

b. *Technical measures*; increased mesh sizes, or the use of the square mesh with some compensation for loss of earnings. This was considered better than a complete ban on fishing or complete compensation or social payments; closures: spawning seasons should be protected and most fishermen were in agreement to closures on cod fishing at least for January to March.

c. *Experimental studies*; in order to come up with accurate estimates more gill-netting trials should be undertaken, in close association with the fishermen involved and at the season when the porpoises are present, i.e. in the summer months between May and July.

d. *Technological measures*; the use of pingers and other acoustic deterrent devices (supported by the government or paid for in part at least by fishermen's associations) as well as the use of different nets.

e. *Responsible fishing practices*; however incomplete the knowledge about the levels of porpoise by-catch in the fisheries in Denmark or the UK sector of the North Sea, it is important to accept this and adopt the practice of responsible fishing. In the case of the gill-net fisheries, this may well mean fishing with shorter nets, restricting the number of nets, reducing the total soak time to less than 24 h, as in Norway, minimising encounter rates with high aggregations of porpoise by switching fishing areas at different times of the year in relation to information and knowledge about porpoise breeding and nursery areas and their patterns of behaviour.

f. *Market mechanisms*; wider use of ITQs; establishment of a European Registry of vessels to prevent quota hopping; requirement to land high-quality fish; abolition of the withdrawal scheme; fish landed should not go for pet food; possible use of eco-labelling to encourage consumer-based price increases.

g. *Institutional mechanisms*; there should be extensive consultations with fishermen and the industry as a whole; self-governing bodies should be established and recognised by the regulatory authorities to enable the fishermen to patrol themselves.

Overall, it was clear from the comments in the interviews that the fishermen wished to become far more actively involved in the decision-making at a regional as well as at a local level.

It is critically important that industry is more actively involved in the arrangements for monitoring and to some extent self-regulation, because, without more consistent financial backing from governments, there is little chance that extensive surveys and observer programmes can be undertaken on a sufficient scale. To ensure that the participation by industry is effective, appropriate incentive structures need to be sought. These could include deferred quota, trading of other resources and price enhancement through product-labelling schemes. These could also include management processes that would promote adaptability to change (Hanna, 1997). In addition, surrogate measures and simple indicators of sustainability and performance should be implemented within a social and political framework.

Acknowledgements

Special thanks to all the fishermen in Grimsby, Esbjerg, Hvide Sande and Thorsminde who graciously participated in the project, and to Roger Proctor and Jason Holt of the CCMS-Proudman Oceanographic Laboratory for providing as yet unpublished results. Additional funding for this report was provided by the WWF-UK. Thanks also to Brian Rackham at the CEFAS laboratory, Lowestoft for data manipulation and extraction.

References

Van Aken, H. G., van Heijst & Maas, L.R. (1987) Observations of fronts in the North Sea. *Journal Marine Research*, **45**, 579–600.

Andersen, S.H. (1975) Changes of migratory behaviour in the harbour porpoise, *Phocoena phocoena*, illustrated by catch statistics from 1834–1944.In: *UN Food and Agricultural Organisation ACMRR Marine Mammal Symposium*, ACMRR/MM/EC/32.

Andersen, S.H. *et al.* (1997) A combined DNA-microsatellite and isozyme analysis of harbour porpoises in Danish waters and west Greenland. *Heredity*, **78**, 270–6.

Barlow, J.S., Swartz, T., Eagle, T. & Wade, P. (1995) *US marine mammal stock assessments: guidelines for preparation, background and a summary of the 1995 assessments. Technical Memorandum* NMFS-OPR-6. National Oceanic and Atmospheric Administration (NOAA) – National Marine Fisheries Service (NMFS).

Berrow, S.T.,Tregenza, N.J.C. & Hammond, P.S. (1994) *Marine mammal bycatch on the Celtic Shelf. EU Final Contract Report* DGXIV/C/1 92/3503.

Bjørge, A. & Øien, N. (1990) *Distribution and abundance of harbour porpoise Phocoena phocoena in Norwegian waters. Scientific Committee of the International Whaling Commission (IWC)* SC/42/SM3.

Bjørge, A.H. Aarefjord, S. Kaarstad, L. Kleivane & Øien, N. (1991) Harbour porpoise *Phocoena phocoena* in Norwegian waters. *International Council for the Exploration of the Sea* C.M. 1991/M:10.

Bowman, M. & Esaias, W. (eds) (1978) *Oceanic Fronts in Coastal Processes. Proceedings of a Workshop at the Marine Sciences Research Center*, 1977, 114 pp. State University, New York.

Camphuysen, C.J. & Leopold, M.F. (1993) The harbour porpoise *Phocoena phocoena* in the southern North Sea, particularly the Dutch sector. *Lutra*, **36**, 1–24.

Charnock, H., Dyer, K.R., Huthnance, J.M., Liss, P.S., Simpson, J.H. & Tett, P.B. (1994) *Understanding the North Sea System*. The Royal Society/Chapman & Hall, London.

Clausen, B. & Andersen, S. (1988) Evaluation of bycatch and health status of the harbour porpoise (*Phocoena phocoena*) in Danish waters. *Danish Reviews in Game Biology*, **13**, 1–24.

Costanza, R., Arger, R., de Groot, S., Farber, S., Grasso, M., Hannon, B., Limburg, K., Naeem, S., O'Neill, R., Paruelo, J., Raskin, R., Sutton, P. & van den Belt, M. (1998) Principles for sustainable governance of the oceans. *Science*, **281**, 198–9.

FAO (1995) The precautionary approach to capture fisheries and species introductions. *Guidelines for Responsible Fisheries*, **2**, 54 pp. Food and Agricultural Organisation.

Fisher, H.D. & Harrison, R.J. (1970) Reproduction in the common porpoise (*Phocoena phocoena*) of the North Atlantic. *Journal of Zoology, London*, **161**, 471–86.

Flintegard, H. (1986) *Fiskeri med garn*. Nordscentertrykkeriet, Hirtshals (in Danish).

Gehringer, J. (1976) Part 216, Regulations governing the taking and importing of marine mammals. *Federal Register* **41**, 52 pp.

Hall, M. (1996) On bycatches. *Reviews in Fish Biology and Fisheries*, **6**, 319–52.

Hall, M. (1998) An ecological view of the tuna–dolphin problem: impacts and trade-offs. *Reviews in Fish Biology and Fisheries*, **8**, 23–34.

Hallenstvedt, A. (1993) Resource management and control. In: *Nordiske Seminar- og Arbejdsrapporter*, **583**. Nordick Ministerråd.

Hammond, P.S., Benke, H., Berggren, P., Borchers, D.L., Buckland, S.T., Collet, A., Heide-Jorgensen, M.P., Heimlich-Boran, S., Hiby, A.R., Leopold, M. & Øien, N. (1995a) *Distribution and abundance of the harbour porpoise and other small cetaceans in the North Sea and adjacent waters. SCANS Final Report*, LIFE 92–2/UK/027.

Hammond, P., Benke, H., Berggren, P., Collet, A., Heimlich-Boran, S., Leopold, M. & Øien, N. (1995b) The distribution and abundance of harbour porpoises and other small cetaceans in the North Sea and adjacent waters. *International Council for the Exploration of the Sea*, C.M. 1995/ N:10.

Hanna, S.S. (1997). The new frontier of American fisheries governance. *Ecological Economics*, **20**, 221–33.

Holt, J. & James, I.J. (1999) A simulation of the Southern North Sea in comparison with measurements from the North Sea Project. *Continental Shelf Research*, **11**.

IWC (1994) The revised management procedure (RMP) for baleen whales. *Report of the International Whaling Commission*, **44**, 145–52.

Kinze, C.C. (1994) Incidental catches of harbour porpoise (*Phococena phocoena*) in Danish waters, 1986089. *Report of the International Whaling Commission*, Special Issue, **15**, 183–7.

Lien, J., Stenson, G.B., Carver, S. & Chardine, J. (1994) How many did you catch? The effect of methodology on bycatch reports obtained from fishermen. *Report of the International Whaling Commission*, Special Issue, **15**, 535–45.

Lowry, N. & Teilmann, J. (1994) Bycatch and bycatch reduction of the harbour porpoise (*Phocoena phocoena*) in Danish waters. *Report of the International Whaling Commission*, Special Issue, **15**, 203–10.

Matthews, J., Fox, A.D. & Prandle, D. (1993) Radar observation of an along-front jet and transverse flow convergence associated with a North Sea Front. *Continental Shelf Research*, **13**, 109–30.

Northridge, S. (1988) *Marine mammal conflicts with fishing gears in Britain. Report to the Seal Group, UK Wildlife Link*. 121 pp.

Northridge, S.P., Tasker, M.L., Webb, A. & Williams, J.M. (1995) Distribution and relative abundance of harbour porpoises (*Phocoena phocoena* L.), white-beaked dolphins (*Lagenorhynchus albirostris* Gray), and minke whales (*Balaenoptera acutorostrata* Lacepede) around the British Isles. *ICES Journal of Marine Science*, **52**, 55–66.

Palka, D. (ed.) (1994) *Results of a scientific workshop to evaluate the status of harbor porpoises* (Phocoena phocoena*) in the western North Atlantic. National Oceanic and Atmospheric Administration (NOAA)-National marine Fisheries Service (NMFS)*, NEFSC Ref. Doc. 94–09.

Pauly, D., Trites, A., Capuli, E. & Christensen, V. (1995) Diet composition and trophic levels of marine mammals. *International Council for the Exploration of the Sea* C.M. 1990/G:66.

Pedersen, Fl. Bo (1994) The oceanographic and biological tidal cycle succession in shallow sea fronts in the North Sea and the English Channel. *Estuarine, Coastal and Shelf Science*. **38**, 249–69.

Pingree, R.D., Holligan, P. & Mardell, G.T. (1978) The effects of vertical stability on phytoplankton distribution in the summer on the north west European shelf. *Deep-Sea Research*, **25**, 1011–28.

Purdom, C.E. & Garrod, D.J. (1990) Fisheries on the Dogger Bank. *International Council for the Exploration of the Sea* C.M. 1990/.

Rhodes, R.A.W. (1996) The new governance: governing without government. *Political Studies*, **44**, 652–65.

De Rougemont, D. (1983) *The Future is Within Us*. Pergamon, Oxford.

Santos, M.B., Pierce, G.J., Ross, H.M., Reid, R. & Wilson, B. (1994) Diets of small cetaceans from the Scottish coast. *International Council for the Exploration of the Sea*, C.M. 1993/N:11.

Simpson, J.H. & Bowers, D. (1981) Models of stratification and frontal movements in shelf seas. *Deep-Sea Research*, **28**, 727–38.

Sonntag, R., Benke, H., Hiby, A.R. & Lick, R. (1998) Breeding ground of harbour porpoises (*Phocoena phocoena*) in the North Sea off Schleswig–Holstein (Germany) and its implication for management. In: *World Marine Mammal Science Conference*, Monaco, 1998, p. 126.

Teilmann, J., Larsen, F. & Desportes, G. (1998) Remote sensing of harbour porpoise behaviour in relation to gillnetting activity in Danish waters. In: *World Marine Mammal Science Conference*, Monaco, 1998, p. 133.

Vedsmand, T. & Nielsen, J.R. (1995) Devolved fisheries management systems: a discussion on implementation of alternative fisheries co-management models in Denmark. *International Council for the Exploration of the Sea* C.M. 1995/S:1.

Vedsmand, T., Nielsen, J.R. & Friis, P. (1995) Decision-making processes in Danish fisheries management: capabilities and aspirations of Danish fishermen's organisations. Working Paper, North Atlantic Regional Studies, Roskilde University.

Vinther, M. (1994) *Investigations on the North Sea gillnet fisheries. DFU Report* 485-95, 26 pp. Landbrugs- og Fiskeriministeriet. Danmarks Fiskeriundersogelser.

Vinther, M. (1995) Incidental catch of the harbour porpoise (*Phocoena phocoena*) in the Danish North Sea gill-net fisheries: preliminary results. In: *Proceedings of the Scientific Symposium in the North Sea, Quality Status Report*, Ebeltoft, 1994.

Chapter 23
Economic and sociocultural priorities for marine conservation

P.J.S. JONES

Institute for Environmental Policy, School of Public Policy, University College London, 5 Gower Street, London, WC1E 6HA, UK

Summary

1. Much of the marine environment is relatively natural and is generally regarded as wilderness, whereas it is in fact significantly impacted by fishing activities. However, the alien, remote and hidden nature of the marine environment reduces the priority that society attaches to these impacts, even though they can damage sources of sociocultural and economic value.

2. A variety of important resources are derived from the marine environment, which could be damaged by the incidental impacts of fishing. There is a risk that fish stocks themselves could potentially be damaged through trophic feedback mechanisms at considerable cost to society. However, concern over this potential risk is largely restricted to ecologically enlightened fisheries managers and the relevant research community. Marine wildlife of importance for coastal tourism could be damaged. With the exception of marine mammals, this is a relatively low priority as the attraction of coasts to tourists is largely related to coastal landscapes, sandy beaches and clean bathing waters, rather than healthy marine wildlife. Populations of marine organisms that might yield biotechnologically important compounds could also be damaged, but this is a very low priority as the degree of this potential is indefinable.

3. The marine environment is an important provider of ecosystem services, but whether the impacts of fishing will endanger the delivery of these services at a given regional scale is uncertain. It could be argued that there is a risk that any significant impacts will reduce the resilience of marine ecosystems and could lead to major ecological perturbations, but it is difficult to predict the outcome of fishing impacts owing to the complexity of marine trophic relations. Again, such risks are a relatively low societal priority.

4. The value that society attaches to the existence of certain marine species and of marine wilderness is a fairly high priority, and has been used to great effect in certain publicity campaigns. The decision not to dump the *Brent Spar* in the deep sea west of the Hebrides was driven by the societal value attached to this marine wilderness, but the fact that nearby areas were subject to increasing fishing exploitation was largely ignored. The European Commission's decision to ban drift nets was largely driven by the impacts of these 'walls of death' on intrinsically appealing species such as seabirds, dolphins and turtles. There is a danger that focusing on the conservation of such flagship species can neglect the wider, ecosystem impacts of fishing.

5. In order to raise the priority that society attaches to the non-target impacts of fishing, it is argued that: (a) greater efforts need to be focused on raising public awareness of the risk that the non-target impacts of fishing could lead to potentially serious ecosystem perturbations; (b) more resources and efforts need to be focused on reducing the uncertainty concerning the functional value of habitats and species that are regionally endangered by fishing activities; and (c) relevant scientists should

become more involved in wider deliberations concerning the non-target impacts of fishing in order to guide their research efforts in a manner that increases the potential potency of their findings, and to maximise the extent to which their findings constructively inform such deliberations.

Keywords: marine ecosystem, ecosystem services, marine resources, flagship species, existence values.

Introduction

Marine habitats and communities are relatively natural in that they are rarely the result of positive anthropogenic intervention, whereas terrestrial environments are, to a degree, intentionally developed and modified. Even those terrestrial areas considered to be of high conservation value, e.g. woodlands, moors, pastures and meadows, are generally semi-natural in that positive management through the maintenance of certain human activities is required to preserve them. Fishing in the marine environment is an exceptional activity in the modern world in that it involves the capture of uncultivated populations and is sustained by naturally productive and diverse ecosystems. In contrast, terrestrial agriculture relies on the enhanced productivity of ecologically and genetically modified, low-diversity systems (Cole-King, 1994). Marine environments are subject to direct human impacts that result from activities such as waste disposal and fishing that lead to significantly modified habitats. However, it is very rare for arguments to be made that such activities should continue in certain marine areas because the impacted habitats are considered to be of importance and value in themselves.

Despite the wide-scale nature of fishing and pollution impacts, a large proportion of the marine environment remains relatively unaffected. As such the sea is perceived by many people as being the last wilderness in their region. However, neither fishing nor its impacts are readily visible, i.e. they are 'out of sight, out of mind'. If they were more apparent, it might be argued that there would be greater pressure to establish areas closed to fishing to match nature reserves that exist on land that are closed to development, farming, etc.

The alien and mysterious nature of the marine environment may be its undoing in that it is perceived as an adversary and the cold-blooded, slimy life that inhabits it does not evoke public inspiration and empathy in the same way as many charismatic terrestrial animals (Kenchington, 1990). Furthermore, conservation priorities in Britain, for example, have historically been focused on the aesthetication or taming of terrestrial nature (Lowe, 1983), in a manner that might be termed 'conservation by gardening'. This technique is clearly not appropriate to the marine environment. These factors reduce the priority that society attaches to the impacts of fishing. However, there are a number of economic and sociocultural priorities that society attaches to marine conservation that are particularly relevant in the context of the impacts of fishing on non-target species and habitats. These relate to the value of marine resources for fishing, tourism and as sources of biotechnological compounds, the service values of marine ecosystems and the existence values of marine wilderness and of certain intrinsically appealing species. Each of these values will be explored in this chapter, and issues concerning the relationship between science and societal concerns about marine conservation will be considered.

Marine resource values

Impacts of fishing on fisheries through ecosystem perturbations

In keeping with the observational, alienation and perceptual hurdles discussed above, it has been argued that society's relationship to the sea is largely defined in terms of the resources it provides, particularly in relation to the importance of sea fish (Cole-King, 1995). Fishing is of very high economic importance and provides many livelihoods, particularly in rural coastal communities. Internationally, humans rely on fish for 16% of their animal protein needs, but in certain countries such as North and South Korea, Ghana, Indonesia, Congo, Japan, Malawi and the Philippines, this figure rises to over 50% (McGinn, 1998). Society's relation to the sea is also defined in terms of the services it provides, particularly as a place to dump solid wastes, dilute and disperse liquid wastes, and undertake marine navigation. By contrast, the land is conceived as a tangible entity in itself with perceivable uses such as the set-aside of areas for landscape and nature conservation (Cole-King, 1995).

Traditionally marine environmental concerns have been focused on high-profile sectoral issues such as the overexploitation of whale stocks, the use of seas as a dumping ground for industrial wastes, the impacts of oil spills as a result of tanker accidents and the direct impacts of fishing in terms of the exhaustion of target fish stocks. More recently, concern has been expressed over the impacts of fishing on non-target species. The repercussions of these impacts can feedback to the fish stocks in question via food chain, habitat and ecosystem perturbations. These concerns have led to increased calls for an ecosystem-based approach to fisheries research and management rather than purely stock-based approaches. However, the understanding of the indirect impacts of fishing on trophic relations in the marine ecosystem remains poor, detailed studies being limited to a few well-documented cases such as coral reefs and areas such as the North Sea (Jennings & Kaiser, 1998). At present, it could be argued that this increasing concern is largely confined to ecologically enlightened fisheries managers, the relevant research community and their funding agencies, and certain environmental campaign groups.

There are examples of communities where this concern is a high priority, particularly those dependent on coral reef fisheries that have already suffered as a result of such indirect impacts, on which the majority of papers reviewing the ecosystem impacts of fishing have been focused. Although these cases are few and their applicability to temperate fish stocks is debatable, they might be regarded as cautionary warnings. Efforts have been made recently to raise societal awareness of the indirect impacts of fishing and to promote support for proposals to set aside areas of the sea to protect both marine ecosystems and fish stocks. Such calls have included an international campaign by marine researchers to set aside 20% of the world's seas as areas closed to fishing (Holmes, 1997).

Fishing regulations often include partially closed areas that are critical to specific fish stocks at certain times of the year, and these are relatively acceptable to fishermen, as the stock conservation benefits are relatively clear. However, gaining political acceptance for the closure of fishing areas on the basis of potential benefits that are relatively unclear from a fish-stock conservation perspective can be extremely difficult, as fishing interests constitute a powerful political lobby. There is, therefore, a strong presumption that, even in areas designated as marine nature reserves, fishing will continue to be allowed in most if not all areas, despite the fact that this undermines many of the objectives of such reserves (Jones, 1994). Gaining political acceptance for closed areas on the basis of the potential for ecosystem impacts to feedback to fish stocks is particularly difficult. Societal awareness about this issue is very low and it remains a relatively low priority, although it is potentially of very real concern.

Impacts of fishing on tourism

Coasts are a very important attraction for tourists, therefore any potential for the non-target impacts of fishing to reduce the tourism earning potential of coastal seas could have significant economic consequences. However, most studies have been confined to the impacts of tourism on tropical ecosystems, particularly the impacts of recreational SCUBA diving on coral reefs, which is a rapidly growing component of the tourism industry (Davis & Tisdell, 1995).

Whilst coastal tourism in Europe is not dependent on healthy marine wildlife to the same extent as diving-based tourism in coral reefs areas, there has been considerable growth in recent years in marine-wildlife-oriented ecotourism. This is related to activities such as whale, dolphin, seal, seabird and wader watching, and SCUBA diving. The marine wildlife on which these activities depend is potentially vulnerable to the impacts of fishing. A few research projects have been undertaken on the impacts of fishing on, for example, vulnerable rocky reef populations, and the potential impacts on marine-wildlife-oriented tourism (e.g. Rayment, 1998). However, suffice it to say that whilst the potential exists for fishing to impact on coastal tourism, this potential is relatively small, as the majority of temperate coastal tourism is dependent more on coastal landscapes, beaches and clean bathing waters, with only a minority sector being dependent upon healthy marine wildlife. Therefore, once again, this issue is currently a relatively low priority in the context of societal concerns about marine conservation.

Impacts of fishing on potential sources of biotechnological compounds

Marine organisms are recognised as potential sources of compounds that could be of biotechnological and pharmaceutical use, and increasing attention has been focused on the marine environment as the search for useful compounds from terrestrial species has become less rewarding (Grant & Mackie, 1977). This attention is particularly focused on biodiversity 'hotspots' in tropical seas, where intense competition among species has led to the evolution of complex toxins that may be of pharmaceutical use, for instance, in fighting cancer. However, colder-water species are also screened and have yielded potentially useful compounds. For instance, promising anticancer compounds have been isolated from a newly discovered species of the temperate sponge *Lissodendoryx* spp., found on deep rocky reefs off the coast of New Zealand (Pain, 1996). The temperate bryozoan *Bugula neritina*, found from California to Europe, yields a drug that could eventually form the basis of a market worth $1 billion a year (Pain, 1998). It has also been argued that deep bathyal seabed environments, such as those in the north-east Atlantic, may contain huge genetic resources that might be of interest to drug companies (John Lambshead quoted in Pearce, 1995).

Clearly marine species have the potential to yield useful compounds, although the degree of this potential is very uncertain and the risk of its loss due to fishing impacts is a relatively low societal concern. However, any irreversible losses caused by the impacts of fishing on the future but, as yet unknown, value of this potential resource could have considerable economic impacts through drug revenues foregone as well as considerable impacts on humanity through the loss of potential medical treatments.

Ecosystem service values

As this book illustrates, there is increasing evidence that fishing has wide indirect impacts on marine ecosystems. However, there is considerable uncertainty surrounding the significance of these impacts in terms of the reduction of the potential of marine ecosystems to provide certain 'ecosystem services' (Table 23.1). It has been estimated that the total global value of such services from marine ecosystems is approximately $20\,949 \times 10^9$ year^{-1}, where the total value from terrestrial ecosystems is $12\,319 \times 10^9$ year^{-1}, and that coastal ecosystems where fisheries are concentrated have the highest total global value of all at $12\,568 \times 10^9$ year^{-1} (Costanza *et al.*, 1997). Whilst it is accepted that these are crude estimates, they do give a good indication of the relatively high importance of coastal ecosystems in providing services on which humans depend.

A key issue is whether the impacts of fishing will undermine the potential of marine ecosystems to deliver these vitally important services. At a wider trophic level, there are a number of indirect impacts that have raised concerns that fishing may have important impacts on ecosystem structure and function (Jennings & Kaiser, 1998). Pauly *et al.* (1998) have highlighted the somewhat worrying trend that the fishing industry is targeting trophically lower levels in marine food webs as fish stocks at higher trophic levels become depleted. The significance of these impacts in terms of the extent to which they might reduce the delivery of vital ecosystem services is uncertain.

Table 23.1 The various ecosystem services provided by the marine environment. Adapted from Costanza *et al.* (1997)

Ecosystem service	Ecosystem function	Examples
Disturbance regulation	Capacitance, damping, resilience and integrity of ecosystem in response to environmental fluctuations	Coastal defence value of marine habitats, particularly in the face of climate change
Nutrient cycling	Storage, internal cycling, processing and acquisition of nutrients	N and P and other elemental or nutrient cycles
Biological control	Trophic–dynamic regulations of populations	Keystone predator control of prey species
Habitat/refugia	Habitat for resident and transient populations	Fish nurseries, habitats for migratory waterfowl and harvested fish

There are also growing concerns amongst conservation biologists that any significant reductions in biodiversity could undermine the stability (Polis, 1998), predictability (McGrady-Steed *et al.*, 1997) and reliability (Naeem & Li, 1997) of ecosystems. Clearly, this is a potential concern where non-target marine species may be threatened with extinction by fishing impacts. There are two schools of thought in relation to such concerns. Some argue on a precautionary basis that it would be unwise to render any species extinct and thus reduce ecosystem resilience, especially in the light of uncertainty about the degree of redundancy of species and the size of future stresses. Others argue that efforts should be focused on conserving critical functional groups of 'driver' species that underpin ecosystem structure rather than trying to conserve all species, many of which are redundant in terms of their contribution to maintaining ecosystem functions and might thus be regarded as 'passengers' that are expendable.

Ehrlich & Walker (1998) point out that these views are not necessarily diametrically opposed. Those that argue for a precautionary approach recognise that not all species are equally important in maintaining ecosystem functioning, but they emphasise our ignorance about which species are redundant. Those that argue on the basis of species redundancy recognise that the extinction of a given species would often not lead to observable negative impacts on the delivery of ecosystem services, and that it makes sense that current conservation initiatives should place the highest priority on those species that are known to be the sole representatives of their functional group. However, they also argue that redundancy is likely to be important in the long term in the face of emerging ecosystem stresses, such as climate change. A species that is presently regarded as redundant might well be the only one in its group that is able to adapt to new environmental conditions and may thus become a driver species. Conserving redundant species is therefore important, as it contributes to ecosystem resilience. Ehrlich & Walker thus conclude that, considering the uncertainties and complexities between biodiversity and ecosystem services, policy decisions should take a precautionary approach to biodiversity protection in order to try to increase, or at least maintain, redundancy in ecosystems and thus maximise ecosystem resilience.

The significance of fisheries impacts on marine biodiversity is not confined to whether species, be they drivers or passengers, are rendered extinct, which is unlikely given the wide distribution of many species and the patchy nature of fishing. It is also important to consider whether regionally important populations of species might be wiped out, undermining the delivery of services at that geographic scale. Thus, the disturbance of ecosystem dynamics at a regional scale by the non-target impacts of fishing could significantly reduce the potential delivery of services in that region without the species in question ever being globally endangered.

It is also important to remember that it is not only the impacts on non-target species that need to be considered. The depletion of target fish stocks can also have impacts on ecosystem functions that may be of significance, but the pathways and outcomes of which are difficult to predict, owing to the complex and non-linear relationships between species, particularly if important top predator species are involved. For instance, it has been reported that overfishing off the coast of Alaska was one of the principal factors that led to the collapse of populations of Steller sea lions and harbour seals, and that these were an important food source for killer whales. These in turn have been forced to seek alternative prey, including sea otters around the Aleutian Islands. The resulting decline in sea otters has set off a cascade of effects. Populations of sea urchins on which sea otters normally prey have exploded, leading to the overgrazing and collapse of kelp beds, which are at the base of the coastal food web and provide an important habitat for many marine species (Estes *et al.*, 1998).

Overall, it is clear that fishing can directly and indirectly lead to ecosystem perturbations that could undermine the potential of these ecosystems to deliver important services at a considerable cost to society. However, wider societal perception of this risk is arguably relatively low, as is the priority that is attached to it. This is related to the fact that, amongst relevant research communities, there is a great deal of uncertainty concerning the risk that these perturbations will occur and the severity of their impacts. Consequently, many argue for a precautionary approach that provides society with sufficient environmental insurance against the impacts of overfishing through measures such as the designation of areas closed to fishing (Lauck *et al.*, 1998; Lindeboom, this volume, Chapter 19). However, for the reasons previously discussed, gaining acceptance for such closed areas is extremely difficult.

Existence values

Existence values are derived from an appreciation of simply knowing that undisturbed natural areas and the species they support are there, and are considered by economists to provide one of the main justifications for wilderness preservation

(Krutilla, 1967). This might be considered in terms of vicarious or symbolic existence values.

Vicarious existence values are based on the 'second-hand' aesthetic appreciation of the marine environment through wildlife documentaries, magazine articles, books, etc., this being particularly important in the marine environment owing to the extreme difficulties in observing it first-hand. They can also be based on the altruistic appreciation that other people can derive values from preserved marine areas.

Symbolic existence values derive from the 'moral conviction that it is right' to preserve natural areas and the species they support (Leopold, 1964) on the basis of their intrinsic value, regardless of any present or future use values. Pearsall (1984) considers that the likelihood of use values, such as the resource and ecosystem service values discussed above, accruing from any given species is very small and that public support for species protection must therefore be largely founded on such moral convictions of which preserved natural areas are an important symbol.

Existence value of wilderness: the **Brent Spar** case

The priority that is attached to the existence value of marine wilderness in relation to the non-target impacts of fishing is arguably relatively low owing to the alien, remote and hidden nature of the marine environment. However, such existence value in relation to other activities can be evoked with potent effect, as the high profile public controversy over the decision to dump the decommissioned *Brent Spar* structure illustrated. Shell's proposal to dump this structure in one of three deep-water locations west of the Hebrides, which had been approved by the British Government, caused widespread public protests in 1995 stimulated by Greenpeace's publicity campaign. This was centred on the argument that the deep-sea marine wilderness should not become a dumping ground for redundant oil industry structures, and the protests led to a dramatic U-turn by Shell. The options appraisal stage was re-opened and a wide consensus-building exercise was initiated in order to identify an option that was acceptable to environmentalists and the wider public.

A notable feature of this case was that the deep-sea 'wilderness' to the east of the dump sites has been subject to increasing exploitation by deep-sea trawlers, particularly on the continental slope at depths of 200–1800 m (B. Bett, Southampton Oceanography Centre, personal communication). However, no major campaign has been mounted by Greenpeace over this exploitation of sensitive deep-sea species. It is uncertain whether this was because the fishing industry is relatively diffuse and therefore less of a target for campaigns, whereas the oil industry is seen as a 'Goliath' against which Greenpeace wished to pitch itself, or because it was considered that the public would be less sympathetic towards an oil company than they might be towards fishermen. What is clear is that Greenpeace's campaign evoked great public concern, which stemmed from the high priority that was attached to the existence value of this deep-sea wilderness, whilst the much wider impacts of deep-sea fishing were not considered to be a priority in this 'landmark' campaign. This also illustrates the malleability of such societal priorities in the face of campaigns whose profile is not necessarily proportionate to the environmental impacts of the activity in question. In this respect, Side (1996) comments that it is unfortunate that it is the marketing of stakeholder goals in such symbolic contests that provides the vehicle for society to form its values, rather than such values being formed through a closer relationship with the natural world.

Existence value of certain species: the 'walls of death' case

The existence values that society holds for certain intrinsically appealing marine species have had a potent effect on many fisheries policies, relating to the by-catch of seabirds, dolphins, turtles, etc. (Hall, 1996). For instance, the European Commission's decision to ban drift nets from most of its waters by the year 2002 was largely driven by concerns about the impacts of drift nets on marine wildlife. Greenpeace's campaign on this occasion highlighted the thousands of dolphins, whales, turtles, etc., that were killed in these 'walls of death' every year. Clearly, the existence value that society holds for such 'charismatic megafauna' can be evoked by arousing public interest and sympathy, and can have a potent effect on fishing policies. However, it has been argued that such flagship species may not be the most appropriate indicator or umbrella species by which wider ecosystem health can be measured or ensured, and that it may be more appropriate to move towards an ecosystem or 'landscape' approach to conservation rather than one based on species (Simberloff, 1998). There is thus a danger that focusing campaigns to reduce the impacts of fishing on flagship species can neglect the more important wider ecosystem impacts of fishing, although it is understandable why campaigns focus on species the existence value of which is relatively easy to evoke.

Discussion

It is clear from the above review that the non-target impacts of fishing can damage the ability of the marine environment to be of value to society through damage to existing or potential resources, to ecological service provision, or to wilderness areas or intrinsically appealing species. Whilst there is a growing scientific awareness of the potential significance of the impacts of fishing on marine ecosystems, there is a strong tendency for publicity campaigns to be focused on flagship species as a means of supporting specific campaigns. Such campaigns may miss important wider ecosystem issues. Public awareness of the wider ecosystem impacts of fishing, particularly with regard to the potentially serious consequences of ecosystem perturbations, should be increased with a view to promoting general interest in the marine environment as a means of raising the priority that society attaches to such impacts.

The relatively low priority that society attaches to the non-target impacts of fishing is also attributable to the considerable scientific uncertainty concerning the present and potential significance of such impacts. Many 'scientific' claims about the potential pharmaceutical or ecosystem service values of species and habitats that may be endangered by fishing at a certain regional scale are often extremely difficult to back up with rigorous scientific evidence, and may thus be regarded more as scientifically informed opinions, wider public support for which is based more on belief than fact-driven concern. This tendency is not restricted to the issue under discussion, as the fact that environmental decisions must often be made in the face of considerable uncertainty is becoming accepted in the context of many pressing issues. However, it is clear that the scientific rationale behind such decisions must be maximised. Whilst the scale, connectivity and complexity of marine ecosystems presents major obstacles for robust empirical research concerning the ecological service value of endangered species and habitats, it is clear that more resources and effort needs to be focused on overcoming these obstacles and increasing the scientific basis of arguments to reduce the non-target impacts of fishing.

Decisions concerning measures to reduce fisheries impacts will often involve major socio-economic trade-offs, and are unlikely to be based solely on scientific rationale as appropriately sound analytical foundations rarely exist. More often they will largely be based on deliberations that are not restricted to scientists, but also include political actors and other stakeholders that act as much on the basis of values as on science (Dietz & Stern, 1998). As such, there is considerable scope for increasing the constructive involvement of scientists, who it must be accepted are also motivated by values and judgements, in broadly based deliberations. This will serve to guide their research efforts in a manner which increases the potential potency of their findings, and to maximise the degree to which their findings constructively inform these deliberations.

However, increasing the profile of scientists in such debates and in raising public awareness of the ecosystem impacts of fishing does raise the critical issue of whether the role of scientists is one of communicating value-free scientific evidence, or whether it should be extended to being advocates or even campaigners for increased marine conservation measures. This is a particularly difficult issue given the problems of achieving scientific certainty about wider impacts on marine ecosystems. In the face of uncertainty, marine scientists may have to go beyond what they can empirically prove when arguing for increased marine conservation measures, but this leaves them open to criticisms that they have abandoned science in favour of values. For instance, one of the authors of a paper (Elliot & Norse, 1998) that compared bottom trawling to the clearcutting of forests and strongly argued for the precautionary establishment of closed areas, has been accused of having 'left his scientific credibility at the door' and of publishing controversial one-sided statements (Oceanspace, 1999).

In the face of the generally low priority that society attaches to the ecosystem effects of fishing, and the problems of proving their potentially major significance, it could be argued that concerned scientists have a legitimate right to adopt the role of advocates or even campaigners. Of course, the quest for scientific evidence to underpin fisheries management must continue, as this book demonstrates. However, if we are to progress from a stock-based to an ecosystem-based fisheries management approach, it must be accepted that a balance will need to be achieved between scientific opinions to inform management decisions and scientific certainty. Concerned scientists may wish become involved in efforts to raise societal awareness about societal awareness about fisheries impacts, which this chapter indicates is relatively low, and it is argued that this can be achieved in a manner that does not necessarily involve abandoning science. To this end, it is important that the distinction is made clear between scientifically provable statements and scientifically informed opinions.

References

Cole-King, A. (1994) Coastal and marine conservation in Britain: ecology and aesthetics, land and sea. *Environmental Values*, **3**, 139–53.

Cole-King, A. (1995) Marine Protected Areas in Britain: a conceptual problem? *Ocean and Coastal Management*, **27**, 109–27.

Costanza, R., d'Arge, R. & van den Belt, M. (1997) The value of the world's ecosystem services and natural capital. *Nature*, **387**, 253–60.

Davis, D. & Tisdell, C. (1995) Recreational scuba-diving and carrying capacity in marine protected areas. *Ocean and Coastal Management*, **26**, 19–40.

Dietz, T. & Stern, P.C. (1998) Science, values and biodiversity. *Bioscience*, **48**, 441–4.

Ehrlich, P. & Walker, B. (1998) Rivets and redundancy. *Bioscience*, **48**, 387.

Estes, J.A., Tinker, M.T., Williams, T.M. & Doak, D.F. (1998) Killer whale predation on sea otters linking oceanic and nearshore ecosystems. *Science*, **282**, 473–6.

Grant, P.T. & Mackie, A.M. (1977) Drugs from the sea: fact or fantasy? *Nature*, **267**, 786–8.

Hall, M.A. (1996) On bycatches. *Reviews in Fish Biology and Fisheries*, **6**, 319–52.

Holmes, B. (1997) Dreams as big as an ocean. *New Scientist*, **155** (2097), 20.

Jennings, S. & Kaiser, M.J. (1998) The effects of fishing on marine ecosystems. *Advances in Marine Biology*, **34**, 201–351.

Jones, P.J.S. (1994) A review and analysis of the objectives of marine nature reserves. *Ocean and Coastal Management*, **24**, 149–78.

Kenchington, R.A. (1990) *Managing Marine Environments*, pp 37–8. Taylor & Francis, London.

Krutilla, J. (1967) Conservation Reconsidered. *American Economic Review*, **57**, 777–86.

Lauck, T., Clark, C.W., Mangel, M. & Munro, G.R. (1998) Implementing the precautionary principle in fisheries management through marine reserves. *Ecological Applications*, **8**, S72-8.

Leopold, A.S. (1964) Wilderness and culture. In: *Wildlands in Our Civilisation* (ed. D. Brower). Sierra Club, San Francisco.

Lowe, P. (1983) Values and Institutions in the history of British nature conservation, In: *Conservation in Perspective* (eds A. Warren & F.B. Goldsmith), pp 329–52. John Wiley, Chichester.

McGinn, A.P. (1998) Rocking the boat: conserving fisheries and protecting jobs. *Worldwatch Paper* 142. Worldwatch Institute, Washington, DC.

McGrady-Steed, J., Harris, P.M. & Morin, P.J. (1997) Biodiversity regulates ecosystem predictability. *Nature*, **390**, 162–65.

Naeem, S. & Li, S. (1997) Biodiversity enhances ecosystem reliability. *Nature*, **390**, 507–9.

Oceanspace (1999) *Oceanspace Electronic Newsletter*, Issue 101, Article 4.

Pain, S. (1996) Hostages of the deep. *New Scientist*, **151** (2047), 38–2.

Pain, S. (1998) Foul medicine. *New Scientist*, **157** (2120), 11.

Pearce, F. (1995) Rockall mud richer than rainforest. *New Scientist*, **147** (1995), 8.

Pearsall, S.H. (1984) *In absentia* benefits of nature preserves: a review. *Environmental Conservation*, **11**, 3–10.

Polis, G.A. (1998) Stability is woven by complex webs. *Nature*, **395**, 744–5.

Pauly, D., Christensen, V., Dalsgaard, J., Froese, R. & Torres, F. (1998) Fishing down marine food webs. *Science*, **279**, 860–3.

Rayment, W. (1998) The conservation of the Lyme Bay reefs. Unpublished M.Sc thesis, Department of Geography, University College London.

Simberloff, D. (1998) Flagships, umbrellas and keystones: is single-species management passé in the landscape era? *Biological Conservation*, **38**, 247–57.

Watling, L. & Norse, E.A. (1998) Disturbance of the seabed by mobile fishing gear: a comparison to forest clearcutting. *Conservation Biology*, **12**, 1180–97.

Chapter 24

Integrated management: the implications of an ecosystem approach to fisheries management

D. SYMES

Department of Geography, University of Hull, HU6 7RX, UK

Summary

1. The aim of the paper is to evaluate some of the opportunities and constraints confronting the incorporation of an ecosystem approach into fisheries management.
2. Its incorporation will complicate an already complex, highly centralised and politicised decision-making process.
3. Full incorporation – in effect, the substitution of fish-stock management by ecosystem management – will prove too radical an agenda; an incremental approach – herein identified as integrated fisheries management (IFM) – is more likely.
4. IFM combines the principles of fisheries and ecosystem management so as to conserve natural biological diversity and ecosystem integrity while supporting a sustainable level of human use.
5. IFM will require a significant reorientation of underlying value systems, scientific information and management procedures, together with a rebalancing of user group interests.
6. The precautionary principle, as the central tenet of IFM, requires more explicit definition and well-defined operational procedures that can be applied to fish stocks and other elements of the marine ecosystem.
7. IFM will require a shift in emphasis from 'numerical management' based on TACs and quotas to 'parametric management', which pays close attention to the physical and biological parameters of the fishery.
8. New institutional structures for IFM will require a 'federalist' approach and greater participative management.
9. International institutions have not encountered great success in the past; three new examples of international collaboration involving an ecosystem approach (Wadden Sea, North Sea Conference, Baltic 21) are outlined but do not appear to break free of the constraints of the past.
10. Review of the CFP (2002) provides a further opportunity for radical reform: three scenarios are considered – *greater centralisation*, *status quo* and *regionalisation*; only the latter offers an appropriate framework for an ecosystem approach, but is unlikely to succeed.
11. Prospects for the adoption of IFM and an ecosystem approach to Europe's fisheries are not very bright, except perhaps in inshore waters.

Keywords: integrated fisheries management, governance, Common Fisheries Policy, ecosystem approach.

Introduction

Concern for the ecosystem effect of fishing activity simply adds to the current dilemmas confronting fisheries policies throughout Europe. Ecologists and conservationists may be forgiven for believing that, when given a clearly defined prior objective of ecosystem sustainability and faced with mounting evidence that certain types of fishing activity continue to exert adverse effects on the diversity and integrity of the marine ecosystem, it will be easy to devise and implement appropriate regulatory measures to halt the degradation of the ecosystem. But the painfully obvious truth is that it is a far from simple task. Overall, the impact of fisheries management on the sustainability of the resource base has been negligible, yet the economic cost in public expenditure, administrative effort and loss of employment has been quite high. Explanations for the failure of fisheries management are legion (Symes, 1996). The underlying problems and their technical solutions are, in fact, rarely simple, incontestable and non-controversial, especially when refracted through the opinions of different user groups. The science of fisheries management is increasingly exposed as insecure and non-stochastic (Smith, 1990; Wilson & Kleban, 1992). Even when the problem can be isolated and reduced to its basic elements, there are several intervening interests – including economic, social, cultural, institutional and political – which have to be accommodated within the decision-making process. It is the task of the management system to balance these interests. Maybe, in attempting to appease so many different interest groups, the system of management has become overly complex, inflexible and therefore ineffective.

  If it is the case that this summary analysis holds true for a discrete, sectoral fisheries policy, how much more difficult will it be to develop a system of integrated management in response to the growing demands for fisheries to be set within a broader framework of ecosystem management? In attempting to answer this question, at least in outline, this chapter will indicate the general nature of the existing decision-making system, identify some of the opportunities and constraints of integrated management and, finally, relocate some of the key issues in the context of the 2002 review of the Common Fisheries Policy (CFP). The aim is not to analyse specific examples of the ecosystem effects of fishing activity, but to examine, from a social science perspective, the general nature of the challenge to management systems posed by the need to incorporate ecosystem objectives within fisheries management.

The policy-making system

Viewed from almost any angle, fisheries management is a complex task (Fig. 24.1). It is clear that the policy framework cannot embrace all aspects of the fishery. Inevitably some areas will be exposed to external, macro-level influences originating in environmental change, the impacts of globalisation or shifting value systems. Within existing fisheries policies, the conservation of marine resources represents only part of the equation. There may well be circumstances in which the thrust of conservation strategy is moderated by other elements of fisheries policy, such as the structural development of the fishing fleet or the regulation of the market, or by concerns relating to other policy areas outwith the scope of fisheries management *per se* (employment, social welfare, regional development, *inter alia*). Indeed in the prevailing political climate of the free market, it can be posited that economic considerations will normally take precedence over socio-cultural and environmental issues.

The policy system is further complicated by the fact that decisions in relation to fisheries management will be made at several different levels by authorities with varying degrees of competence and different areas of application. Within the European Community (EC), three or four levels of political decision-making can be identified: the Community, national, regional and/or local. Although in theory the CFP should establish clear and uniform guidelines and so ensure a harmonisation of policy measures, in reality there is considerable scope for variation of policy implementation at the level of the Member State and, thus, for policy dissonance. In its application, therefore, the CFP can scarcely be regarded as a model of consistency. National interests still prevail, subverting the attempts to create a single market for fisheries in which all Member States stand on an equal footing (Fraga Estevez, 1999). As far as the EC and the CFP are concerned, the whole is not yet equal to the sum of its parts.

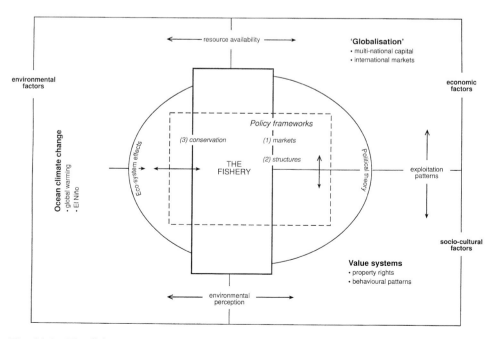

Fig. 24.1 The fishery system.

Although the advent of a common fisheries policy nominally established a framework for the harmonisation of regulation across an ever-widening area of the north-east Atlantic and its tributary seas, it also ensured that fisheries management became a more overtly political process – especially following the enlargement of the Community in the mid-1980s. The increasing politicisation of fisheries management also coincided with an erosion of confidence in the scientific basis of the management system itself. Increasing awareness among scientists, administrators and fishermen alike of the unpredictable behaviour of the ocean environment, as evidenced by episodes such as the collapse of the Barents Sea and Northern cod stocks, led to challenges to the established procedures for stock assessment and scientific advice (Finlayson, 1994). Moreover, there have been calls for participative management, involving the greater incorporation of user group organisations in the management process (Jentoft, 1989; Jentoft & McCay, 1995; Sen & Raakjaer Nielsen, 1996). The new focus on issues relating to ecosystem management simply adds to the destabilisation of established forms of fisheries management.

Fisheries management is, in fact, a relatively recent area of government responsibility. Faced with a growing list of overexploited stocks and the contagious spread of 200-mile Exclusive Economic Zones in the mid-1970s, the state was reluctantly compelled to assume control of their management. It led, not surprisingly, to a highly centralised system of decision making and a hierarchical, top-down mode of delivery. The emergence of new problems and, more especially, the mounting evidence of policy failure, were met by a simplistic response of increasing levels of regulatory control. The fishing industry is now held in a straight-jacket of bureaucratic control and burdened by increasing regulations. This tendency is perhaps unique to fisheries. It stands in marked contrast to developments in other sectors of the economy and in defiance of the prevailing theories of governance (Symes, 1997), where a gradual disengagement of the state from direct intervention has been taking place through deregulation, together with decentralisation and delegation of management responsibility. The reasons for this contrast are not altogether clear, although they may relate, at least in part, to the concept of fisheries as a common good (*res communis* or *res publica*), the alleged absence of clearly defined property rights and the presumed absence of responsible user-group organisations able to shoulder the burden of co-management.

Integrated management

The ecosystem approach

In policy terms the impetus for integrated management was provided by three key events in the 1990s: the Convention on Biological Diversity (1992), publication of the FAO's *Code of Conduct for Responsible Fisheries* (FAO, 1995) and the Intermediate Ministerial Meeting on the Integration of Fisheries and Environmental Issues, held in Bergen in 1997. The concept is clearly in its infancy: the gap between theory and practice is huge, with broadly agreed general principles not yet translated into clearly defined operational systems.

Integrated management may be described as the combining of the principles of fisheries and ecosystem management in a way that conserves natural biological diversity and ecosystem integrity, while supporting a sustainable level of human use (Symes, 1998a). It implies a shift in management practice from one that is firmly rooted in single-stock management to one where the conservation of the whole of the marine ecosystem is recognised as the organising principle. The aim of ecosystem management is not the preservation of the marine ecosystem in some 'ideal' static state, but a guarantee of the resilience and dynamic ability of the physical and biological systems to adapt to change (Hanna & Munasinghe, 1995). According to the FAO's *Code of Conduct,* which comes closest to setting down operational guidelines, it means that 'Management measures should not only ensure the conservation of target species, but also of species belonging to or associated with or dependent on the target species' (FAO, 1995). Even in these terms, it remains a very fisheries centred approach to the concept.

Whereas the ecosystem approach, by definition, stresses the importance of maintaining the diversity, productivity and integrity of natural ecosystems, the Rio Declaration (1992) rightly insists that the concept must also embrace the economic, social and cultural needs of human populations. And, as Garcia & Newton (1994) observed, the ecosystem approach means the incorporation of fisheries management within 'an ever more complex and less mature legal and institutional framework'.

Is, then, the ecosystem approach too radical or merely too vague for effective incorporation within established systems of fisheries management? Adoption of the ecosystem approach should radically alter the current spatial and temporal perspectives of fisheries management (Schramm & Hubert, 1996). Paradoxically, while the areal scale will probably be reduced in line with regional and local ecosystems, the time scale will need to be greatly extended from the present short-term (annual) perspectives of total allowable catches (TACs) and quota-setting to much longer-term (intergenerational) horizons. Logically, the ascendancy of ecosystem management should lead to the abandonment of separate policies for the marine environment and its resources; fisheries management would lose its status as an autonomous policy area and become embedded in a broader and longer-term strategy for the marine ecosystem as a whole (Symes, 1998a).

At present this would seem to be too radical an agenda – too big a step to be credible to fishing interests or bureaucrats alike. What is outlined in the following sections is a more modest approach involving the internalisation of the basic principles of ecosystem management within existing fisheries policies. Even this limited agenda has to confront two particular problems: operationalisation of the defining principles of ecosystem management and creation of appropriate institutional frameworks for integrated fisheries management.

The precautionary principle

The precautionary principle provides the central tenet of ecosystem management. It was formally enunciated in the UN Convention on Environment and Development (1992) in the now almost infamous words:

> *In order to protect the environment, the precautionary approach shall be widely applied by States according to their capabilities. Where there are threats of serious or irreversible damage to the environment lack of full scientific certainty should not be used as a reason for postponing cost effective measures to prevent degradation.*

The principle lacks a stable definition – its meaning and therefore its policy implications vary according to who is actually using the term.

As applied to fisheries management and elaborated in the FAO's *Code of Conduct*, the precautionary principle implies a much more comprehensive approach to fish stock management. Hitherto, the management process has required only the specification of a single target point, the TAC, but no clear specifications of limit reference points – even though these may be inferred from the notion of minimum biologically accepted level (MBAL) applied to spawning stocks – and no indication of the remedial action required should the reference points be breached. In discussing the ecosystem effects of individual fishing for sandeels, Gislason & Kirkegaard (1998) outline a more comprehensive procedure for fish-stock management, including the development of management plans for individual species (or for a complex of species in a mixed fishery), the setting of target and limit reference points and the detailing of contingency action plans should sustainability of the stock(s) be threatened.

Seen in this light, the application of the precautionary principle to fish-stock management is perfectly feasible, but there is, as yet, little parallel development in the operationalisation of the principle with regard to the ecosystem as a whole or to its non-commercial components and their habitats. In the absence of well-developed parameters for precautionary management, ecologists and conservationists have tended to rely upon largely unsubstantiated arguments in favour of 'marine protection areas' (MPAs) or 'no-take zones' (NTZs) applied to large areas of commercial fishing grounds as a catch-all device for implementing the precautionary principle (Gubbay, 1996; Pullen, 1997; Roberts, 1998). While this approach may work for some species, it is likely that in areas like the North Sea, where the target species such as cod are widely dispersed and highly mobile, the closure zones would need to be so extensive as to make them impracticable. It is in this area that progress is most urgently needed if ecosystem management is to make a significant impact on fisheries management.

Integrated fisheries management

Having ruled out a more radical 'big bang' approach of ecosystem management in favour of a more modest, incremental strategy – largely on the grounds of administrative feasibility – it is important to avoid a fragmented, *ad hoc* and essentially piecemeal approach to the issue. Integrated fisheries management, which shifts the emphasis from a system designed to ensure sustainability of commercially important fish stocks to one that combines this objective with that of a sustainable ecosystem, will still require a significant reorientation of the underlying value systems, scientific information and management procedures, and a rebalancing of user-group interests in the quest for a consistent means of ameliorating the effects of fishing activity on the marine ecosystem. It is unlikely that integrated fisheries management will involve a significantly different 'menu' of regulatory measures, but the emphasis is likely to switch from output limitation, sometimes alluded to as 'numerical management', to the more sophisticated deployment of technical conservation measures or 'parametric management' which ensures that the right precautionary measures are adopted in the right places at the right time. Parametric management argues that the decline in fish stocks cannot be explained simply in terms of higher mortality induced through 'overfishing', but rather that

> *fishing activity has led to a degradation of parts of the biotic or physical environment of desirable species, reducing opportunities for growth, reproduction and survival ... [which alters] ... the capacity of the whole system to maintain the partitioning of the energy flows on which the fishery has depended'* (Wilson & Dickie, 1995, p. 154).

Parametric management involves a hierarchical system of regulation including general rules designed to restrain total fishing effort; species-specific rules related to life-cycle behaviours (e.g. seasonal closures); and local environmental rules (e.g. closed areas, gear regulations). Such rules need to be robust in design, specific in their application and stringent in their enforcement.

The existing regulatory framework, intended primarily to limit fishing effort, is based on three types of instrument: output management (TACs and catch quotas), supported by structural measures to reduce the harvesting capacity of the fishing fleet and technical conservation measures in the form of gear regulations and access restrictions. TACs and catch quotas have proved a blunt instrument. They lack regional sensitivity, constrain the flexibility of fishing strategies, induce a mind-set focusing on competition, capitalisation and cost-effectiveness, and also give rise to the misreporting of catches, blackfish landings and discards – although the extent to which discards are a direct reflection of quota management is debatable.

Whether or not the quota system can be improved is a key issue in fisheries politics. If the assumptions concerning the significance of the parametric effects of fishing are correct, altering the system of quota management is not an answer. It addresses the wrong question. The current debate surrounds the introduction of individual transferable quotas (ITQs), which confer quasi-property rights on their owners. According to Hanna *et al.* (1995), property rights are a necessary but not a sufficient condition for resource stability. In the context of the present discussions, the real question is whether ITQs would assist or hinder the development of integrated fisheries management. Suffice it to say that there is surely a fundamental and potentially dangerous contradiction where one key element of the marine ecosystem is treated as 'private property' but all other elements are regarded as '*res nullius*', belonging to no one. It seems fairly predictable that, if the value of the ITQ were to be threatened by other forms of regulation constraining the exercise of fishing rights, the industry would vigorously oppose their introduction. An alternative approach to the problems posed by catch quotas would be to switch the allocation of fishing rights to one based on effort quotas, combined with the banning of discards. This should at least result in a reduction of current malpractices (under-reporting, blackfish landings and discards) and regenerate flexibility in the fishing strategies of individual enterprises, although it would certainly not address the issues of regional sensitivity and the technological treadmill.

In outlining the 'content' of integrated fisheries management, the chosen instruments are intended to form a coherent package of measures to contain fishing effort and reduce the impacts of fishing activity on marine habitats and their ecosystems to acceptable levels, while still laying the basis for a viable fishing industry. The proposals should be viewed more in the nature of a *table d'hôte* than an *à la carte* menu. Integrated fisheries management would therefore:

1. introduce a restrictive licensing system to achieve and maintain fishing capacity at levels commensurate with health of the ecosystem;
2. replace the discredited system of catch quotas by transferable effort quotas, allocated on a community or individual basis;
3. introduce a discard ban, following the Norwegian example;
4. devise regionally specific gear regulations, including mesh sizes, the mandatory introduction of more selective gears and restrictions in the use of certain forms of fishing gear;
5. introduce a more extensive and coordinated network of seasonal closures to protect nursery grounds;
6. introduce emergency closures where the proportion of juvenile or immature fish exceeds a given level; and
7. introduce a limited number of NTZs for the protection of designated habitats and populations of species at risk from serious and irreversible damage.

Apart from the sensitive issue of effort quotas, the only really controversial element of these proposals concerns the limited introduction of NTZs, defined, in this instance, as involving a permanent, long-term closure of the designated area to all forms of extractive use. The list of potential ecosystem benefits accruing to NTZs is undoubtedly impressive (Roberts, 1998), but so too are the costs to those deprived of the opportunity to earn their living as a consequence of the closure of the fishery. There is no guarantee that the benefits claimed in respect of their introduction in tropical reef fisheries would be repeated in the case of mixed demersal fisheries in areas like the North Sea. To be effective in such conditions, their size will have to be very much larger, accounting for as much as 20% or 25% of the total area of the North Sea (Garrod, 1997). Were NTZs to be introduced on a trial basis, it would be important to ensure very careful monitoring of the effects in terms of fish-stock renewal, habitat and ecosystem recovery or preservation and economic costs in order to determine the true benefits of closure. Ideally NTZs should be established in areas where there is little or no existing damage from fishing activity. For much of the north-east Atlantic, this would be a forlorn quest.

At a time when agriculture in the EC is being weaned away from the idea of subsidies to maintain profitability in farming and the retention of farming populations in less-favoured areas, it makes little sense to argue for the introduction of subsidy payments in fisheries. Might it be possible, however, to take a leaf out of recent reforms in agriculture where compensation payments are made to farmers who reintroduce less intensive farming methods in environmentally sensitive areas (ESAs)? As an alternative to NTZs, one might contemplate the designation of environmentally sensitive fishing zones accessible on a carefully regulated basis only to vessels using small-scale, non-destructive fishing gears.

Such schemes are clearly more likely to prove feasible in inshore waters, subject to national rather than Community jurisdiction. Here lies not only the greatest scope for integrated fisheries management but also some of the most diverse and sensitive marine ecosystems and the greatest pressures from fishing and other marine resource use. Yet in the UK to date, only three statutory Marine Nature Reserves have been created since the enabling legislation – the *Wildlife and Countryside Act* – was introduced in 1981. Marine Consultation Areas in Scotland and Sensitive Marine Areas in England, although more numerous, afford no statutory protection. More may be expected of the 35 proposed marine special areas of conservation (SACs) created under the EC's *Habitats Directive* (92/43/EEC). SACs are not expected to interfere significantly with existing fishing practice, although they may limit the opportunity to target new species or introduce new fishing techniques. The appropriate deployment of the precautionary principle, applied to both commercial fish stocks and the protection of the habitat and/or the endangered non-target species, in the management plans for the individual SACs will be crucial in ensuring an acceptable balance between commercial fishing activity and wildlife conservation. While the number of designations may seem quite impressive, the ecosystem benefits will be very site specific and, in total, the SACs will embrace only a very small fragment of the total inshore ecosystem. More generalised benefit may derive from the regulation of inshore fishing undertaken by the 12 Sea Fisheries Committees in England and Wales which, since 1992, have a duty of care for the marine environment in the exercise of their regulatory functions.

Beyond the 6- and 12-n mile limits, marine ecosystem conservation is much more problematic. Within the Community's 200-mile limits, conservation issues can only be addressed directly through fisheries legislation. The rights to manage all other aspects of the marine environment within the coastal state's 200-mile Exclusive Economic Zone were not ceded to the EC and so remain the responsibility of the individual coastal state. Therefore, effective management of the marine ecosystem in an area like the North Sea requires the framing of bilateral or unilateral agreements between the relevant coastal states.

Institutional arrangements

Integrated fisheries management requires a 'federalist' structure with different levels of administrative authority – international, national, regional and local – each dealing with an appropriate area of responsibility. 'General rules' would be decided at the international and/or national level, while species-specific rules and local environmental rules would be set at the regional or local level, making full use of local experiential knowledge of habitats, ecosystems and behavioural patterns of the target species. The integrated approach also implies 'participative management', whereby formulation and implementation of the management objectives and the rules by which these are to be achieved are determined not by bureaucracies but through the involvement of relevant, responsible interest groups.

Clearly the most difficult task, but one which is central to the success of integrated management, lies in translating the good intentions of international accords into a convincing programme of action applicable not only within the territorial waters of individual coastal states but also across the boundaries of national fishing zones (EEZs) and into the high seas. The success of international commissions (NAFO, NEAFC) or supranational administrations (EC), even when dealing with narrow sectoral questions, has not been conspicuous in the past. Within western Europe today, there are three examples of international collaboration in the field of ecosystem management.

The oldest is the Joint Declaration on the Protection of the Wadden Sea signed by the governments of Denmark, Germany and The Netherlands in Copenhagen in 1982 to coordinate activity in implementing international legal instruments relating to nature protection. The subsequent Esbjerg Declaration (1991) elaborated the guiding principles for trilateral cooperation and outlined common management principles and objectives for human use in this area of high conservation value. Developed within a framework of integrated coastal zone management, the realisation of the specific objectives is left to the individual states working within their own institutional and legislative structures.

The series of North Sea Ministerial Conferences was initiated in 1984 in response to the growing concern among the coastal states for the deterioration in the quality of the marine environment as a consequence of both accidental and planned discharges of waste materials into the North Sea. Issues relating to the integration of environmental and fisheries policies were highlighted at the fourth conference in Esbjerg (1995) leading to the setting up of an Intermediate Ministerial Meeting in Bergen in 1997. The North Sea Conferences provide a political framework for the 'comprehensive assessment of the measures needed to protect the North Sea environment' (Svelle *et al.*, 1997). The resulting binding declarations (Anon., 1997), approved by the Ministers no doubt suffer in some measure from the inevitability of compromise in the setting of objectives and targets.

The most recent development in the field of international collaboration in ecosystem management is the Baltic 21 agenda agreed in June 1998, which takes the form of a non-binding action programme for the achievement of sustainable development for the Baltic Sea region by the year 2030. Much broader in scope than the other two schemes – because it is not concerned solely with the marine environment – it focuses on a series of 'joint actions' involving two or more key sectors, 'priority actions' by the seven key sectors, including fisheries, and 'actions for spatial planning'. For fisheries, the International Baltic Sea Fisheries Commission as the lead agency is embarking on long-term strategies for major fish stocks (cod, salmon, herring and sprat), applying the precautionary approach and setting biological reference points to ensure harvesting within safe biological limits.

Leaving aside the Wadden Sea initiative, which is exclusively concerned with territorial waters, both the North Sea Conference and Baltic 21 share a number of characteristics which suggest that they have not freed themselves from the historical baggage of traditional fisheries management. Both concentrate on fish stocks rather than the ecosystem as the focal topic of management and both retain faith in TACs and catch quotas as the principal management tools. It is unlikely that the reform of the CFP will offer a more radical approach to integrated fisheries management.

Redeeming the Common Fisheries Policy

Some doubt already exists over the willingness and ability of the CFP to accommodate the basic precepts of integrated fisheries management. The Policy is essentially a productivist, sectoral policy, similar in style to the Common Agricultural Policy of the 1970s and early 1980s. It is concerned with maximising output, employment and income from the fishery within the constraints imposed by the biological state of the stocks. As a scientifically based, technocratic policy designed by the Commission, it is tempered through the deliberations of the Council of Ministers by a more pragmatic, political concern for the welfare of the fishing industry. Although Article 2 of Regulation 3760/92, updating the CFP, pays lip service to the aims 'to protect and conserve available and accessible living marine aquatic resources ... taking account of ... implications for the marine ecosystem', in reality there is little direct evidence of environmental concern except perhaps in the funding of scientific research. The recent decision to ban drift-net fishing for tuna and related species in Community waters in order to protect non-target marine wildlife from risks of incidental capture is an isolated example. So far, the CFP has studiously ignored the rhetoric of ecosystem management and the precautionary principle, despite the admission of the principle in the terms of the Treaty of Maastricht 1992, Article 130(2). However, in fulfilment of its obligations under the UN's Convention on Biological Diversity (1992), the EU is to draw up a Fisheries

Action Plan by February 2000, which will consider *inter alia* the means for reducing the impacts of fishing activities on non-target species and on marine and coastal ecosystems. If properly integrated into the CFP, this could lay a much firmer foundation for an ecosystem approach to fisheries management in Community waters.

Perhaps the greatest single source of instability in fisheries management in Europe is the prospect of a major reform of the policy systems in 2002. In theory, it could provide a crucial opportunity to redeem the CFP from its history of failure and its reputation as an environmentally unaware policy area. But it would be unwise to expect too much. The legal obligation to review and modify the CFP applies only to a very limited list of topics (Farnell, 1997) – principally the derogation allowing coastal state jurisdiction over the 6- and 12-n mile territorial limits and the designation of the Shetland Box, together with the restrictions on access to Community waters by the new Member States. The Commission has, however, opened the doors to discussion of a much broader and more fundamental political agenda, concerning the tensions between the seemingly contradictory 'principles' embedded in the CFP – 'equal access' and 'relative stability'. The equal access argument derives from the basic tenet of the Treaty of Rome, namely non-discrimination or the concept of the level playing field. Negotiation of the original CFP in the years between 1976 and 1982, involved a number of derogations from this principle as temporary measures to allow different Member States to adjust to the idea of a common system of management applied throughout Community waters. One important negotiated departure from the principle of equal access was the notion of 'relative stability' as applied to the system of national catch quotas, whereby each Member State was guaranteed a fixed share of the TACs for all regulated stocks roughly in line with its historic track record in the 1970s. Some marginal adjustments were made to take account of previous dependence on distant water grounds from which the Community's fleets were now largely excluded and also the level of regional dependence on fishing. The effect was to guarantee the *status quo* and so reduce the scope for radical change.

To date there is no indication that the issues of ecosystem management or integrated fisheries management will feature prominently in the discourse leading up to 2002. Nevertheless, the outcomes of the main debate could have an important bearing on the extent to which the ecosystem approach to fisheries management is adopted in the revised Policy. Broadly three alternative scenarios can be contemplated: centralisation, *status quo* and regionalisation.

1. *Greater centralisation of decision making*, an approach advocated particularly by the Spanish, implies a firm endorsement of the defining principle of the Treaty of Rome – non-discrimination or equal access – and the realisation of a genuine common or single market for fisheries (Fraga Estevez, 1999) with a concomitant weakening of the influence of the individual Member State in their ability to vary the implementation of the CFP. Such a 'solution' to the dilemmas of the CFP would tend towards greater harmonisation or standardisation of

regulations, less regional sensitivity and would take fisheries policy further away from the ideal of an ecosystem approach in both scale and orientation.

2. *The status quo* option is the most likely outcome not so much because of a strong measure of satisfaction with the existing system, but rather because it would represent a middle-of-the-road compromise between the two more extreme solutions. It would imply the retention of 'relative stability' and its paraphernalia of catch quotas, etc., and offer little scope for a significant switch to integrated fisheries management.

3. *Regionalisation of the common fisheries policy* is based in part on the subsidiarity principle agreed in the Treaty of Maastricht, 1992; thereby fisheries management would be devolved to a series of regional councils covering clearly defined regional seas. Each regional council would exercise a considerable degree of autonomy in determining the detailed management strategy for the area in question, subject to scrutiny from the Commission and the full Council of Ministers only in respect of compliance with the broad principles of the EC and the compatibility of neighbouring regimes (Symes, 1998b). In as much as several of the regional seas would be roughly coincident with meso-level marine ecosystems (e.g. the Baltic Sea, North Sea, Mediterranean, *inter alia*), this third scenario would be more likely to facilitate the implementation of an ecosystem approach to fisheries management. At the very least, it would guarantee a much greater level of regional sensitivity than is experienced at present or hinted at in any other proposal for reform. So far, the closest political initiative comes from the joint proposal by the Scottish Fishermen's Federation (SFF) and the National Federation of Fishermen's Organisations (NFFO) covering England and Wales, for 'interlocking coastal state management' which, while seeking to reassert national authority for fisheries management, also advocates the setting up of 'zonal management committees' made up of countries with TAC allocations in the area and responsible for 'establishing minimum fisheries conservation criteria' (SFF and NFFO, 1998)

The regionalisation option is probably the least likely to find favour among the bureaucrats and politicians who will ultimately decide the fate of the CFP after 2002. It is too radical a proposal which, for all its compliance with the concept of subsidiarity, will be judged as certain to undermine the unity of the single market and the authority of the Commission and the Council of Ministers. Little seems likely to change after 2002.

Conclusions

Prospects for the incorporation of an ecosystem approach within European fisheries management are therefore not very sanguine. A radical agenda of change seems to be ruled out by the very nature of the political decision-making process, which errs on the side of caution and compromise. Even a more cautious, incremental approach will struggle to make significant headway in a situation where the user groups are suspicious of the motives and outcomes of an ecosystem approach and where a defensively minded, technocratic management system – such as the CFP – is as much concerned to protect the status of its key actors (fisheries scientists, economists and administrators) as it is to promote a more enlightened approach to the management of the marine ecosystem and its resources.

The prospects may be somewhat brighter in the context of inshore waters, where both the Commission (DG XI) and individual Member State governments appear somewhat more aware of their responsibilities to ensure that resource exploitation does not occur at the cost of ecosystem degradation, and where the fisheries interests themselves may be more alert to the need to minimise the direct and indirect impacts of fishing activity on the marine ecosystem. For all its compelling logic, integrated fisheries management faces an uphill struggle to supplant the mechanistic and productivist ambitions of current policy. Yet, the basic ingredients of integrated fisheries management, as outlined in this chapter, are quite clear. They include an appropriate interpretation of the precautionary principle; the replacement of an authoritarian style of management by a more participative approach; a switch in emphasis from 'numerical' to 'parametric' management systems; and, finally the regionalisation of fisheries policy. Regionalisation probably provides the key to unlocking the other essential ingredients.

References

Anon. (1997) *Statement of Conclusions, Intermediate Ministerial Meeting on the Integration of Fisheries and Environmental Issues.* Ministry of Environment, Oslo.

Farnell, J. (1997) Looking forward to the Common Fisheries Policy after 2002: the Commission's perspective. Unpublished paper presented to the *Greenwich Forum Conference on Rethinking the Common Fisheries Policy – Looking towards 2002*, London, 16 October 1997.

Finlayson, C. (1994) *Fishing for Truth: A Sociological Analysis of Northern Cod Assessment 1977–1990.* Institute of Social and Economic Research Press, Newfoundland.

Food and Agriculture Organisation (1995) *Code of Conduct for Responsible Fisheries.* FAO, Rome.

Fraga Estevez, C. (1999) The inadequacies and ambiguities of the Common Fisheries Policy. In: *Alternative Management Systems for Fisheries* (ed. D. Symes), pp. 21–7. Blackwell Science, Oxford.

Garcia, S.M. & Newton, C.H. (1994) Responsible fisheries: an overview of FAO policy developments (1945–1994). *Marine Pollution Bulletin,* **29**, 528–36.

Garrod, D.J. (1997) Where do we go now to conserve fish stocks? Unpublished paper tabled at the *Greenwich Forum Conference on Rethinking the Common Fisheries Policy – Looking Towards 2002*, London, 16 October 1997.

Gislason, H. & Kirkegaard, E (1998) Is the industrial fishing in the North Sea sustainable? In: *Northern Waters: Management Issues and Practice* (ed. D. Symes), pp. 195–207. Blackwell Science, Oxford.

Gubbay, S. (1996) *Marine Refuges: The Next Step for Nature Conservation and Fisheries Management in the North-East Atlantic.* World Wide Fund for Nature, London.

Hanna, S. & Munasinhge, M. (1995) An introduction to property rights in a social and ecological context. In: *Property Rights in a Social and Ecological Context* (eds S. Hanna and M.

Munasinghe), pp. 3–11. The Beijer International Institute of Ecological Economics and The World Bank, Washington, DC.

Hanna, S., Folke, C. & Maher, K.G. (1995) Property rights and environmental resources. In: *Property Rights and the Environment* (eds S. Hanna and M. Munasinghe). The Beijer International Institute of Ecological Economics and the World Bank, Washington, DC.

Jentoft, S. (1989) Fisheries co-management: delegating government responsibility to fishermen's organizations. *Marine Policy*, **13**, 137–54.

Jentoft, S. & McCay, B. (1995) User participation in fisheries management: lessons drawn from international experiences. *Marine Policy*, **19**, 227–46.

Pullen, S. (1997) The role of Marine Protected Areas and fisheries refuges. *Marine Update*, **28**, 1–4.

Roberts, C. (1998) No-take marine reserves: briefing notes. Paper tabled at the *WWF Workshop on No-take Zones: The Way Forward*, London, 7–8 April 1998.

Schramm H.L. & Hubert, A. (1996) Ecosystem management: implications for fisheries management. *Fisheries*, **21**, 6–11.

Scottish Fishermen's Federation and National Federation of Fishermen's Organisations (1998) *Joint Position Paper on an Alternative Approach to the Present Common Fisheries Policy and the Post-2002 European Fisheries Regime*. Aberdeen and Grimsby.

Sen, S. & Raakjaer Nielsen, J. (1996) Fisheries co-management: a comparative analysis. *Marine Policy*, **20**, 405–18.

Smith, M.E. (1990) Chaos in fisheries management. *Marine Anthropological Studies*, **3**, 1–13.

Svelle, M., Aarefjord, H., Heir, H.T. & Øverland, S. (1997) *Assessment Report on Fisheries and Fisheries Related Species and Habitat Issues*. Ministry of Environment, Oslo.

Symes, D. (1996) Fishing in troubled waters. In: *Fisheries Management in Crisis* (eds K. Crean & D. Symes), pp. 3–18. Blackwell Science, Oxford.

Symes, D. (1997) Fisheries management: in search of good governance. *Fisheries Research*, **32**, 107–14.

Symes, D. (1998a) *The Integration of Fisheries Management and Marine Wildlife Conservation*, Final Report for Joint Nature Conservation Committee, Peterborough, UK.

Symes, D. (1998b) Towards 2002: subsidiarity and the regionalisation of the Common Fisheries Policy. In: *The Politics of Fishing* (ed. T.S. Gray), pp. 176–93. Macmillan, London.

Wilson, J.A. & Kleban, P. (1992) Practical implications of chaos in fisheries. *Marine Anthropological Studies*, **5**, 67–75.

Wilson, J.A. & Dickie, L.M. (1995) Parametric management in fisheries: an ecosystem – social approach. In: *Property Rights in a Social and Ecological Context* (eds S. Hanna and M. Munasinghe), pp. 153–167. The Beijer International Institute of Ecological Economics and The World Bank, Washington, DC.

Part 7
Workshop conclusions

Chapter 25
The implications of the effects of fishing on non-target species and habitats

M.J. KAISER

School of Ocean Sciences, University of Wales – Bangor, Menai Bridge, Gwynedd, LL59 5EY, UK

Introduction

Our knowledge of the effects of fishing on marine ecosystems and their fauna and flora has increased considerably over the last decade. We have now moved beyond the first stage of quantifying the immediate changes that occur as a result of benthic disturbance, or calculating the numbers of fish, birds or mammals caught as by-catch. Although there is still some work required in these areas (Tregenza, Chapter 17), we have now advanced to a position where we are able to calculate the effects of fishing in terms of population change (Fonds & Groenewold, Chapter 9). Such information has enabled us to predict the likely consequences of modifying fishing practices for populations of species that depend, to some extent, on fishing activities as a source of food (Camphuysen & Garthe, Chapter 11). The studies gathered together in this volume are representative of our current level of understanding of the effects of fishing in European waters. Encouragingly, similar studies and conclusions are beginning to emerge elsewhere in the world (see Jennings & Kaiser, 1998, for review).

Distribution of fishing effort and physical interaction with the seabed

While it is important to be able to quantify the direct effects of fishing disturbance on the seabed, it is equally important to know the location of the main areas of the sea that have been and are presently affected by fishing activity. Prior to 1993, the only indication of the distribution of fishing effort was based on the data collected for each International Council for the Exploration of the Sea (ICES) statistical rectangle. While these data have been collected for many years and might permit the construction of historical patterns of disturbance for specific areas of the seabed, these data have always been treated with care due to problems associated with misreporting. In addition, these data are collected at such large scales that they lack the resolution relevant at the scale of benthic habitats. Furthermore, individual national datasets are limited in their use when many areas of the North Sea are fished by several countries, yet there is still no official scheme systematically to compile international data. Then, in 1993, black box recorders were fitted to a proportion of the Dutch beam-trawl fleet. The data yielded from this study have demonstrated how patchily fishing effort is distributed at a scale as small as 9 km^2

and that it is only truly representative of a homogeneous distribution of effort at a scale of 1 km^2 (Rijnsdorp *et al.*, 1998). However, while these data have been a great insight into present-day fishing effort patterns, they are limited in the extent to which they can assist our understanding of the distribution of historical fishing effort. Jennings *et al.* (Chapter 1) have used overflight data collected by government agencies, which have recorded the precise location of fishing vessels while working at sea. The overflight data suggest that fishing effort data collected for ICES rectangles give a good representation of the general distribution of fishing activities, but, as reported by Rijnsdorp *et al.* (1998), give no indication of the micro-scale distribution of fishing within each area. While the large-scale distribution of fishing effort may be suitable for studies of changes in highly mobile species such as fish, they lack resolution for the finer-scale analyses relevant to the study of benthic biota (Ramsay *et al.*, Chapter 10; Craeymeersch *et al.*, Chapter 12).

Bottom-fishing trawls are designed to remain in close contact with the seabed, and an inevitable consequence of their design is the penetration and resuspension of the seabed to some degree. While it appears possible to reduce the direct physical forces exerted on the seabed by modifying fishing practices (e.g. towing with or against the tide), the net benefits are questionable and catches of commercial species would almost certainly suffer (Fonteyne, Chapter 2). In the case of beam trawls, there is no doubt that the tickler chains or chain mats are the part of the gear responsible for causing the majority of physical disturbance and damage to fauna. Van Marlen (Chapter 16) offers some insights into the possible methods for replacing tickler chains with supposedly less-damaging devices such as an electronic tickler chain. Nevertheless, as van Marlen acknowledges, such gear adjustments are likely to be of limited benefit to benthic fauna and habitats, although their contribution should be seen as a positive step. Despite attempts to improve gear design, as long as we continue to pursue bottom-dwelling species using towed fishing gears, there will be inevitable sediment resuspension. The biological and geochemical implications of the resuspension of sediments caused by fishing activities is poorly understood but may be significant (Churchill, 1989). One likely consequence of increasing the suspended fine particles in the water column is the reduction of light available for photosynthetic organisms. Increased turbidity will have greater ecological significance in waters that are normally relatively clear (e.g. sea lochs, fjords or open coasts) compared with shallow areas that are highly perturbed by physical forces and subject to riverine discharges (e.g. the southern North Sea). Ardizzone *et al.* (Chapter 3) investigated whether this might be one of the causes of seagrass regression in the Mediterranean Sea. Their results indicated that increased turbidity increases seagrass regression close to riverine discharges. However, there was no evidence of increased sediment resuspension in areas trawled illegally, and regression occurred patchily within the seagrass bed and was not related to depth. Hence, in this case, it was concluded that the direct physical disruption of the seagrass on the seabed and not sediment resuspension was the cause of regression.

Effects of fishing on benthic fauna and habitats

Previous short-term experimental studies of the effects of fishing on seabed communities have been limited to quantifying immediate or short-term changes. Yet the fundamentally important question is whether the populations of the affected organisms can sustain current levels of disturbance, or whether they are likely to suffer population declines. Bergman & van Santbrink (Chapter 4) are the first to have quantified the absolute mortality rate of benthic biota that encounter bottom trawls. They found that the annual mortality rate of some species is up to 39%, which is less than the annual fishing mortality of plaice in the North Sea. However, our knowledge of the biology of many of the benthic biota is limited and, in many cases, we can only estimate whether or not a species is likely to suffer critical fishing mortality rates (MacDonald *et al.*, 1996). Perhaps the best example of such a species is the common whelk. Whelks mature at an age of about 5 years, lay their eggs in batches on the seabed on hard substrata and are more vulnerable to predation after they have been in physical contact with a trawl (Ramsay & Kaiser, 1998; Mensink *et al.*, in press). Whelks have been absent from the Dutch Wadden Sea for many years, and the evidence published to date suggests that bottom fishing is almost certainly responsible.

Many of the previous studies of the effects of fishing on benthic biota in the North Sea have been hampered by the lack of suitable control areas that have remained unfished. The value of a closed area as a comparative tool for such studies is illustrated by Bradshaw *et al.* (Chapter 6) in their study of the effects of scallop dredging off the Isle of Man. Similarly, Ball *et al.* (Chapter 5) have used wreck sites as reference points against which the fauna of heavily fished areas were compared. Both of these studies provide strong evidence that chronic fishing disturbance has altered benthic communities with time. The closed area off the Isle of Man has permitted Bradshaw *et al.* (Chapter 6) to observe changes after the closed area was instigated. It appears that the benthic communities in the closed area became more heterogeneous after the cessation of fishing. They were then able to conduct an experiment demonstrating that it was possible to reverse this trend and drive the community within the closed area back to a condition found in heavily fished areas. The habitats examined in these two studies are either mud or coarse sediments that are eventually restored with time. Habitats constructed by or composed of living organisms are likely to take much longer to recover. These biogenic habitats are unusual and have limited distribution. Hall-Spencer & Moore (Chapter 7) draw our attention to the effects of scallop dredging on maerl beds that are composed of the thalli of very slow-growing (<1 mm year^{-1}) plants. Scallop dredging is highly efficient in these habitats, so almost the entire stock of scallops is removed after one or two episodes of fishing. The authors clearly demonstrate that recolonisation by the biota and regrowth of the plants in this habitat will take >5 years as a result of just one passage of a gang of scallop dredges. It is quite clear from this study that the short-term gain derived from fishing such a habitat does not justify the environmental damage caused. This is the clearest case of a marine habitat that should be protected immediately from the effects of towed bottom-fishing gear. While we still do not understand the functional significance of high diversity habitats such as maerl beds, it may be necessary to compensate fishers to refrain from fishing them and perhaps engage fishers to protect these habitats long enough for us to understand their ecological role in more detail.

Fishing as a source of energy subsidies

Bottom-fishing directly alters the composition and structure of seabed communities and habitats. In addition, fishing redirects energy within the marine ecosystem via two routes. Firstly, animals damaged, killed or displaced on the seabed become available for consumption by benthic and demersal predators and scavengers. Secondly, material hauled on-board fishing vessels and then discarded overboard is available to avian and midwater predators and scavengers. Demestre *et al.* (Chapter 8) demonstrate that benthic and fish scavengers aggregate in areas of trawl disturbance on a muddy Mediterranean seabed, and as in previous studies (Ramsay *et al.*, 1996, 1997) all the available food appears to be consumed within 72 h. Fonds & Groenewold (Chapter 9) have taken data from Bergman & van Santbrink (Chapter 4) to calculate the total amount of food potentially generated by trawling in the southern North Sea. Using data derived for the energetic requirements of selected scavenger species at different times of the year, they have calculated that the additional energy generated by fishing activities is only enough to provide 7% of the maximum food demand of the entire scavenger population in the Dutch sector of the southern North Sea. These data suggest that, on average, there is insufficient material to stimulate an overall population expansion in scavenging species. Such signals, if they occur, might be manifested most clearly in populations of starfish that are highly resilient to the adverse effects of fishing and are able to eat a wide range of different food-types. Ramsay *et al.* (Chapter 10) related starfish abundance in the southern North Sea and English Channel to trawling effort. In both cases, the response of the starfish populations is similar, i.e. at relatively low to medium levels of fishing disturbance, starfish populations tend to increase, whereas at the highest levels of fishing disturbance, the negative effects of fishing (i.e. starfish mortality) outweigh the benefits derived from energy subsidies. What is clear from this study is that, while energy subsidies may affect starfish populations, there are many other factors that have much stronger influences on the population fluctuations of this species.

It was thought previously that the increases in the population size of many avian scavengers was largely related to increases in the amount of fisheries-generated waste. However, Camphuysen & Garthe (Chapter 11) emphasise that seabirds remain reliant upon natural sources of food that are vulnerable to overexploitation by humans or periodic population collapse due to unforeseen natural phenomena. Fisheries wastes may have aided the expansion in the range of some species by providing a guaranteed source of food when alternatives were unavailable. While seabirds are not entirely reliant on fisheries waste as a source of food, sudden changes in the distribution of fishing effort (that might relate to the instigation of areas closed to fishing) or fisheries policy (relating to mesh-size regulations) could have unexpected side effects for seabird populations.

Long-term changes associated with fishing

We have seen that fishing alters the seabed habitat, directly affects the animals living within or on the substratum and may indirectly influence populations of scavengers. In Chapter 12, Craeymeersch *et al.* have investigated the cumulative effects of many years of fishing disturbance by comparing the composition of benthic fauna in areas subjected to different intensities of fishing effort. While the distribution of fauna was correlated with the latter, fishing effort was only one of many variables that explained the observed patterns of community data. Nevertheless, fishing effort was highly correlated with the occurrence of opportunistic spionid worms that were most abundant in heavily fished areas.

One of the problems faced by any scientist working in the North Sea is that this area has been subjected to human impacts for many hundreds of years, hence the present-day fauna is likely to be the product of human interference and environmental changes. This point is emphasised by Frid & Clark (Chapter 13), who have used historical records of benthic community data to tease apart the different influences of environmental and fishing effects. As they point out, the greatest changes associated with bottom fishing in the benthic fauna of the North Sea probably occurred some years ago with the removal of long-lived and reef-forming fauna. These organisms are intrinsically the most vulnerable to physical disturbance and are likely to be the first to disappear from a physically perturbed environment (see Chapters 6 and 7).

Greenstreet & Rogers (Chapter 14) develop these ideas further for non-target species of fish. As for benthic species, current applied theory indicates those specific life-history characteristics likely to make a species vulnerable to fishing disturbance. These life-history characteristics include large ultimate size, slow growth rate and greater age at maturity. Greenstreet & Rogers examine elasmobranchs in detail, as they are a group of species that have such life-history characteristics likely to render them susceptible to fishing disturbance. In general, trends in the abundance of the different shark, skate and ray species in the North Sea can be attributed to fishing mortality, since they follow predictions based on the life-history characteristics of each species (Jennings *et al.*, 1998). This contrasts with the case of the Georges Bank, where skate and dogfish abundance has actually increased, probably because they were always discarded, and likely to have a high survival rate following discarding. Whilst fishing undoubtedly causes increased mortality for many non-target species, in some cases, it may also increase the scope for population growth through scavenging and reduced predation and competition. It is clear from the studies reported at the workshop that long-term changes in populations of animals are much more clearly seen in large-bodied organisms such as birds and certain fish species and become gradually less clear at lower trophic levels and for smaller-body-sized organisms. The links between fish and invertebrate diversity and the stability and productivity of marine communities are unknown. Hence, we need to improve our understanding of the pattern and processes involved with marine ecosystems so that we can better determine fishing effects. The use of diversity indices may not reveal the subtle shifts associated with the effects of fishing, which are more readily determined using multivariate techniques that have the ability to highlight changes in indicator species or functional groups of organisms (Jennings & Reynolds, Chapter 15).

Conservation methods, issues and implications for biodiversity

Improvement of gear technology certainly has a role to play in the conservation of both target and non-target species. Improvements in mesh configurations and gear design and the use of pingers show promising signs of alleviating some of the by-catch problems that are exhibited throughout the fishing industry (van Marlen, Chapter 16; Tregenza, Chapter 17). Nevertheless, improvements in gear design have only limited potential to reduce by-catch problems associated with catches of benthic invertebrates and may yield conflicting outcomes. For example, while gill nets are highly selective for certain size-classes of fish, they are associated with undesirable by-catches of cetaceans (Tregenza, Chapter 17). No improvement in gear designs will address the problem of seasonality and locality of by-catches. Tregenza (Chapter 17) suggests that some cetacean species are more vulnerable to interactions with fishing gear in certain seasons. Hence, in addition to the use of pingers on set nets, by-catches might be reduced even more effectively if certain areas of the sea were free of set gears at certain times of the year. The behaviour of cetaceans in respect to nets is complex and needs more detailed study to improve our ability to conserve these species in an environment that supports set net fisheries. While by-catches of non-target species are perceived as 'wasteful', fishing across all trophic levels may help to maintain ecosystem stability, whereas biasing catches towards single species may lead to greater shifts in populations of predators and their prey.

Tasker *et al.* (Chapter 18) highlight key conservation issues that pertain to the wide-scale effects of fishing on the marine environment. There is now an agreed strategy among North Sea nations to ensure sustainable and healthy ecosystems in the North Sea (Anon., 1993). This might be achieved by defining best current fishing practices, or improving existing techniques or perhaps vetting proposed new fishing methodologies (Fonteyne, Chapter 2; van Marlen, Chapter 16). In addition, Tasker *et al.* suggest the need for the definition of appropriate levels of fishing mortality for non-target species and acceptable levels of habitat disturbance. In the case of maerl communities, we would conclude that towed fishing disturbance is unacceptable, whereas highly perturbed sandy seabeds would be expected to be more tolerant of increased perturbation. Finally, they suggest the establishment of no-take zones (NTZs) to ensure ecosystem integrity, a suggestion echoed by Frid & Clark (Chapter

13) and Lindeboom (Chapter 19). While there is little evidence to suggest that NTZs are a realistic mechanism to conserve commercial fish stocks in temperate waters (Horwood, Chapter 20), they are unlikely to worsen the current situation. They may have some benefit for more sedentary species that remain confined within the limits of the NTZ (e.g. scallops). Nevertheless, their value in terms of preserving examples of different habitats that may be valuable as breeding or nursery grounds is widely supported; indeed, the UK currently has many such closed areas (although these are not true NTZs). Our current ability to disentangle the effects of fishing from environmental changes that have occurred in European waters is partly attributable to a historical lack of NTZs that would provide true control areas for experimental and comparative studies (Lindeboom, Chapter 19). The scientific value of having areas closed to towed fishing activity is clearly demonstrated by Bradshaw *et al.*'s study (Chapter 6). Not all methods of fishing are equally damaging, hence there is scope for a suite of zoning schemes that permit use of certain gears but not others (Tasker *et al.*, Chapter 18). It should also be recognised that the essential aims of a more environmentally friendly management of fisheries is to ensure a long-term future of the fishing industry. However, as experience has shown, regardless of the good intentions of these objectives, they are unlikely to be realised if we do not begin to develop NTZs and similar schemes with the full participation of the fishing industry (Lindeboom, Chapter 19).

Socio-economic implications and mechanisms for reducing fisheries impacts

The emphasis of the workshop and book thus far has been centred on the repercussions of fishing activities for marine habitats and biota and the mechanisms of minimising these effects. As a marine ecologist, I have been aware that it is easy for those of us that study the biological aspects of fishing activities to become so engrossed in this pursuit that we are guilty of forgetting the social implications of some of our recommendations. It is easy to argue the creation of a NTZ on biological grounds without considering the impacts that this may have for fishers that currently exploit this area. Similarly, imposing bans on certain types of gear will have drastic economic consequences for those fishers that have invested heavily in this particular fishing sector. This final section of the workshop and book was designed to highlight the complexity of implementing marine conservation objectives when this is likely to cause immediate and perhaps indefinite reductions in the income of the industry.

The main objective of fisheries management is to ensure the sustainability of the commercial stock, and while many methods have been tried, none has proved entirely successful. Individual transferable quotas (ITQs) were designed to alleviate problems associated with discarding. However, Pascoe (Chapter 21) demonstrates that, in many scenarios, ITQs can result in an increased tendency to discard fish that are above the minimum legal landing size. However, under an ITQ system, fishers are better able to plan their harvesting strategy. This could result in a fishing pattern that lowers discards, but this depends on the spatial distribution of the stock. In addition, the incentives to adopt more selective gear are also increased. Hence, while a greater proportion of the catch of small fish may be discarded, fewer small fish may be caught and, hence, overall discards may be lower. Even in unmanaged systems, there is a tendency to discard fish owing to constraints such as a vessel's fish-hold capacity. Fishers are sophisticated predators that alter their behaviour to maximise their net financial gain. The techniques employed by Pascoe gives us a useful insight into the likely outcome of adopting one of a number of conservation strategies and how fishers might adopt their behaviour to compensate the effects of legislative restrictions. The techniques used by Pascoe in Chapter 21 have great potential if we are to evaluate the financial incentives necessary to encourage fishers to support the implementation of NTZs and how this might alter their subsequent fishing behaviour.

McGlade & Metuzals (Chapter 22) report a detailed case study of the governance required and options that might be implemented to reduce by-catches of harbour porpoises. They have supplemented existing information on porpoise catches by interviewing fishers involved in set net fisheries, which proved to be an extremely cost-effective method of assessing levels of by-catch. They were then able to undertake a spatio-temporal analysis of fishing effort by fleet and for individual vessels, in relation to oceanographic features, the information gathered from interviews and previously published studies. Their results demonstrated a high coincidence of porpoise by-catch with seasonal patterns of fishing effort for cod that are associated with fronts in the southern and central North Sea and with tidal-mixing in the summer along the inner waters of the Danish coast. Technical factors specific to the fishing operations, such as the height of nets and long soak times in deeper waters, were also important factors that influenced by-catch. This detailed understanding of the interplay between the environment and the behaviour of fishers made it possible to analyse the governing needs required to support a Code of Conduct for Responsible Fisheries. It was concluded that fishing practices, price competition and a lack of participation in decision-making often led to situations where by-catches occurred. Changes to the administrative and market structures were considered necessary for these fisheries to remain sustainable. When interviewed, fishers recognised the need for technical measures (as outlined in Chapters 17 and 21) and responsible fishing practices, and suggested that these should be policed by a self-governing body. These net fisheries are essential to the survival of a number of local communities, and it is important that we seek to implement management systems that continue to permit the fishers to pursue their living, while seeking to minimise any associated adverse effects. However, many of the measures suggested would seem to erode current profitability margins by causing fishers to alter gear and fishing practices. Given current consumer awareness and preference for 'dolphin-friendly' tuna products, it is to be hoped that the deficit in profitability could be subsidised by consumer demand.

Jones (Chapter 23) discusses the problems associated with public (societal) perception of the marine environment and its inhabitants. While flagship species such as cetaceans and turtles evoke public condemnation of activities that might threaten these animals, they have perhaps deflected attention away from other priorities, such as the changes that have occurred lower down the food chain (Pauly *et al.*, 1998). Scientists need to become more actively involved in the education of the public and decision-making processes that would help to focus research and increase its general impact. The uncertainties surrounding our current knowledge and ability to predict the widescale effects of fishing on the marine environment do not enhance our scientific credibility. There is clear scientific justification to set up areas closed to fishing to protect certain vulnerable habitats such as maerl beds (Chapter 7). We should now present this information to the fishing industry and, in joint consultation, publicly present a policy on the agreed management practices for this particular habitat. From the fishers' perspective, this may seem yet another attempt to reduce their access to areas of the sea, yet it must be emphasised that the essential aims of a more environmentally friendly management of fisheries is to ensure a long-term future of the fishing industry. Furthermore, these vulnerable habitats are by definition rare and, as such, would not involve any great loss of accessible area to towed gear fisheries and can still be exploited by fishers using less intrusive gears such as pots and set nets.

Throughout this book, researchers have presented data that demonstrate the undoubted influence of fishing on our marine ecosystem. They have suggested mechanisms or strategies that might be implemented to reduce these effects while still exploiting marine species. However, unless public concerns or fishers demand that action is taken, there is little incentive for governments to act. Research may become the alibi of government procrastination, as a cynic would allude to the current research project culture that runs for 3 years with a further extension of another 3 years, by which time the problem has become that of the next political party in office. With this in mind, Symes (Chapter 24) looks forward to the review of the Common Fisheries Policy in 2002 and has evaluated some of the opportunities and constraints confronting the incorporation of an ecosystem approach into fisheries management. He points out that an ecosystem approach will complicate an already complex, highly centralised and politicised decision-making process, and the option of dispensing with fish-stock management is too radical an agenda to contemplate. A shift to integrated fisheries management (IFM) that combines the principles of fisheries and ecosystem management so as to conserve ecosystem integrity, while supporting a sustainable level of human use, will require a significant reorientation of underlying value systems, scientific information and management procedures, together with a rebalancing of user-group interests. The precautionary principle, as the central tenet of IFM, requires more explicit definition and well-defined operational procedures that can be applied to fish stocks and other elements of the marine ecosystem. While this is an attainable goal for fish stocks, it will be far more difficult to achieve for other elements of the ecosystem whose functions we still do not understand fully. The implementation of IFM would shift the emphasis from numerical management based on TACs and quotas to parametric management that pays close attention to the physical and biological parameters of the fishery. The three current examples of international collaboration involving an ecosystem approach (Wadden Sea, North Sea Conference, Baltic 21) do not appear to have broken free of the constraints of past institutional systems. Symes concludes by suggesting three possible scenarios at the review of the CFP: greater centralisation, *status quo*, and regionalisation; only the latter offers an appropriate framework for an ecosystem approach, but is unlikely to succeed. While the predicted prospects for the adoption of IFM and an ecosystem approach to European fisheries are not very bright, the best chance of success is in inshore waters over which each nation has greater potential for localised management.

The research presented within this book clearly demonstrates the wide-ranging effects of fishing on the marine ecosystem in European waters. Symes' prognosis for the uptake of the information and views presented herein is somewhat pessimistic (Chapter 24). Nevertheless, publicising this information represents a positive step towards increasing the awareness of the general public, government, non-governmental organisations and the fishing industry as to the importance of considering fisheries management within a framework that promotes a healthy and functional marine ecosystem.

References

Anon. (1993) *North Sea Quality Status Report 1993*, 132 pp. Oslo and Paris Commissions, London; Olsen & Olsen, Fredensborg, Denmark.

Churchill, J. (1989) The effect of commercial trawling on sediment resuspension and transport over the Middle Atlantic Bight continental shelf. *Continental Shelf Research*, **9**, 841–64.

Jennings, S. & Kaiser, M.J. (1998) The effects of fishing on marine ecosystems. *Advances in Marine Biology*, **34**, 201–352.

Jennings, S., Reynolds, J.D. & Mills, S.C. (1998) Life history correlates of responses to fisheries exploitation. *Proceedings of the Royal Society of London*, **B265**, 333–9.

MacDonald, D.S., Little, M., Eno, N.C. & Hiscock, K. (1996) Towards assessing the sensitivity of benthic species and biotopes in relation to fishing activities. *Aquatic Conservation: Marine and Freshwater Ecosystems*, **6**, 257–68.

Mensink, B.P., Fischer, C.V., Cadée, G.C., Fonds, M., ten Hallers-Tjabbes, C.C. & Boon, J.P. (in press) *Journal of Sea Research*.

Pauly, D., Christensen, V., Dalsgaard, J., Froese, R. & Torres, F. Jr (1998) Fishing down marine food webs. *Science*, **279**, 860–63.

Ramsay, K. & Kaiser, M.J. (1998) Demersal fishing increases predation risk for whelks *Buccinum undatum* (L.). *Journal of Sea Research*, **39**, 299–304.

Rijnsdorp, A.D., Buijs, A.M., Storbeck, F. & Visser, E. (1998) Micro-scale distribution of beam trawl effort in the southern North Sea between 1993 and 1996 in relation to the trawling frequency of the sea bed and the impact on benthic organisms. *ICES Journal of Marine Science*, **55**, 403–19.

Glossary

Beam trawl
The horizontal opening of this trawl is provided by a beam, made of metal or wood (old fashioned), which may be between 3 and 12 m long. Beam trawls are used mainly to catch flatfish and shrimp fishing.

Benthos
Animals living in or on the seabed.

Biomass
The total weight of organic matter.

Bottom-fishing gear
Towed fishing gear fished in close contact with the seafloor.

Catchability
The capability of a fishing gear to catch fish or other biota during a fishing operation.

Catch efficiency
The ratio between caught animals and those actually present in the path of the gear. It shows the efficiency of a sampling gear, e.g. trawl, grab or core.

Codend
The rearmost part of the trawl where the catch accumulates.

Demersal
Found near the sea bed.

DGPS
Differential Global Positioning System. A world-wide positioning system to be used on land, in air and at sea giving the position by determining its relative position to geostationary satellites.

Direct mortality
Mortality that occurs within 0–3 days due to trawling with a particular gear.

Discard mortality
Mortality within 3 h (immediate) to 3 days (secondary) among discarded animals.

Epifauna
Bottom-dwelling animals that live on the surface of the sea floor, e.g. crabs, shrimps, starfish.

Fishing effort
A measure of the activity of fishing boats. Fishing effort is strictly defined in terms of 'total standard hours fishing per year', but is often described less rigorously in terms of numbers of vessels, fishing time or fishing power.

Fishing intensity
Fishing effort per unit area.

Fishing mortality
Total direct mortality in the population of a species generated by a trawl fishery over a certain time period and expressed as a percentage of the initial population.

Groundrope
Rope or bobbin rope attached to the front of the belly of a net used to help the net over obstacles on the seafloor.

GRT
Gross registered tonnage.

hp
Horse power; 1 hp = 0.7355 kW.

ICES
International Council for Exploration of the Sea (founded 1902, Copenhagen). All nations bordering the North Atlantic are members.

ICES statistical rectangle	A rectangular grid of approximately 30×30 n miles used by the ICES in their study area.
Macrofauna	Bottom-living organisms retained on a 1-mm meshed sieve.
***Nephrops* trawl**	Demersal trawl specially designed for the *Nephrops* (Norway lobster) fishery. *Nephrops* trawls often have separator panels to reduce unwanted by-catches.
Non-target species	All animals and plants not directly fished for. Those species of no commercial value will become discards.
Otter trawl	A large conical net supplied with two otter boards that keep the mouth at the net open horizontally.
Side-scan sonar	An acoustic imaging device to provide pictures of the sea bottom.
Target species	All the animal or plant species a fishing vessel tries to collect in as high numbers as possible. Gears are adapted specifically for the various target species, e.g. flatfish trawl, shrimp trawl, *Nephrops* trawl, oyster dredge, cockle dredge.
Tickler chain	A chain rigged in front of the groundrope of a beam trawl to disturb flatfish from the bottom and to increase the fishing efficiency.
Total direct mortality	Sum of discard mortality and mortality of animals in the trawl path due to the passage of trawl, expressed as a percentage of initial density in the seabed.
Wayline	The navigational course steered by a vessel between two points plotted on a navigational plotter.

Index